Advanced Organic Chemistry: Reaction Mechanisms

Advanced Organic Chemistry: Reaction Mechanisms

Contributors

Yutaka Okada, Hiroshi Huruya et al.

AURIS
Reference

www.aurisreference.com

Advanced Organic Chemistry: Reaction Mechanisms

Contributors: Yutaka Okada, Hiroshi Huruya et al.

Published by Auris Reference Limited

www.aurisreference.com

United Kingdom

Advanced Organic Chemistry: Reaction Mechanisms

ISBN: 978-1-78154-882-0

British Library Cataloguing in Publication Data
A CIP record for this book is available from the British Library

Printed in the United Kingdom

Exclusively distributed by CBS Publishers & Distributors Pvt. Ltd.

Sales & Distribution Rights only for India, Pakistan, Bangladesh, Sri Lanka, Nepal and Bhutan. This book is not to be sold outside these territories.

Contents

List of Abbreviations

GC	Gas chromatography
HCMV	Human cytomegalovirus
THF	tetrahydrofuran
DHFR	dihydrofolate reductase
IR	Infrared
DMF	dimethylformamide
DMFDMA	dimethylformamide dimethyl acetal
LLDPE	Linear-low-density polyethylene
RSM	Response surface methodology
RSM	Response Surface Methodology
ITO	Indium tin oxide
NHS	N-hydroxysuccinimide
EIS	Electrochemical impedance spectroscopy
SEM	Scanning electron microscope
PGs	Prostaglandins
TG	Triglyceride
BDF	Biodiesel fuel

List of Contributors

Yutaka Okada
Department of Applied Chemistry, Ritsumeikan University, Kusatsu-Shi, Japan

Hiroshi Huruya
Department of Applied Chemistry, Ritsumeikan University, Kusatsu-Shi, Japan

Yasuhiro Imori
Department of Applied Chemistry, Ritsumeikan University, Kusatsu-Shi, Japan

Thangavelu Saravanan
Laboratory of Bioorganic Chemistry, Department of Biotechnology, Indian Institute of Technology Madras, Chennai, India

Anju Chadha
Laboratory of Bioorganic Chemistry, Department of Biotechnology, Indian Institute of Technology Madras, Chennai, India
National Center for Catalysis Research, Indian Institute of Technology Madras, Chennai, India

Tarur Konikkaledom Dinesh
Department of Inorganic Chemistry, University of Madras, Guindy Campus, Chennai, India
ITC Life Sciences and Technology Centre, ITC Limited, Peenya Industrial Area, Phase I, Bangalore, India

Namasivayam Palani
ITC Life Sciences and Technology Centre, ITC Limited, Peenya Industrial Area, Phase I, Bangalore, India

Sengottuvelan Balasubramanian[3]
Department of Inorganic Chemistry, University of Madras, Guindy Campus, Chennai, India

Azizollah Habibi
Faculty of Chemistry, Kharazmi University, Tehran, Iran

Yousef Valizadeh
Faculty of Chemistry, Kharazmi University, Tehran, Iran

Marjan Mollazadeh
School of Chemistry, College of Science, University of Tehran, Tehran, Iran

Abdolali Alizadeh
Department of Chemistry, Tarbiat Modares University, Tehran, Iran

Shohei Sanada
Graduate School of Science and Engineering, Yamaguchi University, Tokiwa-dai, Japan

Michinori Sumimoto
Graduate School of Science and Engineering, Yamaguchi University, Tokiwa-dai, Japan

Kenji Hori
Graduate School of Science and Engineering, Yamaguchi University, Tokiwa-dai, Japan

Sarkar M. A. Kawsar
Laboratory of Carbohydrate and Protein Chemistry, Department of Chemistry, Faculty of Science, University of Chittagong, Chittagong, Bangladesh

Hamida A. Ara
Laboratory of Carbohydrate and Protein Chemistry, Department of Chemistry, Faculty of Science, University of Chittagong, Chittagong, Bangladesh

Sheikh Aftab Uddin
The Institute of Marine Science, Faculty of Science, University of Chittagong, Chittagong, Bangladesh

Mohammed K. Hossain
Department of Pharmacy, Faculty of Biological Science, University of Chittagong, Chittagong, Bangladesh

Shagir A. Chowdhury
Laboratory of Carbohydrate and Protein Chemistry, Department of Chemistry, Faculty of Science, University of Chittagong, Chittagong, Bangladesh

Abul F. M. Sanaullah
Laboratory of Carbohydrate and Protein Chemistry, Department of Chemistry, Faculty of Science, University of Chittagong, Chittagong, Bangladesh

Mohammad A. Manchur
Department of Microbiology, Faculty of Biological Science, University of Chittagong, Chittagong, Bangladesh

Imtiaj Hasan
Laboratory of Glycobiology and Marine Biochemistry, Department of Life and Environmental System Science, Graduate School of NanoBio Sciences, Yokohama City University, Yokohama, Japan

Yukiko Ogawa
Divisions of Microbiology, Graduate School of Pharmaceutical Science, Nagasaki International University, Nagasaki, Japan

Yuki Fujii
Divisions of Microbiology, Graduate School of Pharmaceutical Science, Nagasaki International University, Nagasaki, Japan

Yasuhiro Koide
Laboratory of Glycobiology and Marine Biochemistry, Department of Life and Environmental System Science, Graduate School of NanoBio Sciences, Yokohama City University, Yokohama, Japan

Yasuhiro Ozeki
Laboratory of Glycobiology and Marine Biochemistry, Department of Life and Environmental System Science, Graduate School of NanoBio Sciences, Yokohama City University, Yokohama, Japan

Bothaina A. Mousa
[1]Department of Organic Chemistry, Faculty of Pharmacy, Cairo University, Giza, Egypt

Makarem M. Korraa
Department of Organic Chemistry, Faculty of Pharmacy, Cairo University, Giza, Egypt

Mohamed G. Assy
Department of Organic Chemistry, Faculty of Science, Zagazig University, Zagazig, Egypt

Samar A. El-Kalyoubi
Department of Organic Chemistry, Faculty of Pharmacy (Girls), Al-Azhar University, Cairo, Egypt

Asmaa S. Salman
Department of Chemistry, Faculty of Science, Al-Azhar University, Girls' Branch, Nasr City, Cairo, Egypt

Anhar Abdel-Aziem
Department of Chemistry, Faculty of Science, Al-Azhar University, Girls' Branch, Nasr City, Cairo, Egypt

Marwa J.S. Alkubbat
Department of Chemistry, Faculty of Science, Al-Azhar University, Girls' Branch, Nasr City, Cairo, Egypt

Fathi A. Abu-Shanab
Department of Chemistry, Faculty of Science, Al-Azhar University, Assiut, Egypt
Department of Chemistry, Faculty of Science, Gazan University, Gazan, KSA

Sayed A. S. Mousa
Department of Chemistry, Faculty of Science, Al-Azhar University, Assiut, Egypt

Sherif M. Sherif
Department of Chemistry, Faculty of Science, Cairo University, Giza, Egypt

Mohamed I. Hassan
Department of Chemistry, Faculty of Science, Al-Azhar University, Assiut, Egypt

Karim Bouterfas
Laboratory of Vegetal Biodiversity: Conservation and Valorization, Faculty of Life and Natural Sciences, Djillali Liabes University, Sidi Bel-Abbes, Algeria

Zoheir Mehdadi
Laboratory of Vegetal Biodiversity: Conservation and Valorization, Faculty of Life and Natural Sciences, Djillali Liabes University, Sidi Bel-Abbes, Algeria

Djamel Benmansour[2],
Laboratory of Statistics and Random Model, Faculty of Natural and Universe Sciences, Abou-Bekr Belkaid University, Tlemcen, Algeria

Meghit Boumedien Khaled
Department of Biology, Faculty of Natural and Life Sciences, Université Djillali Liabes, Sidi Bel Abbes, Algeria

Mohamed Bouterfas
Laboratory of Microscopy, Microanalysis of the Matter and Molecular Spectroscopy, Faculty of Exact Sciences, Djillali Liabes University, Sidi Bel-Abbes, Algeria

Ali Latreche
Laboratory of Vegetal Biodiversity: Conservation and Valorization, Faculty of Life and Natural Sciences, Djillali Liabes University, Sidi Bel-Abbes, Algeria

Abdulrahman G. Alshammari
Department of Chemistry, Faculty of Science, Al-Imam Mohammad Ibn Saud Islamic University (IMSIU), Riyadh, KSA

Abdel-Rhman B. A. El-Gazzar
Department of Chemistry, Faculty of Science, Al-Imam Mohammad Ibn Saud Islamic University (IMSIU), Riyadh, KSA
Photochemistry Department, Heterocyclic & Nucleosides Unit, National Research Centre, Cairo, Egypt

K. Singh
Amity School of Engineering and Technology, Amity University, Noida, India

Ruchika Chauhan
Amity Institute of Nano Technology, Amity University, Noida, India

Pratima R. Solanki
Amity Institute of Nano Technology, Amity University, Noida, India

Tinku Basu
Amity Institute of Nano Technology, Amity University, Noida, India

Ahmed Abd El-Hameed Hassan
Chemistry Department, Faculty of Science, Al-Azhar University, Nasr City, Egypt
Chemistry Department, Faculty of Medicine, Jazan University, Jazan, KSA

Yuji Ueki
Environment and Industrial Materials Research Division, Quantum Beam Science Center, Sector of Nuclear Science Research, Japan Atomic Energy Agency, Takasaki, Japan

Seiichi Saiki
Environment and Industrial Materials Research Division, Quantum Beam Science Center, Sector of Nuclear Science Research, Japan Atomic Energy Agency, Takasaki, Japan

Takuya Shibata
Environment and Industrial Materials Research Division, Quantum Beam Science Center, Sector of Nuclear Science Research, Japan Atomic Energy Agency, Takasaki, Japan

Hiroyuki Hoshina
Environment and Industrial Materials Research Division, Quantum Beam Science Center, Sector of Nuclear Science Research, Japan Atomic Energy Agency, Takasaki, Japan

Noboru Kasai
Environment and Industrial Materials Research Division, Quantum Beam Science Center, Sector of Nuclear Science Research, Japan Atomic Energy Agency, Takasaki, Japan

Noriaki Seko
Environment and Industrial Materials Research Division, Quantum Beam Science Center, Sector of Nuclear Science Research, Japan Atomic Energy

Agency, Takasaki, Japan

Pulimamidi Rabindra Reddy
Department of Chemistry, Osmania University, Hyderabad, India

Ravula Chandrashekar
Department of Chemistry, Osmania University, Hyderabad, India

Hussain Shaik
Department of Chemistry, Osmania University, Hyderabad, India

Battu Satyanarayana
Department of Chemistry, Osmania University, Hyderabad, India

Preface

Organic chemistry is a chemistry sub discipline involving the scientific study of the structure, properties, and reactions of organic compounds and organic materials, i.e., matter in its various forms that contain carbon atoms. The text Advanced Organic Chemistry: Reaction Mechanisms explains the theories and examples of organic chemistry, providing the most comprehensive resource about organic chemistry. In first chapter, the reaction with heterocycles containing nitrogen, oxygen or sulfur atoms has been carried out. Chemoenzymatic synthesis of an enantiomerically enriched bicyclic carbocycle using candida parapsilosis ATCC 7330 mediated enantioselective hydrolysis has been presented in second chapter. In third chapter, we report a highly efficient, and environmentally benign procedure for the reaction of 1,2-pheny- lene-diamine derivatives 1 with arylidenemalononitrile 2 in aqueous medium as a green solvent to produce benzimidazole derivatives 3. In fourth chapter, we theoretically investigate the Pd-catalyzed acylation reaction of allylic ester using the DFT method. Fifth chapter focuses on chemically modified uridine molecules incorporating acyl residues to enhance antibacterial and cytotoxic activities. In sixth chapter, a series of new pyrimidine fused ring analogs have been synthesized and their biological effects are determined. Seventh chapter deals with synthesis, spectroscopic characterization and antimicrobial activity of some new 2-substituted imidazole derivatives. In eighth chapter, we report the reaction of acetylacetone 1a with one mole of N,N'-dimetylfor-mamide dimethyl acetal (DMFDMA) in dry dioxane gave the corresponding enamine 2a. The aim of ninth chapter is to optimize the extraction conditions of some phenolic compounds from white horehound leaves. Tenth chapter describes a class of novel 2- and 4-alkylamino-pyrimidines and their nucleosides designed by considering the structural features of the previously reported derivatives proved to be bioactive inhibitor compounds. In eleventh chapter, an attempt has been made to develop PtNP and PPY based electrodeposited tri-layer nanocomposite film and the same has been evaluated as a transducer matrix for total cholesterol estimation using impedance spectroscopy. In twelfth chapter, we report a variety of synthesis of heteroaromatics developed using functionally substituted enaminones as readily obtainable building blocks possessing multiple electrophilic and nucleophilic moieties. The objective of last chapter is to characterize the catalytic properties of the grafted fibrous catalyst and determine the optimal biodiesel fuel (BDF) production conditions using the grafted fibrous catalyst

Chapter 1

LIGAND EXCHANGE REACTION OF FERROCENE WITH HETEROCYCLES

Yutaka Okada, Hiroshi Huruya, Yasuhiro Imori

Department of Applied Chemistry, Ritsumeikan University, Kusatsu-Shi, Japan

ABSTRACT

The ligand exchange reaction with heterocycles containing nitrogen, oxygen or sulfur atoms was carried out. For the reaction with heterocycles, the order of the reactivity was S-heterocycles > N-heterocycles > O-heterocycles. Furthermore, when the results for the heterocycles were compared to those for the corresponding hydrocarbons, the hydrocarbons had a higher reactivity. These results mean that the reactivity would be mainly governed by the electron density of these arenes.

INTRODUCTION

The ligand exchange reaction is one of the typical reactions of ferrocenes. This is the reaction which was siscovered by Nesmeyanov et al. (Scheme 1) [1] . As for this reaction, the Cp ring of the ferrocene is exchanged for the aromatic ring, like benzene, in the presence of AlCl$_3$ and the product is able to be isolated as a stable PF$_6$ salt. Recently, this reaction is used for the preparation of several functional materials containing ferrocene moiety [2] [3] .

This reaction is initiated by pulling the Cp ring by AlCl$_3$ in the vertical direction, so that the ferrocene decomposes into two fragments. The resulting fragment containing an iron atom is coordinated by an aromatic ring (Scheme 2) [4] [5] .

The authors have reported the substituent effects for the reaction between an alkylferrocene and alkylbenzenes [6] - [8] . In this study, the reaction with heterocycles containing nitrogen, oxygen or sulfur atoms was carried out. These results were compared to the reaction with carbocycles.

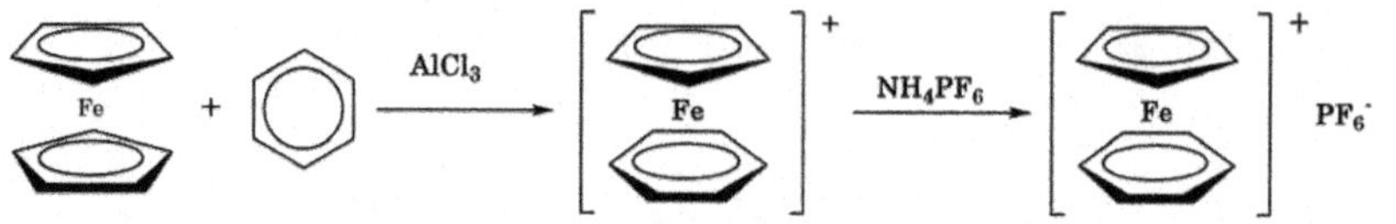

Scheme 1: The ligand exchange reaction.

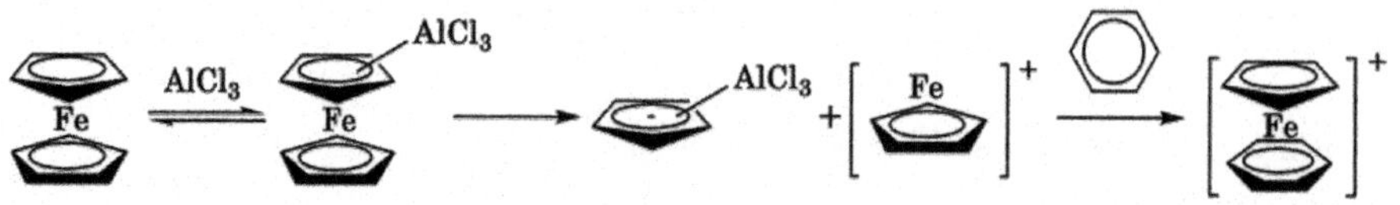

Scheme 2: The resulting fragment.

EXPERIMENTAL

Measurements of Ligand Exchange Reaction

The reaction was carried out in a small vessel, which contained the substrates and the solvent. The vessel was heated to a constant temperature in an oil bath. The reaction mixture was analyzed by HPLC at a specific time [7] [8] .

Measurement Apparatus

The ^{1}H-NMR spectra were measured in chloroform-d at room temperature using a JEOL A-400 spectrometer. The NMR data are shown in Tables 1-4.

The mass spectra were obtained using a Shimadzu LCMS-QP8000. The ionization mode was used for the Atmospheric Pressure Chemical Ionization method.

RESULTS AND DISCUSSION

The Ligand Exchange Reaction with Five-Membered Heterocycles

The yields of the reaction are shown in Table 5. For pyrrole, there are no products under these conditions. For thiophene, a small amount of the ligand exchange product was produced.

Pyrrole and thiophene are analogues containing nitrogen and sulfur, respectively. The difference in the reactivity would be due to the difference in the π-electron density of the aromatic ring. Namely, the electronegativity of nitrogen is 2.6, and that of sulfur is 3.0. Therefore, concerning the deviation of π-electron on the heterocycle, the sulfur analogue is larger than the nitrogen one, so the former would not act as an η^5-ligand,

The yield of the ligand exchange product for thiophene was 3.1% - 4.8%, which was much lower than that of the hydrocarbon mentioned in previous papers [7] [8] . These results are supported by the fact that the product from thiophene shows lower chemical shifts of the Cp ring than that of the other products (Table 6).

The Ligand Exchange Reaction with Fused Five-Membered Heterocycles

As shown in Table 7, the product from benzothiophene was detected, whereas no product was detected from indole and benzofuran. The carbon analogue, indene, produced the exchange product. Furthermore, the NMR data (Table 6) showed that the CpFe moiety coordinates to the benzene ring, not but the five-membered ring.

As mentioned above, the electronegativity of nitrogen is higher than that of sulfur. The electronegativity of oxygen is 3.4 which is higher than that of nitrogen. Therefore, these hetero atoms would produce a decrease in the π-electron density and deviation on the benzene ring. These would decrease the ability to coordinate as the η^5-ligand. The sulfur analogue, benzothiophene, gave the exchange product as well as the carbon analogue, indene, because of the lower electronegativity of the sulfur atom.

However, the yield of the ligand exchange product for benzothiophene was much lower than that of the tricyclic compounds as will be mentioned later.

Table 1: ^{1}H-NMR chemical shifts (p.p.m.) of substrates for the ligand exchange reactions on acetone-d$_6$.

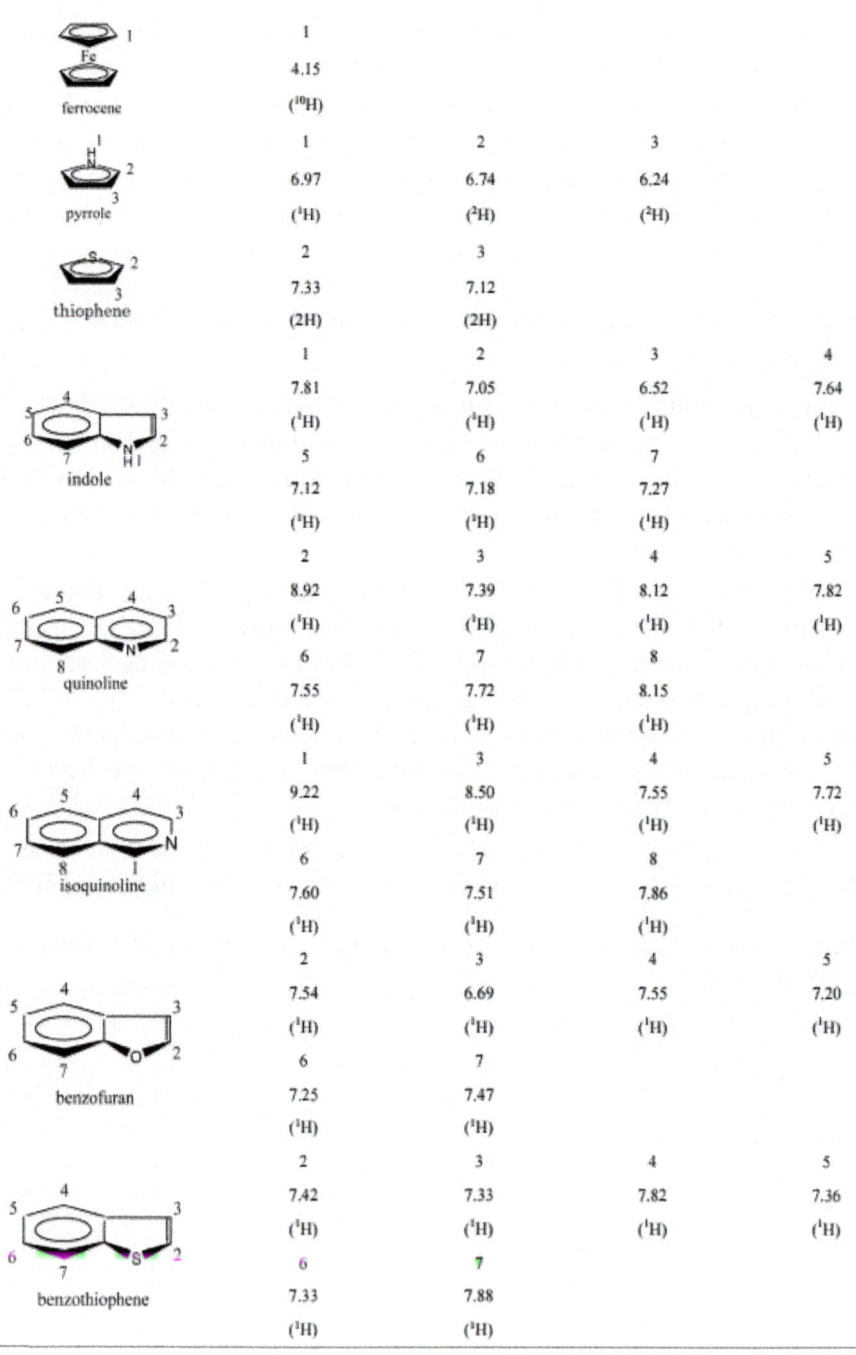

ferrocene

1
4.15
(^{10}H)

pyrrole

1	2	3
6.97	6.74	6.24
(^{1}H)	(^{2}H)	(^{2}H)

thiophene

2	3
7.33	7.12
(2H)	(2H)

indole

1	2	3	4
7.81	7.05	6.52	7.64
(^{1}H)	(^{1}H)	(^{1}H)	(^{1}H)
5	6	7	
7.12	7.18	7.27	
(^{1}H)	(^{1}H)	(^{1}H)	

quinoline

2	3	4	5
8.92	7.39	8.12	7.82
(^{1}H)	(^{1}H)	(^{1}H)	(^{1}H)
6	7	8	
7.55	7.72	8.15	
(^{1}H)	(^{1}H)	(^{1}H)	

isoquinoline

1	3	4	5
9.22	8.50	7.55	7.72
(^{1}H)	(^{1}H)	(^{1}H)	(^{1}H)
6	7	8	
7.60	7.51	7.86	
(^{1}H)	(^{1}H)	(^{1}H)	

benzofuran

2	3	4	5
7.54	6.69	7.55	7.20
(^{1}H)	(^{1}H)	(^{1}H)	(^{1}H)
6	7		
7.25	7.47		
(^{1}H)	(^{1}H)		

benzothiophene

2	3	4	5
7.42	7.33	7.82	7.36
(^{1}H)	(^{1}H)	(^{1}H)	(^{1}H)
6	7		
7.33	7.88		
(^{1}H)	(^{1}H)		

	1	2	3	4
carbazol	7.42 (^{2}H)	7.42 (^{2}H)	7.24 (^{2}H)	8.08 (^{2}H)
dibenzofuran	1	2	3	4
	7.56 (^{2}H)	7.44 (^{2}H)	7.32 (^{2}H)	7.93 (^{2}H)
dibenzothiophene	1	2	3	4
	7.85 (^{2}H)	7.45 (^{2}H)	7.45 (^{2}H)	8.16 (^{2}H)

Table 2: ^{1}H-NMR chemical shifts (p.p.m.) of arene complexes of ferrocene on acetone-d_6.

	2	3	6		
	6.61 (^{2}H)	6.50 (^{2}H)	5.21 (^{5}H)		
	2	3	4	5	6
	8.38 (^{1}H)	7.67 (^{1}H)	7.67 (^{1}H)	6.41 (^{1}H)	6.48 (^{1}H)
	7	8			
	7.37 (^{1}H)	4.85 (^{5}H)			
	1	2	3	4	5
	7.71 (^{1}H)	7.55 (^{1}H)	7.43 (^{1}H)	8.40 (^{1}H)	7.71 (^{1}H)
	6	7	8	9	10
	6.22 (^{1}H)	6.29 (^{1}H)	7.21 (^{1}H)	10.6 (^{1}H)	4.65 (^{5}H)
	1	2	3	4	5
	7.82 (^{1}H)	7.63 (^{1}H)	7.59 (^{1}H)	8.43 (^{1}H)	7.82 (^{1}H)
	6	7	8	10	
	6.43 (^{1}H)	6.50 (^{1}H)	7.45 (^{1}H)	4.91 (^{5}H)	
	1	2	3	4	5
	8.16 (^{1}H)	7.83 (^{1}H)	7.74 (^{1}H)	8.59 (^{1}H)	7.74 (^{1}H)
	6	7	8	10	
	6.57 (^{1}H)	6.63 (^{1}H)	7.61 (^{1}H)	4.85 (^{5}H)	

Table 3: ^{13}C-NMR chemical shifts (p.p.m.) of substrates for the ligand exchange reactions on acetone-d$_6$.

pyrrole	1				
	67.8				
pyrrole	2	3			
	117.6	107.9			
thiophene	2	3			
	125.1	126.8			
indole	2	3	4	5	
	124.3	102.2	120.6	121.9	
	6	7	3a	7a	
	119.7	111.1	127.7	135.7	
quinoline	2	3	4	5	
	150.3	130.0	135.9	127.7	
	6	7	8	4a	8a
	126.4	129.4	125.5	128.2	148.2
isoquinoline	1	3	4	5	6
	152.4	142.9	120.4	126.4	130.2
	7	8	4a	8a	
	127.2	127.5	135.7	128.6	
benzofuran	2	3	4	5	
benzofuran	144.9	106.5	121.2	122.7	
	6	7	3a	7a	
	124.2	111.4	127.5	155.0	
benzothiophene	2	3	4	5	
benzothiophene	126.1	123.7	123.5	124.0	
	6	7	3a	7a	
	124.1	122.4	139.5	139.6	
carbazol	1	2	3	4	
	110.6	125.9	119.5	120.4	
	4a	8b			
	123.4	139.6			
dibenzofuran	1	2	3	4	
	111.6	127.1	122.6	120.6	
	4a	8b			
	124.2	138.1			
dibenzothiophene	1	2	3	4	
	121.5	126.6	124.3	122.7	
	4a	8b			
	135.5	139.4			

Table 4: ^{13}C-NMR chemical shifts (p.p.m.) of arene complexes of ferrocene on acetone-d$_6$.

	1	2	3	4	5
	112.2	130.1	113.0	121.8	79.1
	6	7	8	4a	4b
	71.1	83.4	67.8	122.7	81.2
	8a	8b	10		
	88.0	130.1	76.3		
	1	2	3	4	5
	114.7	133.3	125.4	124.7	81.6
	6	7	8	4a	4b
	86.2	91.7	76.0	127.0	86.2
	8a	8b	10		
	130.9	161.5	79.6		
	1	2	3	4	5
	108.3	132.5	126.1	125.9	84.1
	6	7	8	4a	4b
	86.5	88.9	82.0	126.4	87.1
	8a	8b	10		
	128.4	136.3	80.2		

Table 5: Yields of ligand exchange reactions of ferrocene with five-membered heterocycles.

Entry	Arene	Fc:arene:AlCl$_3$	Solvent	Temperature/°C	Time/hr.	Yield/%
1	Pyrrole	1:1:1	1,2-dichloroethane	80	2	0
2	Pyrrole	1:1:2	1,2-dichloroethane	80	2	0
3	Pyrrole	2:1:2	1,2-dichloroethane	80	2	0
4	Thiophene	1:1:1	1,2-dichloroethane	80	2	3.4
5	Thiophene	1:1:2	1,2-dichloroethane	80	2	4.8
6	Thiophene	1:1:1	1,2-dichloroethane	80	2	3.1

Table 6: The difference of chemical shifts (p.p.m.) between the ligand exchange products and the substrates.

Entry	Arene	$\Delta\delta^{1)}$/p.p.m.	
		Cp ring	Arene
1	Thiophene	+1.06	−0.72 - 0.63
2	Benzothiophene	+0.70	−0.95 - 0.15
3	Carbazol	+0.50	−1.15 - 0.21
4	Dibenzophenone	+0.76	−1.16 - 0.35
5	Dibenzothiophene	+0.70	−0.88 - 0.24

$^{1)}\Delta\delta = \delta$(ligand exchange product) − δ(substrate).

Table 7: Yields of ligand exchange reactions of ferrocene with fused five-membered heterocycles.

Entry	Arene	Fc:arene:AlCl₃	Solvent	Temperature/°C	Time/hr.	Yield/%
1	Indole	1:1:1	1,2-dichloroethane	80	2	0
2	Indole	1:1:1	1,2-dichloroethane	80	4	0
3	Indole	1:1:1	1,2-dichloroethane	80	6	0
4	Indole	1:1:1	decalin	130	4	0
5	Benzofuran	1:1:1	1,2-dichloroethane	80	2	0
6	Benzofuran	1:1:1	1,2-dichloroethane	80	6	0
7	Benzofuran	1:1:1	decalin	130	2	0
8	Benzofuran	1:1:2	1,2-dichloroethane	80	2	0
9	Benzothiophene	1:1:1	1,2-dichloroethane	80	2	13
10	Benzothiophene	1:1:2	1,2-dichloroethane	80	2	10
11	Benzothiophene	2:1:2	1,2-dichloroethane	80	2	40

The Ligand Exchange Reaction with Fused Six-Membered Heterocycles

As shown in Table 8, there are no products from the reaction between ferrocene and fused six-membered heterocycles, quinoline and isoquinoline. For the carbon analogue, naphthalene, the expected product was produced.

These results would be due to the electron withdrawing effect by the hetero atom similar to the case of five- membered heterocycles.

The Ligand Exchange Reaction with Fused Tricyclic Heterocycles

The ligand exchange reaction between ferrocene and the fused tricyclic heterocycles was carried out. As the results, the expected products were formed for all the tested heterocycles; i.e. carbazol, dibenzofuran and dibenzothiophene

(Table 9). The NMR data showed that these products have the CpFe moiety on the end ring. This structure would be more stable than that bearing the CpFe moiety at the central ring due to the aromaticity of the rings.

The yields for these reactions were much higher than those for above-mentioned heterocycles. This would be due to the electron withdrawing effect by the hetero atom is divided into the two end rings, so that the effect is weakened.

Furthermore, the end ring protons bearing no CpFe moiety are shifted to a lower field as the NMR data show in Table 6. These shifts show that electrons are supplied from the end ring to produce the bond between the CpFe and the aromatic ring. On the contrary, this electron transfer caused the lower field shifts of the ring protons bearing the CpFe moiety. This effect, which does not appear for monocyclic and bicyclic heterocycles, would increase the reactivity of these tricyclic heterocycles.

As shown in Table 9, the yields for the carbazol, dibenzofuran and dibenzothiophene were almost equal, so the reactivities of these heterocycles were difficult to compare to each other. Therefore, these three heterocycles were put in a vessel at the same time, and the reaction was completed. The results are shown in Table 10.

From Table 10, for the mixture of carbazol and dibenzofuran, only the product with carbazol was formed (Entry 1). For the mixture of carbazol and dibenzothiophene, the ratio of products was 95:5, and for the mixture of dibenzofuran and dibenzothiophene, the ratio was 36:64 (Entries 2 and 3). These ratios were determined from the relative intensity of the LC-MS's molecular weight related ion. Based on these results, the order of the reactivities of these heterocycles is determined to be carbazol > dibenzothiophene > dibenzofuran.

This order was supported by the NMR data. In general, electrons are supplied from the Cp ring to the Fe atom and the coordinated aromatic ring in these ligand exchange products, so that the chemical shifts of the Cp protons in the products are at a lower field than those in ferrocene. Table 1shows such lower field shifts of the Cp protons. Moreover, the shift amount was the lowest for carbazole, and the highest for dibenzofuran. The fact that the shift amount was low means that the supplied amount of electron is low. This result shows that the intensity

Table 8: Yields of ligand exchange reactions of ferrocene with fused six-membered heterocycles.

Entry	Arene	Fc:arene:AlCl₃	Solvent	Temperature/°C	Time/hr.	Yield/%
1	Quinoline	1:1:1	1,2-dichloroethane	80	2	0
2	Quinoline	1:1:1	1,2-dichloroethane	80	4	0
3	Quinoline	1:1:1	1,2-dichloroethane	80	6	0
4	Quinoline	1:1:2	1,2-dichloroethane	80	2	0
5	Isoquinoline	1:1:1	1,2-dichloroethane	80	2	0
6	Isoquinoline	1:1:1	1,2-dichloroethane	80	4	0
7	Isoquinoline	1:1:1	1,2-dichloroethane	80	6	0
8	Isoquinoline	1:1:2	1,2-dichloroethane	80	2	0

Table 9: Yields of ligand exchange reactions of ferrocene with fused tricyclic heterocycles.

Entry	Arene	Fc:arene:AlCl₃	solvent	Temperature/°C	Time/hr.	Yield/%
1	Carbazol	1:1:1	1,2-dichloroethane	80	2	29
2	Carbazol	1:1:2	1,2-dichloroethane	80	2	37
3	Carbazol	2:1:2	1,2-dichloroethane	80	2	64
4	Dibenzofuran	1:1:1	1,2-dichloroethane	80	2	23
5	Dibenzofuran	1:1:2	1,2-dichloroethane	80	2	38
6	Dibenzofuran	2:1:2	1,2-dichloroethane	80	2	53
7	Dibenzofuran	5:1:12	1,2-dichloroethane	80	2	43
8	Dibenzofuran	5:1:12	decalin	130	4	91
9	Dibenzothiophene	1:1:1	1,2-dichloroethane	80	2	25
10	Dibenzothiophene	1:1:2	1,2-dichloroethane	80	2	38
11	Dibenzothiophene	2:1:2	1,2-dichloroethane	80	2	60

Table 10: Ratio of ligand exchange products of ferrocene with fused tricyclic heterocycles.

Entry	Fc:N:O:S:AlCl₃[1]	Solvent	Temperature/°C	Time/hr.	Yield/%	Ratio of products N:O:S/%
1	1:1:1:0:1	1,2-dichloroethane	80	2	63	100:0:-
2	1:1:0:1:1	1,2-dichloroethane	80	2	32	95:-:5
3	1:0:1:1:1	1,2-dichloroethane	80	2	38	-:36:64

[1]N = carbazol, O = dibenzofuran, S = dibenzothiophene.

of the Cp-Fe bond in the ligand exchange products does not change from that of ferrocene. Therefore, the product would be thermodynamically stable. As mentioned in the Introduction, the ligand exchange is a two-step reaction. The first step is pulling the Cp ring by AlCl₃, and the second step is the coordination of an aromatic ring. Therefore, the reactivity would depend on the second step. The transition state of the second step should be similar to the ligand exchange product, therefore, the order of stability of the transition state is equal to that of the product.

CONCLUSION

For the ligand exchange reaction with heterocycles, the order of the reactivity was S-heterocycles > N-hetero- cycles > O-heterocycles. Furthermore, the reactivity of these heterocycles was lower than those for the corresponding hydrocarbons. The reactivity would be affected by the electron density of these arenes.

CITE THIS PAPER

YutakaOkada,HiroshiHuruya,YasuhiroImori,   (2015)   Ligand   Exchange Reaction of Ferrocene with Heterocycles. *International Journal of Organic Chemistry*,**05**,282-290. doi:10.4236/ijoc.2015.54028

REFERENCES

1.   Nesmeyanov, A.N., Vol'kenau, N.A. and Isaeva, L.S. (1967) Mobility of the Halogen in the Cyclopentadienyl Ring of Chlorocyclopentadienylphenyliron. Doklady Chemistry, 176, 106-109.

2.   Li, F.-C., Tsai, S.-C., Yeh, C.-Y., Yeh, J.-Y., Chou, Y.-S., Ho, J.-R. and Tsiang, R.C.-C. (2014) Organic Thin-Film- Transistor Au/Poly(3-Hexylthiophene)/(Bilayer Die-lectrics)/Si Having Carbon Nanotubes Chemically Bonded to Poly(3-Hexylthiophene) in the Active Layer. Journal of Nano-science and Nanotechnology, 14, 5019-5027. http://dx.doi.org/10.1166/jnn.2014.9259

3.   Cha, I., Yagi, Y., Kawahara, T., Hashimoto, K., Fujiki, K., Tamesue, S., Yamauchi, T. and Tsubokawa, N. (2014) Grafting of Polymers onto Graphene Oxide by Trapping of Polymer Radicals and Ligand-Exchange Reaction of Polymers Bearing Ferrocene Moieties. Colloids and Surfaces, A441, 474-480. http://dx.doi.org/10.1016/j.colsurfa.2013.10.002

4.   Astruc, D. and Dabard, R. (1976) On the Mechanism of Ligand Exchange and Complexation by Aluminum Chloride of Ferrocene and Its Alkylated and Acylated Derivatives. Journal of Organometallic Chemistry, 111, 339-347. http://dx.doi.org/10.1016/S0022-328X(00)98142-7

5.   Astruc, D. and Dabard, R. (1976) Ligand Exchange Reactions between

Arenes and Ferrocene Derivatives Bearing at the α Position One or Two Substituents Complexed by Aluminum Trichloride. Tetrahedron, 32, 245-249. http://dx.doi.org/10.1016/0040-4020(76)87009-3

6.   6. Astruc, D. (1983) Organo-Iron Complexes of Aromatic Compounds. Applications in Synthesis. Tetrahedron, 39, 4027-4095. http://dx.doi.org/10.1016/S0040-4020(01)88627-0

7.   Hayashi, T., Okada, Y. and Shimizu, S. (1996) Substituent Effects for the Disproportionation, Synproportionation and Ligand Exchange Reaction of t-Butylferrocenes. Transition Metal Chemistry, 21, 418-422. http://dx.doi.org/10.1007/BF00140783

8.   Okada, Y., Yoshigami, Y. and Hayashi, T. (2003) Steric Effects on the Ligand Exchange Reactions of Alkylferrocenes. Transition Metal Chemistry, 28, 794-799. http://dx.doi.org/10.1023/A:1026006314010

Chapter 2

CHEMOENZYMATIC SYNTHESIS OF AN ENANTIOMERICALLY ENRICHED BICYCLIC CARBOCYCLE USING CANDIDA PARAPSILOSIS ATCC 7330 MEDIATED ENANTIOSELECTIVE HYDROLYSIS

Thangavelu Saravanan[1], Anju Chadha[1,2], Tarur Konikkaledom Dinesh[3,4], Namasivayam Palani[4], Sengottuvelan Balasubramanian[3]

[1]Laboratory of Bioorganic Chemistry, Department of Biotechnology, Indian Institute of Technology Madras, Chennai, India

[2]National Center for Catalysis Research, Indian Institute of Technology Madras, Chennai, India

[3]Department of Inorganic Chemistry, University of Madras, Guindy Campus, Chennai, India

[4]ITC Life Sciences and Technology Centre, ITC Limited, Peenya Industrial Area, Phase I, Bangalore, India

ABSTRACT

Enantiomerically enriched (R)-1-(2-bromocycloalkenyl)-3-buten-1-ol and its derivatives were obtained via enantioselective hydrolysis [resolution] with good enantioselectivities (E = 31 to >500) using Candida parapsilosis ATCC 7330. The various reaction parameters were optimized for enantioselective hydrolysis to achieve high enantiomeric excess (ee) and conversions. Among the substrates tested, (RS)-1-(2-bromocyclohex-1-en-1-yl) but-3-yn-1-yl acetate was hydrolysed by the biocatalyst in 12 h to the corresponding (R)-alcohol in 49% conversion and >99 ee. The optically pure allylic alcohol thus obtained was used as a chiral starting material for the synthesis of an enantiomerically enriched bicyclic alcohol effectively establishing a chemoenzymatic route.

INTRODUCTION

Enantiomerically enriched allylic and propargylic alcohols are very useful intermediates in the asymmetric synthesis of more complex molecules [1] [2] and are also resourceful synthons for the synthesis of various biologically active compounds [3] [4] . Several methods are available for the synthesis of such systems, e.g. enantioselective addition of chiral allylboranes [5] and allyltitanation [6] to aldehydes, asymmetric organozinc additions to aldehydes [7] [8] and the elimination of chiral vinyl sulfoxides [9] . The widely used methods for the preparation of these intermediates are enantioselective acylation of secondary allylic alcohols an enantioselective hydrolysis of the corresponding esters catalysed by enzymes and nonenzymatic chiral catalysts [1] [10] - [13] . In particular, enantiomerically pure 1-(2'-bromocycloalkenyl)-3-buten-1-ol and its derivatives can be cyclised and utilized for the synthesis of optically active condensed carbocycles (Scheme 1). These bicyclic allylic alcohols can be oxidized to bicyclicenones which are important chiral synthons for the development of biologically active compounds [14] [15] and is an important structural feature of many natural products [16] - [19] . Synthesis of various bicyclicenones by intramolecular cyclization has been described in literature [20] - [22] . Intramolecular tri butyl tin hydride mediated cyclization of optically active 1-(2'-bromocycloalkenyl)-3-buten-1-ols is a potential route for the synthesis of optically pure condensed carbocyles [23] - [25] . However, synthesis of optically pure carbocycles from the corresponding chiral intermediates i.e., enantiopure allylic alcohols has not been reported.

Various chemical [1] [22] and biocatalytic [26] - [29] methods are available for the synthesis of enantiomerically pure aryl/alkyl-allyl alcohols. Lipases, generally obtained from microorganisms, are the most widely used enzymes in organic synthesis, mainly due to their high enantioselectivity [30] - [32] , the possibility of using non- conventional solvents like organic solvents and ionic liquids [33] and are economical, stable, and easy to reprocess. Among the methods reported, only few specific examples are available for the lipase mediated resolution of aliphatic cyclic allylic alcohols. In this context, Mc Cubbin et al. resolved a series of aryl and aliphatic vinyl chlorohydrins by enzymatic kinetic resolution using Amano lipase AK. The resolution of alkyl and cyclic aliphatic vinyl chlorohydrins was a fast reaction but the enantioselectivity was lower as compared to that of aryl substrates [34] . Takabe et al. reported the one-pot acid promoted catalyzed cyclisation of geranyl amine N-oxide to synthesise 1-acetyl-4,4-dimethyl-1-cyclohexene which in turn was reduced and resolved by Amano lipase AK with good enantioselectivity [35] . Similarly Kuriata et al. synthesized chiral bicycle [3.1.0] hexane derivatives via chemoenzymatic synthesis in which the synthesized (+)-1-[(1S,5R)-6,6-dimethylbicyclo[3.1.0] hex-2-en-2- yl)]ethanol was resolved by a transesterification strategy using Amano PS lipase from B. cepacia [36] . However, the enantioselective hydrolysis of the corresponding acetate using various lipases from B. cepacia and C. rugosa was unsuccessful. The versatility of the biocatalyst Candida parapsilosis ATCC 7330 was established in our lab. This yeast is used for the preparation of a variety of enantiopure compounds by deracemization [36] - [39], asymmetric reduction [40] [41] and oxidative kinetic resolution [42].

This biocatalyst is also reported for resolution of N-protected amino acid esters with excellent enantioselectivity (E = 40 to >500) [43]. In continued efforts to expand the scope of the biocatalyst, herein we report the enantioselective hydrolysis of 1-(2-bromocycloalkenyl) but-3-en-1-yl acetates using fermenting cells of C. parapsilosis ATCC 7330. These cyclic acetates with different functionalities (bromo and allyl/propargyl), in addition to yielding important chiral synthons, provide an interesting set of substrates to study hydrolysis using C. parapsilosis ATCC 7330. To the best of our knowledge, this is the first report for the synthesis of enantiomerically enriched 1-(2-bromocycloalkenyl) but-3-en-1-ol and its derivatives by biocatalyst mediated enantioselective hydrolysis. The enantiopure enriched 1-(2-bromocycloalkenyl) but-3-en-1-ol thus obtained was further cyclized to 3-methyl-2,3,4,5,6,7-hexahydro-1-H-inden-1-ol which can be used for the synthesis of optically pure condensed carbocycles.

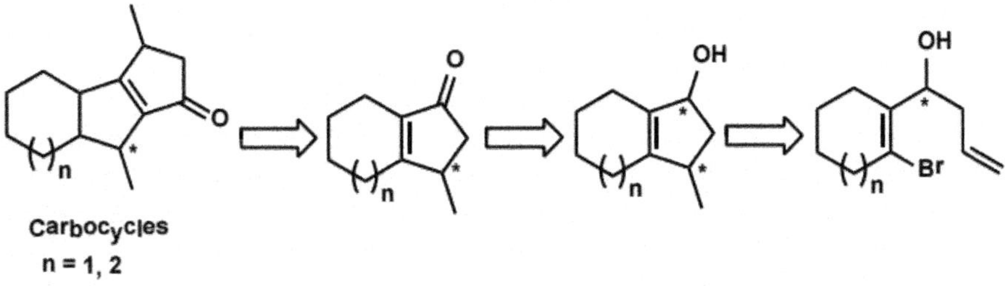

Carbocycles
n = 1, 2

Scheme 1: Retrosynthetic route for carbocycle from bicyclic allylic alcohol.

RESULT AND DISCUSSION

Initially, the model substrate (RS)-1-(2-bromocyclohex-1-en-1-yl) but-3-en-1-yl acetate 1a was subjected to enantioselective hydrolysis with commercially available lipases (Lipase PS, Amano lipase AK and CAL-B) under reported conditions [44]. All the lipases tested gave <5% conversion even after 3 days. Further, enantioselective hydrolysis of (RS)-1-(2-bromocyclohex-1-en-1-yl) but-3-en-1-yl acetate 1a was carried out with the resting cells (14 h culture) of Candida parapsilosis ATCC 7330 using our earlier reported conditions [43]. In this reaction, the substrate 1a (0.2 mmol) was incubated with the resting cells of C. parapsilosis ATCC 7330 suspended in water (20 mL; 6 g wet cell mass from 300 mL culture) using ethanol as a cosolvent (0.5% v/v). The product 2a was obtained with moderate conversion (26%) and only 20% ee as analysed by gas chromatography (GC) using a chiral column (Scheme 2). The poor enantioselectivity could be due to the presence of multiple hydrolases with opposite enantioselectivities for the same substrate. Next, the substrate 1a (0.08 mmol) was dissolved in cosolvent ethanol (0.5% v/v) and directly added in to a 12 h fermenting culture (50 mL) at 25°C and incubation was continued. After 48 h the product was obtained with an improvement in ee (89%) but the conversion was very poor (8%).

To further improve the conversion and enantiomeric excess, the reaction conditions were optimized with 1a as follows.

Optimization of Cosolvents

Reportedly, conversion and enantioselectivity in lipase catalysed hydrolysis can be improved by the addition of a suitable cosolvent [45] . First, the reaction was carried out in the absence of a cosolvent and the product was obtained with 82% ee and 19% conversion. When the reaction was carried out with ethanol, isopropyl ether or acetone as a cosolvent, the conversion dropped from 19 to 8%. Similar observations have been reported by Szymanski et al. for the enzymatic hydrolysis of α-acetoxyamides when acetone was used as a cosolvent [46] . Dioxane and DMSO showed moderate conversion (25% - 37%). All long chain hydrophopic cosolvents (n-octane, isooctane and hexane) gave good conversion (up to 56%). This result is consistent with the literature reports that hydrophobic solvents favor the binding of hydrophobic substrates to lipase [27] . Among the cosolvents tested in this study, only hexane showed reasonable conversion (40%) and good ee (90%) (Figure 1).

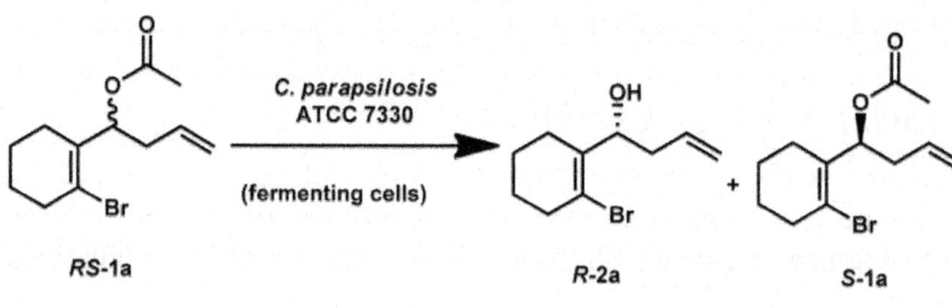

Scheme 2: Enantioselective hydrolysis of (RS)-1a.

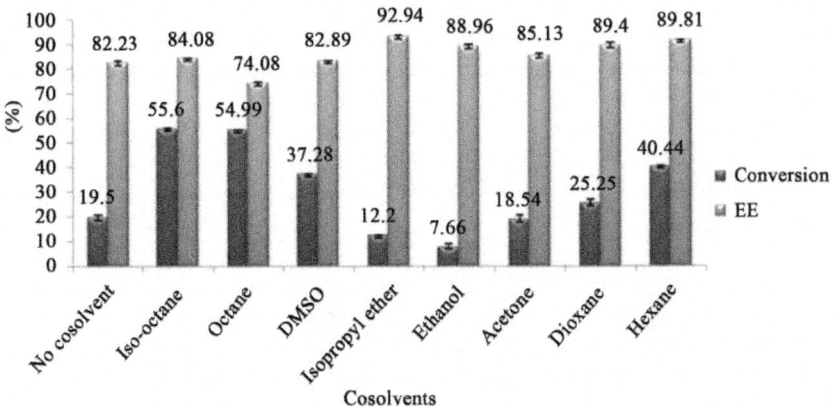

Figure 1: Optimization of cosolvents.

Optimization of Reaction Time

Further, the reaction time was varied from 12 to 60 h using hexane as cosolvent. It was found that 48 h showed improvement in both conversion (40%) and enantiomeric excess (89.6%) (Table 1). Further increase in the reaction time resulted in decrease in ee (82.2%). Hence 48 h was chosen as the optimum incubation time for further experiments.

Effect of Temperature

To study the effect of temperature on conversion and ee of enantioselective hydrolysis, the reaction was performed at various temperatures from 25˚C - 35˚C. The product 2a was obtained with good conversion (40%) and ee (89%) at 25˚C (Table 2). With increasing temperature (30˚C & 35˚C), the conversion also increased (66 & 68% respectively) but the enantiomeric excess reduced sharply (61 & 57% respectively) (Table 2). This Candida grows best at 25˚C [36], which supports the observation that maximum conversion is obtained at 25˚C. All further experiments were carried out at 25˚C.

Optimization of Substrate Concentration

In whole cell biocatalysis, conversion and enantioselectivity can be controlled by optimizing the substrate concentration due to the presence of multiple enzymes with opposite stereo preferences and different Km values for the same substrate [45]. The substrate 1a was added in varied concentrations (1 - 2.5 mM) keeping all the conditions identical (Figure 2). Notably, higher conversion

(53%) was observed with low substrate concentration (1 mM), while the enantiomeric excess was moderate (75%). At a substrate concentration of 1.5 mM, the enantiomeric excess increased to 89% with good conversion (39%).

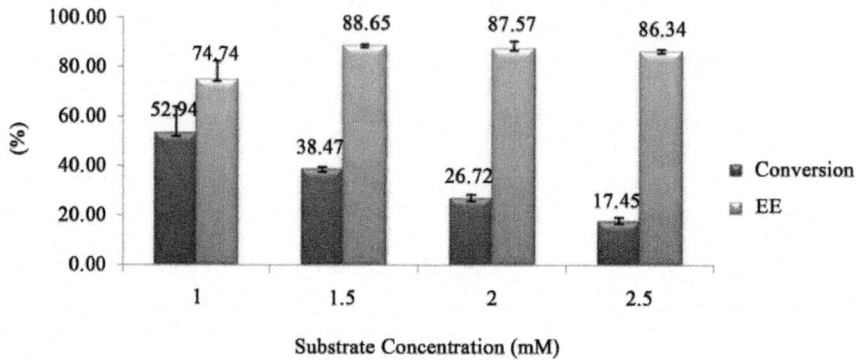

Figure 2: Optimization of substrate concentration (1a).

Table 1: Optimization of reaction time for improved conversion and enantiomeric excess.

S.No.	Time (h)	Conversion (%)	ee (%)
1	12	25	88.3
2	24	32	89.1
3	48	40	89.6
4	60	44	82.2

Table 2: Effect of temperature.

S.No.	Temperature (°C)	Conversion (%)	ee (%)
1	25	40	89.4
2	30	66	61.1
3	35	68	56.8

Further increase in the substrate concentration showed a drop in conversion however the ee was maintained. Hence, 1.5 mM of substrate was chosen as the optimum substrate concentration.

Substrate Scope

Under the optimized reaction conditions, three substrates were used (Scheme 3). The experiment with substrate 1a yielded product (R)-1-(2-bromocyclohex-

1-en-1-yl)but-3-en-1-ol 2a in 33% isolated yield and 89% ee (E = 31) (Table 3, entry 2a). To further explore the substrate scope of this biocatalyst, substrate 1b (Scheme 3) was subjected to hydrolysis using fermenting cells of C. parapsilosis ATCC 7330. Initially, under the optimized reaction time of 48 h, a moderate conversion (20%) was obtained with excellent ee (>99%). The conversion was further improved to 32% by extending the reaction time to 60 h and the product (R)-1-(2-bromocyclohept-1-en-1-yl) but-3-en-1-ol 2b was isolated with moderate yield (29%) and excellent ee (> 99%) (E => 200) (Table 3, entry 2b). The obtained high enantioselectivity suggests that the substrate with the 7-membered ring is seemingly better than the one with the 6-membered ring for the biocatalyst. Another substrate 1c (Scheme 3) with an alkyne chain was subjected for enantioselective hydrolysis and the product (R)-1-(2-bromocyclohex-1-en-1-yl) but-3- yn-1-ol 2c was obtained with excellent ee (>99%) in lesser reaction time (12h) compared to substrates 1a & 1b. Further increase in the reaction time resulted in decrease in ee. The product 2c was isolated in 42% yield and 99% ee (E => 500) (Table 3, entry 2c).

Scheme 3: Enantioselective hydrolysis of (RS)-1a-c.

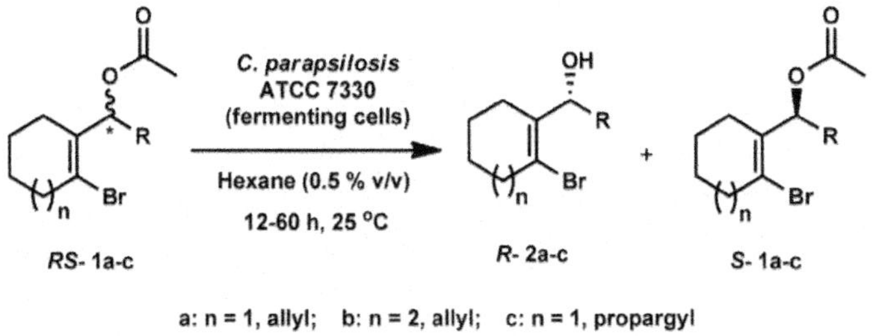

a: n = 1, allyl;   b: n = 2, allyl;   c: n = 1, propargyl

Table 3: Enantioselective hydrolysis of (RS)-1a-1c.

Entry	Product[a]	Reaction time (h)	Conversion (%)[b]	Isolated Yield (%)	ee of alcohol (R-2a-c)[b] (%)	ee of acetate (S-1a-c) (%)[c]	E value[d]
2a	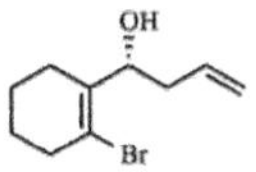	48	40	33	89	59	31
2b	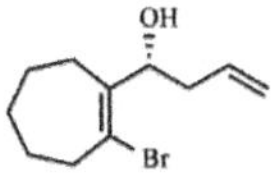	60	32	29	99	46	>200
2c	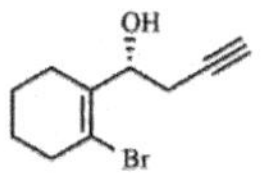	12	49	42	99	99	>500

[a]Absolute configuration was assigned by comparing specific rotation value with the analogue substrate [47] ; [b]Conversion and ee was determined by GC using Chirasil Dex CB chiral column;[c]Enantiomeric excess of the unreacted acetates were determined after hydrolysis using 5% methanolic KOH; [d]Enantioselectivity ratios were calculated according to: E = ln[(eep(1 − ees)/(eep+ees)]/ln[(eep(1 + ees)/(eep + ees) [48] ; E values in the range 200 - 500 are mentioned as >200, more than 500 as >500.

^{1}H NMR (500 MHz,CDCl$_3$) δ:5.86 (dd, J = 8,5.5 Hz, 1H), 5.74 - 5.82 (m, 1H), 5.10 - 5.15 (ddd, J = 17, 3.5, 1.5 Hz, 1H), 5.06 - 5.09 (ddd, J = 9.5, 1.5, 0.5 Hz, 1H), 2.51 - 2.55 (m, 2H), 2.44 - 2.50 (m, 1H), 2.36 - 2.41 (m, 1H), 2.08 - 2.17 (m, 2H), 2.06 (brs, 3H), 1.66 - 1.72 (m, 3H), 1.61 - 1.64 (m, 1H); ^{13}C NMR (125 MHz, CDCl$_3$) δ: 169.9, 133.7, 133.3, 121.3, 117.6, 75.7, 36.7, 36.6, 25.9, 24.5, 21.9, 20.9; HRMS for C$_{12}$H$_{17}$O$_2$Br (Cal: 295.0310 [M + Na]$^+$, Found: 295.0310); ee = 59%, retention times 15.0 (S, major), 15.5 min (R, minor); [α] D^{25} + 1.1 (c 1, CHCl$_3$).

^{1}H NMR (500 MHz, CDCl$_3$) δ: 5.74 - 5.79 (m, 2H), 5.06 - 5.11 (m, 1H), 5.03 - 5.04 (m, 1H), 2.72 - 2.79 (m, 2H), 2.37 - 2.43 (m, 1H), 2.27 - 2.35(m, 1H), 2.19 - 2.26 (dd, J = 6.5, 5 Hz, 2H), 2.02 (brs, 3H), 1.70 - 1.74 (m, 2H), 1.53 - 1.56 (m, 3H), 1.38 - 1.52 (m, 1H); ^{13}C NMR (125 MHz, CDCl$_3$) δ: 169.9, 138.7, 133.2, 125.0, 117.7, 76.7, 41.5, 36.7, 31.6, 27.9, 25.9, 25.1, 20.9; HRMS forC$_{13}$H$_{19}$O$_2$Br (Cal: 309.0466 [M + Na]$^+$, Found: 309.0466); ee = 46%, retention times 20.3 (S, major), 21.5 min (R, minor); [α] D^{25} + 1.7 (c 1, CHCl$_3$).

^{1}H NMR (500 MHz, CDCl3) δ: 5.87 - 5.89 (m, 1H), 2.56 - 2.66 (m, 2H), 2.52 - 2.54 (m, 2H), 2.18 - 2.28 (m, 1H), 2.08 (s, 3H), 1.97 - 1.98 (t, J = 2.5 Hz, 1H), 1.62 - 1.75 (m, 4H), 1.59 (brs, 1H);^{13}CNMR (125 MHz, CDCl$_3$) δ: 169.7, 133.1, 122.1, 79.4, 74.0, 70.2, 36.7, 26.0, 24.4, 22.4, 21.8, 20.8; HRMS for C$_{12}$H$_{15}$O$_2$Br (Cal: 293.0153 [M + Na]$^+$, Found: 293.0154); ee => 99%, retention times19.4 (R, minor), 19.8 min (S, major); [α] D^{25} + 3.2 (c 0.5, CHCl$_3$).

^{1}H NMR (500 MHz,CDCl$_3$) δ: 5.78 - 5.86 (m, 1H), 5.12 - 5.18 (m, 2H), 4.78 - 4.81 (dd = t, J = 6.5 Hz, 1H), 2.50 - 2.52 (m, 2H), 2.28 - 2.34 (m, 3H), 2.04 - 2.10 (m, 1H), 1.63 - 1.74 (m, 5H); ^{13}C NMR (125 MHz, CDCl$_3$) δ:136.9, 134.3, 120.3, 118.0, 73.2, 38.9, 36.8, 25.4, 24.7, 22.1; ee = 89%; retention times 15.0 (S, minor), 15.5 min (R, major); [α]D^{25} + 25.4 (c 0.5, CHCl$_3$).

Lipase PS (from Burkholderia cepacia) and Amano lipase AK (from Pseudomonas fluorescence) were obtained from Amano Pharmaceuticals Co., Nagoya, Japan. CAL B (Candida antarctica lipase B, Novozym 435) was purchased from Sigma. Candida parapsilosis ATCC 7330 was obtained from ATCC (Manassas, VA) and maintained at 4°C in yeast malt agar medium that contained 5 g/L peptic digest of animal tissue, 3 g/L malt extract, 3 g/L yeast extract, 10 g/L dextrose and 20 g/L agar. All chemicals used were of analytical grade and distilled prior to use. TLC was carried out on Kieselgel 60 F254 aluminium sheets (Merck1.05554). ^1H and ^{13}C NMR spectra were recorded in $CDCl_3$ solution on a Bruker AVANCE III 500 MHz spectrometer. Chemical shifts are expressed in ppm values using TMS as an internal standard. The enantiomeric excess was determined by gas chromatography using Perkin Elmer Clarus 600 gas chromatograph and Shimadzu GC 2014 gas chromatograph fitted with flame ionisation detector using VARIAN Chirasil Dex CB chiral column (0.25 μm × 25 mm × 30 m). The injector and detector were set at 220°C, respectively. Nitrogen was used as the carrier gas. Optical rotations were determined using Autopol-IV automatic polarimeter.

General Procedure for the Synthesis of (RS)-1a-1c [23]

Tin powder (1.5 mmol) was suspended in a mixture of water (2 mL) and diethyl ether (2mL). To this mixture allyl bromide (3 mmol) was added and treated with few drops of 48% hydrobromic acid and the mixture was stirred well for 5 min at RT. To this reaction mixture bromoaldehyde (1 mmol) was added and the resultant mixture was stirred for 6 h at RT (reaction was monitored by TLC). After the complete conversion, the reaction was quenched with water (5 ml) and extracted with 20% dichloromethane/hexane mixture (10 mL × 3). The organic layers were combined and dried over anhydrous sodium sulfate and concentrated under reduced pressure. The crude reaction mixture was purified by silica gel column chromatography using hexane/ethyl acetate (9:1) as a mobile phase (Yield: 60% - 80%). The respective bromo alcohol (1 mmol) was stirred with (2 mL) of acetic anhydride containing catalytic amount of sodium acetate for 24 h at room temperature.

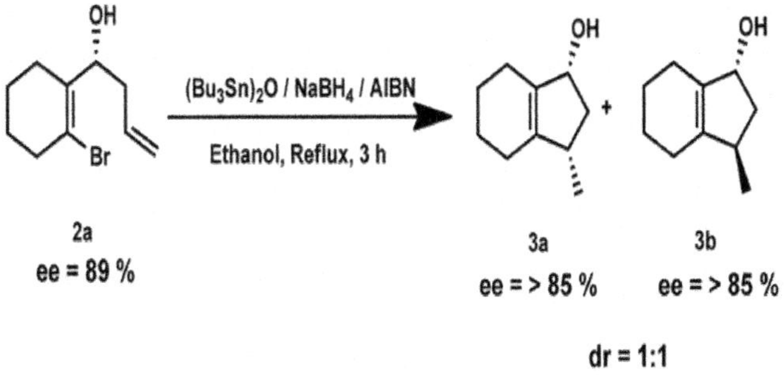

Scheme 4. Radical cyclization of 2a.

The progress of the reaction was monitored by TLC. After the completion of reaction, the reaction mixture was diluted with dichloromethane and washed with water. The organic layer was dried over sodiumsulphate and concentrated under reduced pressure. The crude reaction mixture was purified by column chromatography over silica gel (100 - 200 mesh) using hexane/ethyl acetate mixture (9:1) as a mobile phase to obtain the pure product (Yield: 70% - 80%).

Growth Conditions for Candida parapsilosis ATCC 7330

Candida parapsilosis ATCC 7330 was pre-cultured for 12 h at 25°C with shaking at 200rpm in yeast malt broth medium that contained 5 g/L peptic digest of animal tissue, 3 g/L malt extract, 3 g/L yeast extract and 10 g/L dextrose. The pre-cultured broth, 4 mL [4% (v/v)] was transferred to a 500 mL Erlenmeyer flask that contained 96 mL of yeast malt broth. The culture was grown on rotatory shaker at 25°C and 200 rpm for 12 h.

Typical Procedure for Enantioselective Hydrolysis Using Candida parapsilosis ATCC 7330

To the 500 mL Erlenmeyer flask that contained 12 h cultivated cells (100 mL culture), the substrate (racemic acetate, 1.5 mM) in 500 μL of hexane as co-solvent was added and the incubation was continued at 25°C and 200 rpm. After getting the maximum conversion and enantiomeric excess of the product, the cells were centrifuged at 6000 rpm for 10 min. The formed product and the remaining substrate were extracted from the supernatant using ethylacetate. The organic layer was dried over anhydrous sodium sulphate. The solvent was removed by evaporation and the crude extract was purified by silica gel column chromatography using hexane/ ethyl acetate (95:05) as a mobile phase eluent.

General Procedure for the Synthesis of 3a and 3b [23]

The respective enantiomerically enriched (R)-1-(2-bromocyclohex-1-en-1-yl) but-3-en-1-ol 2a (1 mmol) was dis- solved in ethanol (25 mL). To this bis-tri-n-butyltinoxide (5 mol%), sodium borohydride (1 mol) and catalytic amount of azobisisobutyro nitrile (AIBN) were added. The reaction mixture was refluxed for 3 h at 80°C under N_2 atmosphere and the reaction was monitored by TLC. After the completion of reaction, the reaction mixture was concentrated under reduced pressure. The products were purified by column chromatography over silica gel (100 - 200 mesh) using hexane/ethyl acetate mixture (9:1) as a mobile phase to obtain the cyclized product 3a and 3b (Yield: 70% - 80%).

Spectroscopic Characterization

(S)-1-(2-bromocyclohex-1-enyl) but-3-enyl acetate 1a

^{1}H NMR (500 MHz, CDCl$_3$) δ:5.86 (dd, J = 8,5.5 Hz, 1H), 5.74 - 5.82 (m, 1H),

5.10 - 5.15 (ddd, J = 17, 3.5, 1.5 Hz, 1H), 5.06 - 5.09 (ddd, J = 9.5, 1.5, 0.5 Hz, 1H), 2.51 - 2.55 (m, 2H), 2.44 - 2.50 (m, 1H), 2.36 - 2.41 (m, 1H), 2.08 - 2.17 (m, 2H), 2.06 (brs, 3H), 1.66 - 1.72 (m, 3H), 1.61 - 1.64 (m, 1H); ^{13}C NMR (125 MHz, CDCl$_3$) δ: 169.9, 133.7, 133.3, 121.3, 117.6, 75.7, 36.7, 36.6, 25.9, 24.5, 21.9, 20.9; HRMS for C$_{12}$H$_{17}$O$_2$Br (Cal: 295.0310 [M + Na]$^+$, Found: 295.0310); ee = 59%, retention times 15.0 (S, major), 15.5 min (R, minor); [α] D^{25} + 1.1 (c 1, CHCl$_3$).

GC conditions: Perkin Elmer Clarus 600 gas chromatograph. The initial column temperature of 100°C was held for 1 min, then raised to150°C at a rate of 8°C/min and held at 150°C for 2 min. Further the temperature raised to 180°C at a rate of 1°C/min and finally held at 170°C for 3 min.

(S)-1-(2-bromocyclohept-1-enyl) but-3-enyl acetate 1b

^{1}H NMR (500 MHz, CDCl$_3$) δ: 5.74 - 5.79 (m, 2H), 5.06 - 5.11 (m, 1H), 5.03 - 5.04 (m, 1H), 2.72 - 2.79 (m, 2H), 2.37 - 2.43 (m, 1H), 2.27 - 2.35(m, 1H), 2.19 - 2.26 (dd, J = 6.5, 5 Hz, 2H), 2.02 (brs, 3H), 1.70 - 1.74 (m, 2H), 1.53 - 1.56 (m, 3H), 1.38 - 1.52 (m, 1H); ^{13}C NMR (125 MHz, CDCl$_3$) δ: 169.9, 138.7, 133.2, 125.0, 117.7, 76.7, 41.5, 36.7, 31.6, 27.9, 25.9, 25.1, 20.9; HRMS forC$_{13}$H$_{19}$O$_2$Br (Cal: 309.0466 [M + Na]$^+$, Found: 309.0466); ee = 46%, retention times 20.3 (S, major), 21.5 min (R, minor); [α] D^{25} + 1.7 (c 1, CHCl$_3$).

GC conditions: Shimadzu GC 2014 gas chromatograph. The initial column temperature of 120°C was held for 1 min, then raised to 150°C at a rate of 8°C/min and held at 150°C for 2 min. Further the temperature raised to 170°C at a rate of 1°C/min and finally held at 170°C for 4 min.

(S)-1-(2-bromocyclohex-1-enyl) but-3-ynyl acetate 1c

^{1}H NMR (500 MHz, CDCl3) δ: 5.87 - 5.89 (m, 1H), 2.56 - 2.66 (m, 2H), 2.52 - 2.54 (m, 2H), 2.18 - 2.28 (m, 1H), 2.08 (s, 3H), 1.97 - 1.98 (t, J = 2.5 Hz, 1H), 1.62 - 1.75 (m, 4H), 1.59 (brs, 1H);^{13}CNMR (125 MHz, CDCl$_3$) δ: 169.7, 133.1, 122.1, 79.4, 74.0, 70.2, 36.7, 26.0, 24.4, 22.4, 21.8, 20.8; HRMS for C$_{12}$H$_{15}$O$_2$Br (Cal: 293.0153 [M + Na]$^+$, Found: 293.0154); ee => 99%, retention times19.4 (R, minor), 19.8 min (S, major); [α] D^{25} + 3.2 (c 0.5, CHCl$_3$).

GC conditions: Shimadzu GC 2014 gas chromatograph. The initial column temperature of 120°C was held for 1 min, then raised to 150°C at a rate of 8°C/min and held at 150°C for 2 min. Further the temperature raised to 170°C at a rate of 1°C/min and finally held at 170°C for 4 min.

(R)-1-(2-bromocyclohex-1-enyl) but-3-en-1-ol 2a [20]

^{1}H NMR (500 MHz,CDCl$_3$) δ: 5.78 - 5.86 (m, 1H), 5.12 - 5.18 (m, 2H), 4.78

- 4.81 (dd = t, J = 6.5 Hz, 1H), 2.50 - 2.52 (m, 2H), 2.28 - 2.34 (m, 3H), 2.04 - 2.10 (m, 1H), 1.63 - 1.74 (m, 5H); ^{13}C NMR (125 MHz, CDCl$_3$) δ:136.9, 134.3, 120.3, 118.0, 73.2, 38.9, 36.8, 25.4, 24.7, 22.1; ee = 89%; retention times 15.0 (S, minor), 15.5 min (R, major); [α]D^{25} + 25.4 (c 0.5, CHCl$_3$).

GC conditions: Perkin Elmer Clarus 600 gas chromatograph. The initial column temperature of 100°C was held for 1 min, then raised to 150°C at a rate of 8°C/min and held at 150°C for 2 min. Further the temperature raised to 170°C at a rate of 1°C/min and finally held at 170°C for 3 min. Split Ratio = 1:10. Flow rate = 1.5 ml/min.

(R)-1-(2-bromocyclohept-1-enyl) but-3-en-1-ol 2b [20]

^{1}H NMR (500 MHz, CDCl$_3$) δ: 5.79 - 5.88 (m, 1H), 5.13 - 5.19 (m, 2H), 4.78 - 4.81 (dd = t, J = 7 Hz, 1H), 2.74 - 2.84 (ddd, J = 11.5, 7, 3.5 Hz, 2H), 2.28 - 2.34 (m, 4H), 1.75 - 1.80 (m, 3H), 1.55 - 1.62 (m, 2H), 1.50 - 1.55 (m, 2H); ^{13}C NMR (125 MHz, CDCl$_3$) δ:142.2, 134.3, 123.4, 117.9, 74.7, 41.6, 39.0, 31.6, 27.3, 26.3, 25.3; ee => 99%; retention times 20.3 (S, minor), 21.5 min (R, major); [α] D^{25} + 15.8 (c 0.6, CHCl$_3$).

GC conditions: Shimadzu GC 2014 gas chromatograph. The initial column temperature of 120°C was held for 1 min, then raised to 150°C at a rate of 8°C/min and held at 150°C for 2 min. Further the temperature raised to 170°C at a rate of 1°C/min and finally held at 170°C for 4 min.

(R)-1-(2-bromocyclohex-1-enyl) but-3-yn-1-ol 2c [21]

^{1}H NMR (500MHz,CDCl$_3$) δ: 4.89 - 4.91 (dd = t, J = 6.5 Hz 1H), 2.49 - 2.54 (m, 2H), 2.47 - 2.49 (m, 2H), 2.46 - 2.47 (m, 1H), 2.05 - 2.06 (dd = t, J = 3 Hz, 1H), 1.63 - 1.74 (m, 5H), 1.24 (s, 1H); ^{13}C NMR (125 MHz, CDCl$_3$) δ: 136.1, 121.0, 80.5, 72.3, 70.7, 36.8, 25.3, 24.7, 24.6, 22.0; ee = > 99%, retention times 19.4 (R, major), 19.8 min (S, minor); [α] D^{25} + 12.2 (c 0.5, CHCl$_3$).

GC conditions: Shimadzu GC 2014 gas chromatograph. The initial column temperature of120°C was held for 1 min, then raised to150°C at a rate of 8°C/min and held at 150°C for 2 min. Further the temperature raised to 170°C at a rate of 1°C/min and finally held at 170°C for 4 min.

(1R, 3R)-2,3,4,5,6,7-hexahydro-3-methyl-1-H-inden-1-ol 3a

^{1}H NMR (400 MHz, C$_6$D$_6$) δ:4.46 - 4.47 (d, J = 6.4 Hz 1H); 2.56 (brs, 1H); 2.15 - 2.19 (m, 1H); 1.85 - 1.90 (m, 3H); 1.50 - 1.66 (m, 7H); 0.86 - 0.88 (d, J = 6.8 Hz 3H); ^{13}C NMR (100 MHz, CDCl$_3$) δ: 143.9, 135.6, 78.6, 42.6, 39.6, 23.5, 22.8, 22.7, 22.7, 19.5; HRMS for C$_{10}$H$_{16}$O (Cal: 151.1123 [M]$^+$, Found: 151.1121); (Diasteromeric ratio 1:1) ee = > 88%; retention times 8.52 min (R, major), 8.60 min (S, minor); [α] D25 + 8.3 (c 0.5, CHCl$_3$).

GC conditions: Shimadzu GC 2014 gas chromatograph. The initial column temperature of 120°C was held for 1 min, then raised to 150°C at a rate of 8°C/min and held at 150°C for 2 min. Further the temperature raised to 170°C at a rate of 1°C/min and finally held at 170°C for 4 min.

CONCLUSION

Enantiomerically enriched 1-bromohexa-1,5-dien-3-ols 2a-2c were synthesized with good ee (up to >99%) by enantioselective hydrolysis from the corresponding racemic acetates using fermenting cells of C. parapsilosis and are reported here for the first time. The reaction parameters like cosolvent (hexane), incubation time (12 - 60 h), temperature (25°C) and substrate concentration (1.5 mM) were optimized to obtain maximum conversion and enantiomeric excess. Notably, the product 1-(2-bromocyclohex-1-en-1-yl) but-3-yn-1-ol 2c was obtained with good yield and excellent ee in less reaction time than the other two substrates (E => 500). (R)-1-(2-bromocyclo- hex-1-en-1-yl)but-3-en-1-ol was used as chiral starting material for the first time to prepare enantiomerically enriched (1R, 3R)-3-methyl-2,3,4,5,6,7-hexahydro-1-H-inden-1-ol and (1R, 3S) 3-methyl 2,3,4,5,6,7-hexahydro- 1-H-inden-1-ol with good ee (>85%) via a chemoenzymatic approach.

ACKNOWLEDGEMENTS

One of the authors [TS] thanks UGC, New Delhi for Senior Research Fellowship, SAIF, IIT Madras for instrumentation facilities and Mr. T. Rathnavel Pandian for assistance in optimization study. TKD and NP thank ITC Life Sciences Technology Centre, Bangalore for providing the necessary infrastructure to carry out the experiments. Professor K. K. Balasubramanian, INSA Senior Scientist, Department of Biotechnology, IIT Madras is thanked for reviewing the paper and providing valuable input. Dr. Sowmyalakshmi Venkataraman, IIT Madras for helping with the analysis.

DISCLAIMER

All views expressed herein are authors' views and in no way, expressed or implied, are that of or necessarily represent the positions of ITC Limited, my current employer.

CITE THIS PAPER

ThangaveluSaravanan,AnjuChadha,Tarur Konikkaledom Dinesh,Namasivaya mPalani,SengottuvelanBalasubramanian, (2015) Chemoenzymatic Synthesis of an Enantiomerically Enriched Bicyclic Carbocycle Using *Candida parapsilosis* ATCC 7330 Mediated Enantioselective Hydrolysis. *International Journal of Organic Chemistry,***05**,271-281. doi:10.4236/ijoc.2015.54027

REFERENCES

1.   Bellemin-Laponnaz, S., Tweddell, J., Ruble, J.C., Breitling, F.M. and Fu, G.C. (2000) The Kinetic Resolution of Allylic Alcohols by a Non-Enzymatic Acylation Catalyst; Application to Natural Product Synthesis. Chemical Communications, 1009-1010. http://dx.doi.org/10.1039/b002041i

2.   Hoveyda, A.H., Evans, D.A. and Fu, G.C. (1993) Substrate-Directable Chemical Reactions. Chemical Reviews (Washington DC), 93, 1307-1370. http://dx.doi.org/10.1021/cr00020a002

3.   Fontana, A., d'Ippolito, G., D'Souza, L., Mollo, E. and Parameswaram, P.S. (2001) New Acetogenin Peroxides from the Indian Sponge Acarnus bicladotylota. Journal of Natural Products, 64, 131-133. http://dx.doi.org/10.1021/np0002435

4.   Roush, W.R. and Sciotti, R.J. (1998) Enantioselective Total Synthesis of (-)-Chlorothricolide via the Tandem Inter- and Intramolecular Diels-Alder Reaction of a Hexaenoate Intermediate. Journal of the American Chemical Society, 120, 7411-7419. http://dx.doi.org/10.1021/ja980611f

5.   Marino, J.P., McClure, M.S., Holub, D.P., Comasseto, J.V and Tucci, F.C. (2002) Stereocontrolled Synthesis of (-)- Macrolactin A. Journal of the American Chemical Society, 124, 1664-1668. http://dx.doi.org/10.1021/ja017177t

6.   BouzBouz, S., Pradaux, F., Cossy, J., Ferroud, C. and Falguieres, A. (2000) Enantioselective Synthesis of Propargylic Alcohols by Addition of Enantiopure Cyclopentadienyldialkoxyallyltitanium Complexes to Acetylenic Aldehydes. Tetrahedron Letters, 41, 8877-8880. http://dx.doi.org/10.1016/S0040-4039(00)01597-5

7.   Frantz, D.E., Faessler, R. and Carreira, E.M. (2000) Facile Enantioselective Synthesis of Propargylic Alcohols by Direct Addition of Terminal Alkynes to Aldehydes. Journal of the American Chemical Society, 122, 1806-1807. http://dx.doi.org/10.1021/ja993838z

8.   Pu, L. and Yu, H.-B. (2001) Catalytic Asymmetric Organozinc Additions to Carbonyl Compounds. Chemical Reviews (Washington DC), 101, 757-824. http://dx.doi.org/10.1021/cr000411y

9.   Nakamura, S., Kusuda, S., Kawamura, K. and Toru, T. (2002) Preparation

of Optically Pure Propargylic and Allylic Alcohols from 2-(Trimethylsilyl) vinyl Sulfoxides as a Chiral Ethynyl Anion Synthon: Computational Studies on Elimination Reaction of 2-(Trimethylsilyl)vinyl Sulfoxides. Journal of Organic Chemistry, 67, 640-647. http://dx.doi.org/10.1021/jo0157223

10. Birman, V.B. and Jiang, H. (2005) Kinetic Resolution of Alcohols Using a 1,2-Dihydroimidazo[1,2-a]quinoline Enantioselective Acylation Catalyst. Organic Letters, 7, 3445-3447. http://dx.doi.org/10.1021/ol051063v

11. Rotticci, D., Norin, T. and Hult, K. (2000) Mass Transport Limitations Reduce the Effective Stereospecificity in Enzyme-Catalyzed Kinetic Resolution. Organic Letters, 2, 1373-1376. http://dx.doi.org/10.1021/ol005639m

12. Vedejs, E. and Daugulis, O. (1999) 2-Aryl-4,4,8-trimethyl-2-phosphabicyclo[3.3.0]octanes: Reactive Chiral Phosphine Catalysts for Enantioselective Acylation. Journal of the American Chemical Society, 121, 5813-5814. http://dx.doi.org/10.1021/ja990133o

13. Onaran, M.B. and Seto, C.T. (2003) Using a Lipase as a High-Throughput Screening Method for Measuring the Enantiomeric Excess of Allylic Acetates. Journal of Organic Chemistry, 68, 8136-8141. http://dx.doi.org/10.1021/jo035067u

14. Therasse, P., Arbuck, S.G., Eisenhauer, E.A., Wanders, J. and Kaplan, R.S. (2000) New Guidelines to Evaluate the Response to Treatment in Solid Tumors. European Organization for Research and Treatment of Cancer, National Cancer Institute of the United States, National Cancer Institute of Canada. Journal of the National Cancer Institute, 92, 205-216. http://dx.doi.org/10.1093/jnci/92.3.205

15. Trost, B.M. and Pinkerton, A.B. (2000) A New Strategy for Cyclopentenone Synthesis. Organic Letters, 2, 1601-1603. http://dx.doi.org/10.1021/ol005853a

16. Herrmann, J.L., Richman, J.E. and Schlessinger, R.H. (1973) Novel Linch-Pin Construction of Dihydrojasmone. High Yield Synthesis of Cis-Jasmone. Tetrahedron Letters, 14, 3275-3278. http://dx.doi.org/10.1016/S0040-4039(01)86893-3

17. Mathew, J. and Alink, B. (1990) A Novel Route to Substituted Cyclopent-2-en-1-one; Application to the Synthesis of Cis-Jasmone and Dihydrojasmone. Journal of the Chemical Society, Chemical Communications, No. 9, 684-686. http://dx.doi.org/10.1039/c39900000684

18. Mikolajczyk, M. and Balczewski, P. (1987) Methylenomycin B: A New Synthesis from a β-Keto Phosphonate. Synthesis, 1987, 659-661. http://dx.doi.org/10.1055/s-1987-28041

19. Smith III, A.B., and Boschelli, D. (1983) Stereocontrolled Total Synthesis of (±)-Xanthocidin, Two Diastereomers, (±)-Epixanthocidin and (±)-β-Isoxanthocidin, and (±)-Dedihydroxy-4,5-didehydroxanthocidin, a Likely Biosynthetic Precursor. The Journal of Organic Chemistry, 48, 1217-1226. http://dx.doi.org/10.1021/jo00156a015

20. Ray, D., Mal, S.K. and Ray, J.K. (2005) Palladium-Catalyzed Novel Cycloisomerization: An Unprecedented Domino Oxidative Cyclization towards Substituted Carbocycles. Synlett, 2005, 2135-2140. http://dx.doi.org/10.1055/s-2005-872241

21. Ray, D., Nasima, Y., Sajal, M.K., Ray, P. and Urinda, S. (2013) Palladium-Catalyzed Intramolecular Oxidative Heck Cyclization and Its Application toward a Synthesis of (±)-β-Cuparenone Derivatives Supported by Computational Studies. Synthesis, 45, 1261-1269. http://dx.doi.org/10.1055/s-0032-1316884

22. Vedejs, E. and MacKay, J.A. (2001) Kinetic Resolution of Allylic Alcohols Using a Chiral Phosphine Catalyst. Organic Letters, 3, 535-536. http://dx.doi.org/10.1021/ol006923g

23. Dinesh, T.K., Palani, N. and Balasubramanian, S. (2015) Intramolecular Radical Cyclization of Vinyl, Aryl and Alkyl Halides Using Catalytic Amount of Bis-tri-n-butyltin Oxide/Sodium Borohydride: A Novel Entry to Functionalized Carbocycles. Synlett, 26, 1055-1058. http://dx.doi.org/10.1055/s-0034-1380325

24. Hart, D.J. (1984) Free-Radical Carbon-Carbon Bond Formation in Organic Synthesis. Science, 223, 883-887. http://dx.doi.org/10.1126/science.223.4639.883

25. Stork, G. and Mook Jr., R., (1983) Vinyl Radical Cyclization. 2. Dicyclization via Selective Formation of Unsaturated Vinyl Radicals by Intramolecular Addition to Triple Bonds. Applications to the Synthesis of Butenolides and Furans. Journal of the American Chemical Society, 105, 3720-3722. http://dx.doi.org/10.1021/ja00349a067

26. Lee, D., Huh, E.A., Kim, M.-J., Jung, H.M. and Koh, J.H. (2000) Dynamic Kinetic Resolution of Allylic Alcohols Mediated by Ruthenium- and Lipase-Based Catalysts. Organic Letters, 2, 2377-2379. http://dx.doi.org/10.1021/ol006159y

27. Kadnikova, E.N. and Thakor, V.A. (2008) Enantioselective Hydrolysis of 1-Arylallyl Acetates Catalyzed by Candida antarctica Lipase.

Tetrahedron: Asymmetry, 19, 1053-1058. http://dx.doi.org/10.1016/j. tetasy.2008.04.018

28. Marques, FA., Oliveira, M.A., Frensch, G., Sales Maia, B.H.L.N. and Barison, A. (2011) Highly Efficient Kinetic Resolution of Allylic Alcohols with Terminal Double Bond. Letters in Organic Chemistry, 8, 696-700. http://dx.doi.org/10.2174/157017811799304151

29. Chen, P. and Xiang, P. (2011) Kinetic Resolution of Allylic Alcohols via Stereoselective Acylation Catalyzed by Lipase PS-30. Tetrahedron Letters, 52, 5758-5760. http://dx.doi.org/10.1016/j.tetlet.2011.08.093

30. Lau, R.M., van Rantwijk, F., Seddon, K.R. and Sheldon, R.A. (2000) Lipase-Catalyzed Reactions in Ionic Liquids. Organic Letters, 2, 4189-4191. http://dx.doi.org/10.1021/ol006732d

31. Reetz, M.T. (2002) Lipases as Practical Biocatalysts. Current Opinion in Chemical Biology, 6, 145-150. http://dx.doi.org/10.1016/S1367-5931(02)00297-1

32. Palani, N., Chadha, A. and Balasubramanian, K.K. (1998) Mechanism of Lithium Perchlorate/Diethyl Ether-Catalyzed Rearrangement of α- and β-Endo- and -Exo-Dicyclopentadienyl Vinyl Ethers: Use of Deuterium Labeling and a Chiral Probe. The Journal of Organic Chemistry, 63, 5318-5323. http://dx.doi.org/10.1021/jo970715t

33. Vidya, P. and Chadha, A (2010) Pseudomonas cepacia Lipase Catalyzed Esterification and Transesterification of 3-(Furan-2-yl) Propanoic Acid/ Ethyl Ester: A Comparison in Ionic Liquids vs Hexane. Journal of Molecular Catalysis B: Enzymatic, 65, 68-72. http://dx.doi.org/10.1016/j. molcatb.2010.01.032

34. . McCubbin, J.A., Maddess, M.L. and Lautens, M. (2008) Enzymatic Resolution of Chlorohydrins for the Synthesis of Enantiomerically Enriched 2-Vinyloxiranes. Synlett, 2008, 289-293.

35. Takabe, K., Yamada, T., Miyamoto, T. and Mase, N. (2008) Cyclization of N,N-Diethylgeranylamine N-Oxide in One-Pot Operation: Preparation of Cyclic Terpenoid-Aroma Chemicals. Tetrahedron Letters, 49, 6016-6018. http://dx.doi.org/10.1016/j.tetlet.2008.08.001

36. Chadha, A. and Baskar, B. (2002) Biocatalytic Deracemization of α-Hydroxy Esters: High Yield Preparation of (S)-Ethyl 2-Hydroxy-4-phenylbutanoate from the Racemate. Tetrahedron: Asymmetry, 13, 1461-1464. http://dx.doi.org/10.1016/S0957-4166(02)00403-2

37. Baskar, B., Pandian, N.G., Priya, K. and Chadha, A. (2005) Deracemization of Aryl Substituted α-Hydroxy Esters Using Candida parapsilosis ATCC 7330: Effect of Substrate

Structure and Mechanism. Tetrahedron, 61, 12296-12306. http://dx.doi.org/10.1016/j.tet.2005.09.104

38.   Saravanan, T., Jana, S. and Chadha, A. (2014) Utilization of Whole Cell Mediated Deracemization in a Chemoenzymatic Synthesis of Enantiomerically Enriched Polycyclic Chromeno[4,3-b] Pyrrolidines. Organic & Biomolecular Chemistry, 12, 4682-4690. http://dx.doi.org/10.1039/c4ob00615a

39.   Titu, D. and Chadha, A. (2008) Enantiomerically Pure Allylic Alcohols: Preparation by Candida parapsilosis ATCC 7330 Mediated Deracemisation. Tetrahedron: Asymmetry, 19, 1698-1701. http://dx.doi.org/10.1016/j.tetasy.2008.07.012

40.   Mahajabeen, P. and Chadha, A. (2011) One-Pot Synthesis of Enantiomerically Pure 1,2-Diols: Asymmetric Reduction of Aromatic α-Oxo Aldehydes Catalyzed by Candida parapsilosis ATCC 7330. Tetrahedron: Asymmetry, 22, 2156- 2160. http://dx.doi.org/10.1016/j.tetasy.2011.12.008

41.   Venkataraman, S. and Chadha, A. (2015) Preparation of Enantiomerically Enriched (S)-ethyl 3-Hydroxy 4,4,4-Trifluorobutanoate Using Whole Cells of Candida parapsilosis ATCC 7330. Journal of Fluorine Chemistry, 169, 66-71. http://dx.doi.org/10.1016/j.jfluchem.2014.11.004

42.   Sivakumari, T. and Chadha, A. (2014) Regio- and Enantio-Selective Oxidation of Diols by Candida parapsilosis ATCC 7330. RSC Advances, 4, 60526-60533. http://dx.doi.org/10.1039/C4RA08146C

43.   Stella, S. and Chadha, A. (2010) Resolution of N-Protected Amino Acid Esters Using Whole Cells of Candida parapsilosis ATCC 7330. Tetrahedron: Asymmetry, 21, 457-460. http://dx.doi.org/10.1016/j.tetasy.2010.02.011

44.   Waldinger, C., Schneider, M., Botta, M., Corelli, F. and Summa, V. (1996) Aryl Propargylic Alcohols of High Enantiomeric Purity via Lipase Catalyzed Resolutions. Tetrahedron: Asymmetry, 7, 1485-1488. http://dx.doi.org/10.1016/0957-4166(96)00166-8

45.   Kinoshita, M. and Ohno, A. (1996) Factors Influencing Enantioselectivity of Lipase-Catalyzed Hydrolysis. Tetrahedron, 52, 5397-5406. http://dx.doi.org/10.1016/0040-4020(96)00179-2

46.   Szymanski, W. and Ostaszewski, R. (2007) Chemoenzymatic Synthesis of Enantiomerically Enriched α-Hydroxyamides. Journal of Molecular Catalysis B: Enzymatic, 47, 125-128. http://dx.doi.org/10.1016/j.molcatb.2007.04.007

47. Malkov, A.V., Barlog, M., Jewkes, Y., Mikusek, J. and Kocovsky, P. (2011) Enantioselective Allylation of α,β-Unsaturated Aldehydes with Allyltrichlorosilane Catalyzed by METHOX. The Journal of Organic Chemistry, 76, 4800-4804. http://dx.doi.org/10.1021/jo200712p

48. Rakels, J.L.L., Straathof, A.J.J. and Heijnen, J.J. (1993) A Simple Method to Determine the Enantiomeric Ratio in Enantioselective Biocatalysis. Enzyme and Microbial Technology, 15, 1051-1056. http://dx.doi. org/10.1016/0141-0229(93)90053-5

Chapter 3

GREEN AND HIGH EFFICIENT SYNTHESIS OF 2-ARYL BENZIMIDAZOLES: REACTION OF ARYLIDENE MALONONITRILE AND 1,2-PHENYLENEDIAMINE DERIVATIVES IN WATER OR SOLVENT-FREE CONDITIONS

Azizollah Habibi[1], Yousef Valizadeh[1], Marjan Mollazadeh[2], Abdolali Alizadeh[3]

[1]Faculty of Chemistry, Kharazmi University, Tehran, Iran

[2]School of Chemistry, College of Science, University of Tehran, Tehran, Iran

[3]Department of Chemistry, Tarbiat Modares University, Tehran, Iran

ABSTRACT

A fast, high efficiency and environmentally friendly procedure for the synthesis of 2-aryl benzimidazole derivatives has been reported. Reaction between 1,2-phenylenediamine derivatives and arylidene malononitrile under aqueous media and also solvent-free conditions generates 2-aryl benzimidazole derivatives with a high yield.

INTRODUCTION

In recent years, significant attentions have been considered to the organic reaction under aqueous media, particularly from the viewpoint of green chemistry [1] - [4] . Using water, in contrast to common hazardous organic solvents, offers many advantages such as: simplicity of reaction conditions, ease of work-up and product isolation, increasing the selectivity of a wide variety of organic reactions and accelerating reaction rates [5] [6] . Benzimidazole derivatives are important scaffolds in medicinal chemistry due to their biological and pharmacological activities. These compounds exhibit activity against several viruses include HIV [7] [8] , influenza [9] , herpes (HSV-1) [10] , RNA [11] and human cytomegalovirus (HCMV) [12] . They have also been employed as antihypertensive, antiviral, anticancer, antiulcer, antifungal and untihistamine [13] - [18].

In view of the biological importance of benzimidazoles, there have been growing interests in the development of efficient, fast, simple and environment friendly synthetic methods for the preparation of these molecules. Several procedures have been reported for the synthesis of 2-substituted benzimidazoles: Condensation of 1,2- phenylenediamines with carboxylic acids, acid chlorides, nitriles, imidates and orthoesters under strong acidic conditions, sometimes combined with very high temperatures or useing microwave irradiation, [19] - [22] oxidative cyclodehydrogenation of 1,2-phenylenediamine and aldehydes in the presence of different oxidants [23] - [26] , transition-metal-catalyzed intramolecular cyclization of 2-haloanilides and their analogues [27] - [29] and also the condensation reactions of 1,2-phenylenediamine with β-ketonitriles [30] , β-ketoesters [31] [32] , or β- diketones [33] under microwave radiation and high temperature conditions or in the presence of a catalyst. Although, all of these methods are widely employed, but they have drawbacks such as low yields, the use of expensive and toxic reagents, catalysts and solvents, long reaction times, formation of side-products, tedious work-up procedure, and in some cases, harsh reaction conditions are required. Therefore, development of efficient, economical, and environmentally benign synthetic protocols for their construction is an important goal in diverse areas of chemistry. In addition, a number of other useful green reactions for the synthesis of benzimidazole derivatives have been reported in the literature. For example, Su and co-workers reported synthesis of substituted benzimidazoles from 1,2-phenylenediamine and arylaldehydes or arylmethylenemalononitriles absorbed on silica gel by intermittent grinding or by a microwave-assisted technique under solvent- and catalyst-free conditions [34] Also, Chikashita and co-workers described formation of 2-aryl benzmidazoles with reaction between of arylidenemalononitriles or β-nitrostyrenes with 1,2-phenylenediamine in ethanol at boiling temperature through a simple and efficient transfer-hydrogenation process from the in situ generated benzimidazolines to activate olfines [35] .

In order to further development of synthetic route of benzimidazoles under green reaction conditions, here, we devoted our effort for the synthesis of 2-aryl benzimidazole derivatives in water as a green solvent as well as solvent-free conditions (Scheme 1).

RESULT AND DISCUSSION

In this work, we report a highly efficient, and environmentally benign procedure for the reaction of 1,2-pheny- lenediamine derivatives 1 with arylidenemalononitrile 2 in aqueous medium as a green solvent to produce benzimidazole derivatives 3. Also, in continuation of our goal towards performing of this reaction under another green condition, we have developed reaction between reactants under solvent-free condition using thermal heating method after grinding. Arylidenemalononitrile 2 was reacted with 1,2-phenylenediamine derivatives in the presence of water to produce the related products 3 with excellent yields. We initially employed 1,2-phenyl-nediamine 1a (1 mmol) and arylidenemalononitrile 2a (2 mmol) in water at room temperature as a model reaction. In this condition the reaction wasn't complete after 14 hours (Table 1, entry 1). Therefore, various conditions have been designed to determine the optimized conditions. Different solvents such as water, ethanol, methanol, acetone, dimethylsuloxide, tetrahydrofuran, and chloroform were explored. Also, the reaction was performed under different temperatures such as 25°C, 50°C, 75°C and 90°C. The results are summarized in Table 1. As can be seen, the best result was obtained by the reaction mixture in water at 75°C for 20 min to yield product

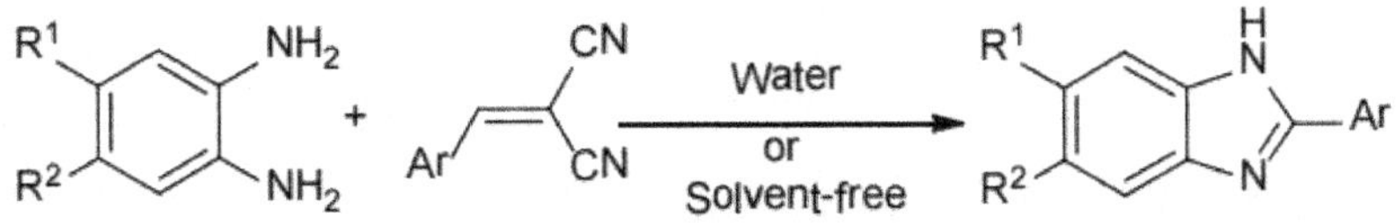

R^1, R^2 = H or Me

Ar = C_6H_5, 2-Cl-C_6H_4, 4-Cl-C_6H_4, 3-NO_2-C_6H_4, 4-NO_2-C_6H_4, 3-OCH_3-C_6H_4, 4-OCH_3-C_6H_4, 4-CH_3-C_6H_4, 4-Br-C_6H_4, 2-thiophenyl

Scheme 1: Synthesis of 2-aryl benzimidazole derivatives.

Table 1: Effect of different reaction conditions for synthesis of product 3a[a].

Entry	Solvent	Tem. [°C]	Time [min]	Yield [%][b]
1	H_2O	25	14h	40[c]
2	H_2O	50	5	75
3	H_2O	50	10	79
4	H_2O	50	15	83
5	H_2O	50	20	83
6	H_2O	75	5	80
7	H_2O	75	10	86
8	H_2O	75	15	89
9	H_2O	75	20	92
10	H_2O	75	30	92
11	H_2O	90	10	88
12	H_2O	90	20	90
13	H_2O	90	30	90
14	EtOH	reflux	20	91
15	MeOH	reflux	20	89
16	Acetone	reflux	20	70
17	DMSO	50	20	85
18	THF	50	20	77
19	$CHCl_3$	reflux	20	65
20	No solvent	75	20	79
21	No solvent	90	20	82
22	No solvent	90	30	87

[a]Reaction condition: 1,2-phenylenediamine 1a (1 mmol) and arylidenemalononitrile 2a (2 mmol);[b]Isolated yield. [c]yield based on TLC analysis.

3a (Table 1, entry 9). Although, the reaction gave high to excellent yields in organic solvents, but using water is the most advantageous to this method (Table 1, entries 14 - 19). After optimizing the reaction condition, to explore the scope and generality, the synthesis of benzimidazole derivatives 3a-p were carried out through the reaction of 1,2-phenylenediamine derivatives and a wide diversity of arylidenemalononitrile in high yields (Table 2). Interestingly, we observed that the position and nature of substitution on the ring of arylidenemalononitrile did not make much difference in reactivity, indicating the wide scope of this methodology. In continuation of this study, we are interested in solvent-free conditions as another green procedure by using grinding method. Thus, we have synthesized a series of 2-substituted benzimidazoles 3a-p by the reaction of reactants under this method on heating. Therefore, the reaction of arylidenemalononitrile 2a and 1,2-phenylne- diamine 1a proceeded successfully in an open vial through grinding of two components together and then heating at 90°C for 30 min. This reaction started immediately after heating, with liquification of the mixture, followed by solidification of the mixture of reaction. By comparing the reaction time and yields of entries 20 to 22 in Table 1, it was found that 30 min and 90°C was best conditions for this reaction. Also, it was found that both

Table 2: Synthesis of 2-aryl benzimidazoles 3a-p.

Compound	Ar	R^1	R^2	M.p. (lit. m. p.)/[°C]	H$_2$O[a] Yield [%][c]	Solvent-free[b] Yield [%][c]	Ref.
3a	C$_6$H$_5$	H	H	287-289 (288-290)	92	87	[36]
3b	2-Cl-C$_6$H$_4$	H	H	231-233 (233-234)	86	87	[41]
3c	4-Cl-C$_6$H$_4$	H	H	291-293 (287-289)	89	88	[37]
3d	3-NO$_2$-C$_6$H$_4$	H	H	204-207 (205)	88	87	[38]
3e	4-NO$_2$-C$_6$H$_4$	H	H	325-327 (328)	90	91	[40]
3f	3-MeO-C$_6$H$_4$	H	H	202-205 (201-204)	86	83	[41]
3g	4-MeO-C$_6$H$_4$	H	H	223-226 (225)	93	90	[38]
3h	4-Me-C$_6$H$_4$	H	H	275-276 (278)	92	90	[38]
3i	4-Br-C$_6$H$_4$	H	H	299-300 (298)	85	87	[40]
3j	2-C$_4$H$_4$S	H	H	330-333 (330)	90	88	[40]
3k	C$_6$H$_5$	Me	H	240-242 (241-242)	93	91	[36]
3l	2-Cl-C$_6$H$_4$	Me	H	104-106 (106-108)	87	88	[36]
3m	4-Cl-C$_6$H$_4$	Me	H	225-227 (224)	90	89	[42]
3n	C$_6$H$_5$	Me	Me	251-252 (244-247)	85	87	[39]
3o	2-C$_4$H$_4$S	Me	Me	240-242	93	91	[43]
3p	3-MeO-C$_6$H$_4$	Me	Me	240-243	93	90	[43]

[a]Reaction condition: 1,2-phenylenediamine derivatives 1 (1 mmol), arylidene malononitrile 2 (2 mmol), water (5 ml) under 75°C and 20 min; [b]Reaction condition: 1,2-phenylenediamine derivatives 1 (1 mmol), arylidene malononitrile 2 (2 mmol), grinding heating at 90°C for 30 min;[c]Isolated yield.

electron-donating and electron-deficient groups were suitable for this reaction because the products were obtained in excellent yields. In addition, a heterocyclic arylidenemalononitrile such as 2-(thiophen-2-ylmethylene) malononitrile could react with 1,2-phenylenediamine and 4,5-dimethyl-1,2-phenylenediamine to afford the corresponding benzimidazole (Table 2). The known compounds were identified by comparison of their melting point with those reported earlier (see references inTable 2). Also, a number of these compounds was characterized by its ¹H-NMR. A plausible mechanism based on reported previous work [35] is proposed in Scheme 2. Initially, Michael addition reaction of 1,2-phenylenediamine 1 with the arilydenemalononitrile 2 gave intermediate 4. The consequent proton transfer results transformation of 4 into 5. Then this intermediate converted to benzimidazoline 7 with leave malononitrile as leaving group perhaps by one of two paths (path a or b). The benzimidazole 3 is formed through a simple and efficient transfer-hydrogenation process from in situ generated benzimidazoline to arylidenemalononitrile.

CONCLUSION

In summary, we have reported green and highly efficient method for the synthesis of 2-aryl benzimidazoles in water as well as under solvent-free and catalyst-free conditions. The main advantages of these procedures are environmentally friendly, the operational simplicity, short reaction times, simple work-up procedures and high yields.

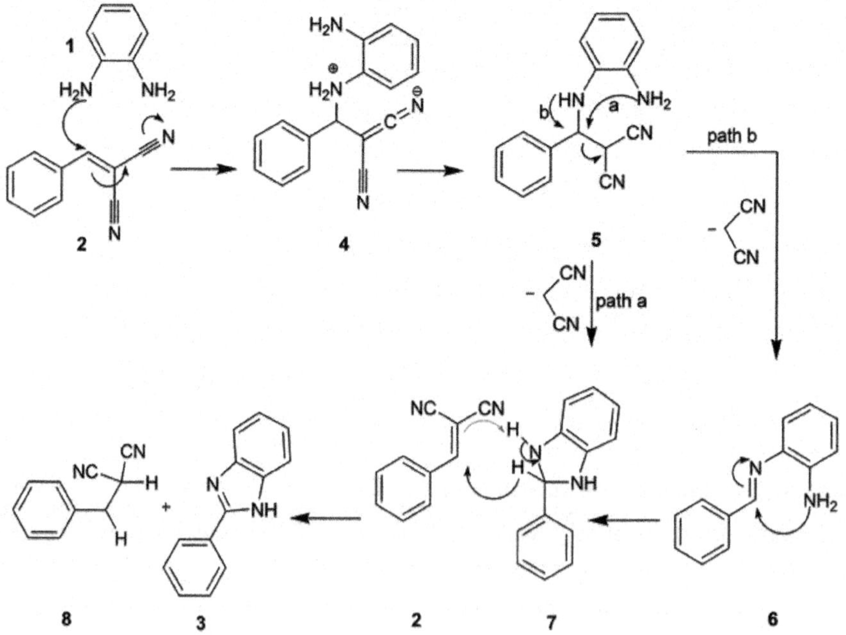

Scheme 2: Plausible mechanism for the formation of products 3a-p.

EXPERIMENTAL

1,2-phenylenediamine derivatives, malononitrile and aldehyde derivatives were purchased from the Merck and Fulka companies and were used without further purification. Melting points were determined on Electrothermal 9100 apparatus. ^{1}H NMR spectra was recorded on a Bruker Avance 300 MHz employing tetramethylsilane as an internal standard.

General Procedure for the Synthesis of 2-Aryl Benzimidazole 3a-p in Water

1,2-phenylenediamine (1 mmol) was dissolved in 5 ml water at 75°C. Then, arylidenemalononitrile (2 mmol) was added to this solution, and immediately the reaction mixture liquefied and resolidified. The reaction was monitored by TLC (petroleum ether: Ethyl Acetate (8:2)) till the disappearance of the starting arylidene malononitrile. After cooling the resultant reaction mixture, recrystalization in ethanol-water and finally pure 2-aryl bezimidazole was filtered out.

General Procedure for the Synthesis of 2-Aryl Benzimidazole 3a-p in Solvent-Free Conditions Using Conventional Heating Method

Arylidenemalononitrile (2 mmol) and 1,2-phenylenediamine (1 mmol) were mixed thoroughly with glass stirrer and heated at 90°C. The reaction mixture liquefied and resolidified in 30 min. Completion of the reaction was checked by TLC (petroleum ether: Ethyl Acetate (8:2)). After cooling the resultant semi-solid reaction mixture, crystallization was performed in ethanol-water and 2-aryl bezimidazole was filtered out.

2-Phenylbenzimidazole [36] : m.p. = 287°C - 289°C, ^{1}H NMR (300 MHz, DMSO-d$_6$): δ = 7.19 - 7.25 (m, 2H, ArH), 7.48 - 7.54 (m, 1H, ArH), 7.62 (d, J = 8.6 Hz, 2H, ArH), 7.67 - 7.73 (m, 2H, ArH), 8.17 (d, J = 8.6 Hz, 2H, ArH), 8.54 (s, 1H, NH).

2-(2-Cholorophenyl) benzimidazole [41] : m.p. = 231°C - 233°C. ^{1}H NMR (300 MHz, DMSO-d$_6$): δ = 7.26 - 7.36 (m, 2H, ArH), 7.38 - 7.44 (m, 2H, ArH), 7.48 - 7.51 (m, 1H, ArH), 7.69 (m, 2H, ArH), 8.41 (m, 1H, ArH), 10.36 (br s, 1H, NH).

2-(4-Chlorophenyl) benzimidazole [37] : m.p. = 291°C - 93°C, ^{1}H NMR (300 MHz, CDCl$_3$): δ = 7.29 - 7.32 (m, 1H, ArH), 7.49 - 7.54 (m, 3H, ArH), 7.73 - 7.88 (m, 3H, ArH), 7.98 - 8.00 (m, 1H, ArH), 9.3 (br s, 1H, NH).

2-(3-Nitrophenyl) benzimidazole [38] : m.p. = 204°C - 207°C, ^{1}H NMR (300

MHz, DMSO-d$_6$): δ = 7.20 - 7.30 (m, 2H, ArH), 7.57 (d, J = 7.3 Hz, 1H, ArH), 7.71 (d, J = 7.8 Hz, 1H, ArH), 7.85 (dd, J$_1$ = 7.9 Hz, J$_2$ = 7.9 Hz, 1H, ArH), 8.32 (dd, J$_1$ = 2.0 Hz, J$_2$ = 7.9 Hz, 1H, ArH), 8.61 (d, J = 7.9 Hz, 1H, ArH), 9.01 (dd, J = 2.0 Hz, J = 2.0 Hz, 1H, ArH), 13.30 (s, 1H, NH).

2-(3-Methoxyphenyl) benzimidazole [41] : m.p. = 202°C - 205°C, ^{1}H NMR (300 MHz, DMSO-d$_6$): δ =3.84 (s, 3H, OMe), 6.95 (d, J = 8.6 Hz, 2H, ArH), 7.00 - 7.92 (m, 4H, ArH), 8.03 (d, J = 8.6 Hz, 2H, ArH).

2-(4-Methoxyphenyl) benzimidazole [38] : m.p. = 223°C - 226°C, ^{1}H NMR (300 MHz, DMSO-d$_6$): δ = 3.83 (s, 3H, OMe), 6.95 (d, J = 8.8 Hz, 2H, ArH), 7.23 (dd, J$_1$ = 3.2 Hz, J$_2$ = 6.0 Hz, 2H, ArH), 7.60 (dd, J$_1$ = 3.2 Hz, J$_2$ = 6.0 Hz, 2H, ArH), 8.04 (d, J = 8.8 Hz, 2H, ArH).

2-(Thiophen-2-yl) benzoimidazole [40] : m.p. = 330°C - 333°C, ^{1}H NMR (300 MHz, CDCl$_3$): δ = 7.15 - 7.18 (m, 1H, ArH), 7.27 - 7.29 (m, 2H, ArH), 7.47 - 7.49 (m, 2H, ArH), 7.61 - 7.62 (m, 1H, ArH), 7.80 - 7.81 (m, 1H, ArH).

5,6-Dimethyl-2-phenylbenzoimidazole [43] : m.p. = 251°C - 252°C, ^{1}H NMR (300 MHz, DMSO-d$_6$): δ = 2.31 (s, 6H, 2Me), 7.34 - 7.54 (m, 4H, ArH), 8.12 (d, J = 8.0 Hz, 2H, ArH), 12.69 (br s, 1H, NH).

2-(3-Methoxyphenyl)-5,6-dimethylbenzoimidazole [43] : m.p. = 240°C - 243°C. ^{1}H NMR (300 MHz, DMSO-d$_6$): δ = 2.31 (s, 6H, 2Me), 3.84 (s, 3H, OMe), 6.99 - 7.03 (m, 2H, ArH), 7.30 - 7.45 (m, 2H, ArH), 7.69 - 7.72 (m, 2H, ArH), 12.60 (1, br s, NH).

CITE THIS PAPER

AzizollahHabibi,   YousefValizadeh,MarjanMollazadeh,AbdolaliAlizad eh, (2015) Green and High Efficient Synthesis of 2-Aryl Benzimidazoles: Reaction of Arylidene Malononitrile and 1,2-Phenylenediamine Derivatives in Water or Solvent-Free Conditions. *International Journal of Organic Chemistry*,**05**,256-263. doi: 10.4236/ijoc.2015.54025

REFERENCES

1.   Lindstrom, U.M. (2007) Organic Reactions in Water: Principles Strategies and Applications. Blackwell, Oxford. http://dx.doi.org/10.1002/9780470988817

2.   Lindstrom, U.M. (2002) Stereoselective Organic Reactions in Water. Chemical Reviews, 102, 2751-2772. http://dx.doi.org/10.1021/cr010122p

3.   Shapiro. N. and Vigalok. A. (2008) Highly Efficient Organic Reactions "on Water", "in Water", and Both. Angewandte Chemie, 120, 2891-2894. http://dx.doi.org/10.1002/ange.200705347

4.   Li, C.-J. (2005) Organic Reactions in Aqueous Media with a Focus on Carbon-Carbon Bond Formations:? A Decade Update. Chemical Reviews, 105, 3095-3166. http://dx.doi.org/10.1021/cr030009u

5.   Grieco, P.A. (1998) Organic Synthesis in Water. Blackie Academic & Professional, London. http://dx.doi.org/10.1007/978-94-011-4950-1

6.   Li, C.-J. and Chan, T.-H. (2007) Comprehensive Organic Reactions in Aqueous Media. Wiley & Sons, New York. http://dx.doi. org/10.1002/9780470131442

7.   Rida, S.M.S., EI-Hawash, A.M., Fahmy, H.T.Y., Hazzaa, A.A. and EI-Meligy, M.M.M. (2006) Synthesis of Novel Benzofuran and Related Benzimidazole Derivatives for Evaluation of in Vitro Anti-HIV-1, Anticancer and Antimicrobial Activities. Archives of Pharmacal Research, 29, 826-833. http://dx.doi.org/10.1007/BF02973901

8.   Roth, T., Morningstar, M.L., Boyer, P.L., Hughes, S.H., Buckheitjr, R.W. and Michejda, C.J. (1997) Synthesis and Biological Activity of Novel Nonnucleoside Inhibitors of HIV-1 Reverse Transcriptase. 2-Aryl-Substituted Benzimidazoles. Journal of Medicinal Chemistry, 40, 4199-4207. http://dx.doi.org/10.1021/jm970096g

9.   Hisano, T., Ichikawa, M., Tsumoto, K. and Tasaki, M. (1982) Synthesis of Benzoxazoles, Benzothiazoles and Benzimidazoles and Evaluation of Their Antifungal, Insecticidal and Herbicidal Activities. Chemical and Pharmaceutical Bulletin, 30, 2996-3004. http://dx.doi.org/10.1248/cpb.30.2996

10.   Migawa, M.T., Girardet, J.-L., Walker, J.A., Koszalka, G.W., Chamberlain, S.D., Drach, J.C. and Townsend, L.B. (1998) Design, Synthesis, and Antiviral Activity of α-Nucleosides:? d- and l-Isomers of Lyxofuranosyl-and (5-Deoxylyxofuranosyl)benzimidazoles. Journal of Medicinal Chemistry, 41, 1242-1251. http://dx.doi.org/10.1021/jm970545c

11.   Dreyer, C. and Hausen, P. (1978) Inhibition of Mammalian RNA Polymerase by 5,6-Dichlororibofuranosylbenzimi- dazole (DRB) and DRB Triphosphate. Nucleic Acids Research, 5, 3325-3335. http://dx.doi. org/10.1093/nar/5.9.3325

12.   Porcari, A.R., Devivar, R.V., Kucera, L.S., Drach, J.C. and Townsend, L.B. (1998) Design, Synthesis, and Antiviral Evaluations of 1-(Substituted benzyl)-2-substituted-5,6-dichlorobenzimidazoles as Nonnucleoside Analogues of 2,5,6- Trichloro-1-(β-d-ribofuranosyl) benzimidazole. Journal of Medicinal Chemistry, 41, 1252-1262.http:// dx.doi.org/10.1021/jm970559i

13.   Erhardt, P.W. (1987) In Search of the Digitalis Replacement. Journal

of Medicinal Chemistry, 30, 231-237.http://dx.doi.org/10.1021/jm00385a001

14. Tomczuk, B.E., Taylor Jr., C.R., Moses, L.M., Sutherland, D.B., Lo, Y.S., Johnson, D.N., Kinnier, W.B. and Kilpatrick, B.F. (1991) 2-Phenyl-3H-imidazo[4,5-b]pyridine-3-acetamides as Non-Benzodiazepine Anticonvulsants and Anxiolytics. Journal of Medicinal Chemistry, 34, 2993-3006. http://dx.doi.org/10.1021/jm00114a007

15. Spasov, A.A., Yozhitsa, I.N. and Bugaeva, L.I. (1999) Benzimidazole Derivatives: Spectrum of Pharmacological Activity and Toxicological Properties (A Review). Pharmaceutical Chemistry Journal, 33, 232-243. http://dx.doi.org/10.1007/bf02510042

16. Gravatt, G.L., Baguley, B.C., Wilson, W.R. and Denny, W.A. (1994) DNA-Directed Alkylating Agents. 6. Synthesis and Antitumor Activity of DNA Minor Groove-Targeted Aniline Mustard Analogs of Pibenzimol (Hoechst 33258). Journal of Medicinal Chemistry, 37, 4338-4345. http://dx.doi.org/10.1021/jm00051a010

17. Horton, D.A., Bourne, G.T. and Smythe, M.L. (2003) The Combinatorial Synthesis of Bicyclic Privileged Structures or Privileged Substructures. Chemical Reviews, 103, 893-930. http://dx.doi.org/10.1021/cr020033s

18. Kim, J.S., Gatto, B., Yu, C., Liu, A., Liu, L.F. and La Voie, E.J. (1996) Substituted 2,5'-Bi-1H-benzimidazoles:? Topoisomerase I Inhibition and Cytotoxicity. Journal of Medicinal Chemistry, 39, 992-998.http://dx.doi.org/10.1021/jm950412w

19. Lu, J., Yang, B. and Bai, Y. (2002) Microwave Irradiation Synthesis of 2-Substituted Benzimidazoles Using PPA as a Catalyst under Solvent-Free Conditions. Synthetic Communications, 32, 3703-3709.http://dx.doi.org/10.1081/SCC-120015381

20. Geratz, J.D., Stevens, F.M., Polakoski, K.L. and Parrish, R.F. (1979) Amidino-Substituted Aromatic Heterocycles as Probes of the Specificity Pocket of Trypsin-Like Proteases. Archives of Biochemistry and Biophysics, 197, 551-559.http://dx.doi.org/10.1016/0003-9861(79)90279-0

21. Tidwell, R.R., Geratz., J.D., Dann, O., Volz, G., Zeh, D. and Loewe, H. (1978) Diarylamidine Derivatives with One or Both of the Aryl Moieties Consisting of an Indole or Indole-Like Ring. Inhibitors of Arginine-Specific Esteroproteases. Journal of Medicinal Chemistry, 21, 613-623. http://dx.doi.org/10.1021/jm00205a005

22. Fairley, T.A., Tidwell, R.R., Donkor, I., Naiman, N.A., Ohemeng, K.A., Lombardy, R.J., Bentley, J.A. and Cory, M. (1993) Structure, DNA

Minor Groove Binding, and Base Pair Specificity of Alkyl- and Aryl-Linked Bis(amidinobenzi- midazoles) and Bis(amidinoindoles). Journal of Medicinal Chemistry, 36, 1746-1753.http://dx.doi.org/10.1021/jm00064a008

23. Riadi, Y., Mamouni, R., Azzalou, R., EI Haddad, M., Routier, S., Guillaumet, G. and Lazar, S. (2011) An Efficient and Reusable Heterogeneous Catalyst Animal Bone Meal for Facile Synthesis of Benzimidazoles, Benzoxazoles, and Benzothiazoles. Tetrahedron Letters, 52, 3492-3495. http://dx.doi.org/10.1016/j.tetlet.2011.04.121

24. Bachhav, H.M., Bhagat, S.B. and Telvekar, V.N. (2011) Efficient Protocol for the Synthesis of Quinoxaline, Benzoxazole and Benzimidazole Derivatives Using Glycerol as Green Solvent. Tetrahedron Letters, 52, 5697-5701.http://dx.doi.org/10.1016/j.tetlet.2011.08.105

25. Blacker, A.J., Farah, M.M., Hall, M.I. Marsden, S.P, Saidi, O. and Williams, J.M. (2009) Synthesis of Benzazoles by Hydrogen-Transfer Catalysis. Organic Letters, 11, 2039-2042. http://dx.doi.org/10.1021/ol900557u

26. Patil, V.D., Patil, J., Rege, P. and Dere, G. (2011) Mild and Efficient Synthesis of Benzimidazole Using Lead Peroxide Under Solvent-Free Conditions. Synthetic Communications, 41, 58-62. http://dx.doi.org/10.1080/00397910903531789

27. Saha, P., Ramana, T., Purkait, N., Ashif, A.M., Paul, R. and Punniyamurthy, T. (2009) Ligand-Free Copper-Catalyzed Synthesis of Substituted Benzimidazoles, 2-Aminobenzimidazoles, 2-Aminobenzothiazoles, and Benzoxazoles. Journal of Organic Chemistry, 74, 8719-8725. http://dx.doi.org/10.1021/jo901813g

28. Evindar, G. and Batey, R.A. (2006) Parallel Synthesis of a Library of Ben-zoxazoles and Benzothiazoles Using Ligand- Accelerated Copper-Catalyzed Cyclizations of Ortho-Halobenzanilides. Journal of Organic Chemistry, 71, 1802-1808.http://dx.doi.org/10.1021/jo051927q

29. Yang, D., Fu, H., Hu, L., Jiang, Y. and Zhao, Y. (2008) Copper-Catalyzed Synthesis of Benzimidazoles via Cascade Reactions of O-Haloacetanilide Derivatives with Amidine Hydrochlorides. Journal of Organic Chemistry, 73, 7841- 7844. http://dx.doi.org/10.1021/jo8014984

30. Kamila, S., Koh, B. and Biehl, E.R. (2006) Microwave-Assisted "Green" Synthesis of 2-Alkyl/Arylbenzothiazoles in One Pot: A Facile Approach to Anti-Tumor Drugs. Journal of Heterocyclic Chemistry, 43, 1609-1612. http://dx.doi.org/10.1002/jhet.5570430627

31. Kamila, S., Zhang, H. and Biehl, E.R. (2005) One-Pot Synthesis of 2-Aryl-

and 2-Alkylbenzothiazoles under Microwave Irradiation. Heterocycles, 65, 2119-2126. http://dx.doi.org/10.3987/COM-05-10466

32. Cai, L., Ji, X., Yao, Z., Xu, F. and Shen, Q. (2011) Efficient Synthesis of Functionalized Benzimidazoles and Perimidines: Ytterbium Chloride Catalyzed C-C Bond Cleavage. Chinese Journal of Chemistry, 29, 1880-1886.http://dx.doi.org/10.1002/cjoc.201180328

33. Wang, Z.-X. and Qin, H.-L. (2005) Reaction of 1,3-Dicarbonyl Compounds with o-Phenylenediamine or 3,3'-Diami nobenzidine in Water or under Solvent-Free Conditions via Microwave Irradiation. Journal of Heterocyclic Chemistry, 42, 1001-1005. http://dx.doi.org/10.1002/jhet.5570420540

34. Yu, C., Guo, P., Jin, C. and Su, W. (2009) The Synthesis of Benzimidazole Derivatives in the Absence of Solvent and Catalys. Journal of Chemical Research, 5, 333-336.

35. Itoh, K., Ishida, H. and Chikashita, H. (1982) The Reactions of Benzylidenmalononitriles β-Nitrostyrenes with o-Phenylenediamine including the New Organic Redox Reactions between the Olefins and 2-Phenylbenzimidazolines. Chemistry Letters, 1117-1118. http://dx.doi.org/10.1246/cl.1982.1117

36. Li, J., Benard, S., Neuville, L. and Zhu, J. (2012) Copper Catalyzed N-Arylation of Amidines with Aryl Boronic Acids and One-Pot Synthesis of Benzimidazoles by a Chan-Lam-Evans N-Arylation and C-H Activation/C-N Bond Forming Process. Organic Letters, 14, 5980-5983. http://dx.doi.org/10.1021/ol3028847

37. Chari, M.A., Shobha, D., Kenawy, E.R., Al-Deyab, S.S., Subba Reddy, B.V. and Vinu, A. (2010) Nanoporous Aluminosilicate Catalyst with 3D Cage-Type Porous Structure as an Efficient Catalyst for the Synthesis of Benzimidazole Derivatives. Tetrahedron Letters, 51, 5195-5199. http://dx.doi.org/10.1016/j.tetlet.2010.07.132

38. Lei, M., Ma, L. and Hu, L. (2012) One-Pot Synthesis of 1H-Benzimidazole Derivatives Using Thiamine Hydrochloride as a Reusable Organocatalyst. Synthetic Communications, 42, 2981-2993. http://dx.doi.org/10.1080/00397911.2011.573610

39. Rostamizadeh, S., Aryan, R. and Ghaieni, H.R. (2011) Aqueous 1 M Glucose Solution as a Novel and Fully Green Reaction Medium and Catalyst for the Oxidant-Free Synthesis of 2-Arylbenzimidazoles. Synthetic Communications, 41, 1794-1804. http://dx.doi.org/10.1080/00397911.2010.492460

40. Saha, D., Saha, A. and Ranu, B.C. (2009) Remarkable Influence of

Substituent in Ionic Liquid in Control of Reaction: Simple, Efficient and Hazardous Organic Solvent Free Procedure for the Synthesis of 2-Aryl Benzimidazoles Promoted by Ionic Liquid, [pmim]BF4. Green Chemistry, 11, 733-737. http://dx.doi.org/10.1039/b823543k

41. Heravi, M.M., Tajbakhsh, M., Ahmadi, A.N. and Mohajerani, B. (2006) Zeolites. Efficient and Eco-Friendly Catalysts for the Synthesis of Benzimidazoles. Monatshefte fur Chemie, 137, 175-179.

42. Cohen, V.I. (1979) A New Method of Synthesis of Some 2-Aryl and 2-Heterocyclic Benzimidazole, Benzox-azole and Benzothiazole Derivatives. Journal of Heterocyclic Chemistry, 16, 13-16. http://dx.doi.org/10.1002/jhet.5570160103

43. Lam, T., Hilgers, M.T., Cunningham, M.L., Kwan, B.P., Nelson, K.J., Brown-Driver, V., Ong, V., Trzoos, M., Hough, G., Joy Shaw, K. and Finn, J. (2014) Structure-Based Design of New Dihy-drofolate Reductase Antibacterial Agents: 7-(Benzimidazol-1-yl)-2,4-diaminoquinazolines. Journal of Medicinal Chemistry, 57, 651-668. http://dx.doi.org/10.1021/jm401204g

Chapter 4

THEORETICAL STUDY ON PD-CATALYZED ACYLATION OF ALLYLIC ESTERS WITH ACYLSILANES AND ACYLSTANNANES

Shohei Sanada, Michinori Sumimoto, Kenji Hori

Graduate School of Science and Engineering, Yamaguchi University, Tokiwadai, Japan

ABSTRACT

Acylation of allylic esters with acylsilanes and acylstannanes in the presence of a palladium complex was investigated theoretically using the DFT (B3PW91) method. We examined along the reaction that was reported by Tsuji's. In this mechanism for generating active species, a Pd dinuclear complex 5 (the reaction of Pd and 2) was produced. Then, 5 is decomposed to two mononuclear complex 6. The reaction of 6 and 1 forms an intermediate 7, which is active species. In catalytic cycle from 7, the O (1) atom of 7 attacks the Si or Sn atom in TS_{7-8} to produce 8. Then, the C(1)-C(2) reductive elimination from 8 occurs through the TS_{8-9} to yield 9. Therefore, 9 decomposed to Pd(0), 3 and 4. However, reaction mechanism from 9 to 6 should be considered because Pd(0) + 3 + 4 are less stable than 9 by 29.2 kcal/mol, 9 does not decompose. We proposed the reaction mechanism from 9, as shown below: 1) 2 attacks 9 to form 10. 2) 10 released 4 to produce a five coordinated intermediate 11. 3) 11 changes its structure to another π-allyl complex 12. 4) The product 3 was released from 12 and 6 formed again for a next catalytic cycle. The rate-determining step of these reaction is nucleophilic attack of carbonyl oxygen to R^A (7 → 8), and the $\Delta G^{\ddagger}$ for I, II and III was calculated to be 27.1, 39.1, and 30.9 kcal/mol, respectively. As a result, we elucidated the reaction mechanism of acylation of allylic esters with acylsilanes and acylstannanes in the presence of a palladium complex.

INTRODUCTION

The palladium-catalyzed acylation of allylic esters has been widely applied in organic synthesis and is one of the most important reactions. Carbanions [1] - [5] , enolates [6] - [10] , and amines [11] - [15] are often employed as the nucleophilic agent of this reaction until now. The acylated products obtained by this reaction are used as materials for constructing natural products, pharmaceutical compounds, low molecular organic compounds, polyesters, and polycarbonates. However, since functional groups that can be directly introduced to the allylic system are still limited, further developments of the research are expected.

In 1993, Tsuji and co-workers found out the palladium-catalyzed silylation reaction of allylic esters with disilanes (Scheme 1) [16] . Additionally, they reported new acylation reaction of allylic ester with acyl silane or acyl stannane based on reactions in the past [17] [18] .

In these reactions, acyl silane and acyl stannane act as the acylating agent in the presence of Pd catalyst. Three kinds of reactions by the difference in the substituents (R^A and R^B) as shown in Scheme 2 were reported by these works. In these three reactions, I and II proceeds, but III does not occur.

Reaction mechanism of this acylation is proposed by Tsuji and co-workers as shown in Scheme 3. The first step of these catalytic cycles is formed a π-aryl complex by Pd complex and 2. After this reaction, a π-aryl complex reacts with 1 to form 4 and a new π-aryl complex. Finally, 3 is generated from a π-aryl complex, a Pd catalyst reproduces. Thus, several experimental works have been well performed, but a reaction mechanism has not been theoretically investigated yet, to our knowledge. In this study, we theoretically investigated the Pd-catalyzed acylation reaction of allylic ester using the DFT method [19] . Here our attentions were focused on clarifying the reaction mechanism and the substituent effect in these reactions.

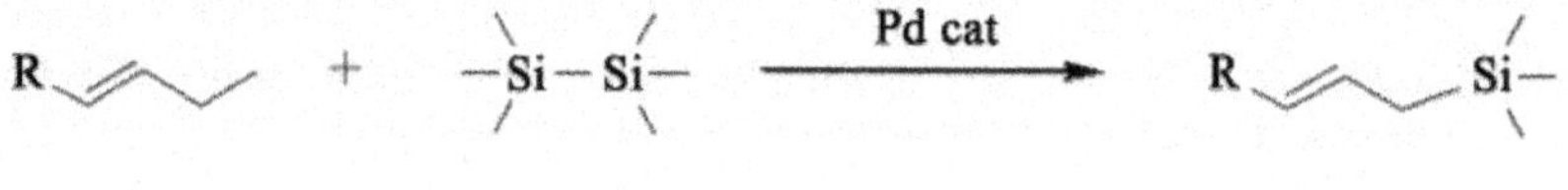

Scheme 1: Silylation reaction reported by Tsuji and co-workers.

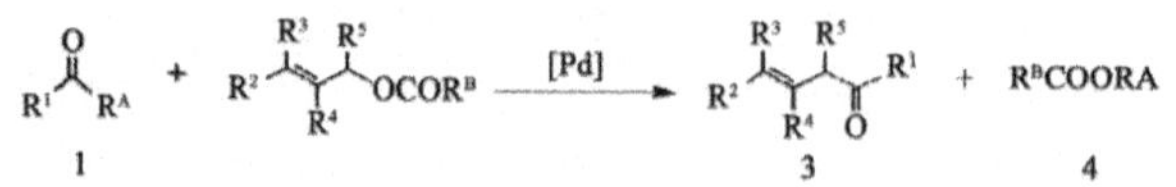

Reaction	R^A	R^B
I	$SnMe_3$	CF_3
II	$SiMe_3$	CF_3
III	$SiMe_3$	CH_3

Scheme 2: Acylation of 2 with 1.

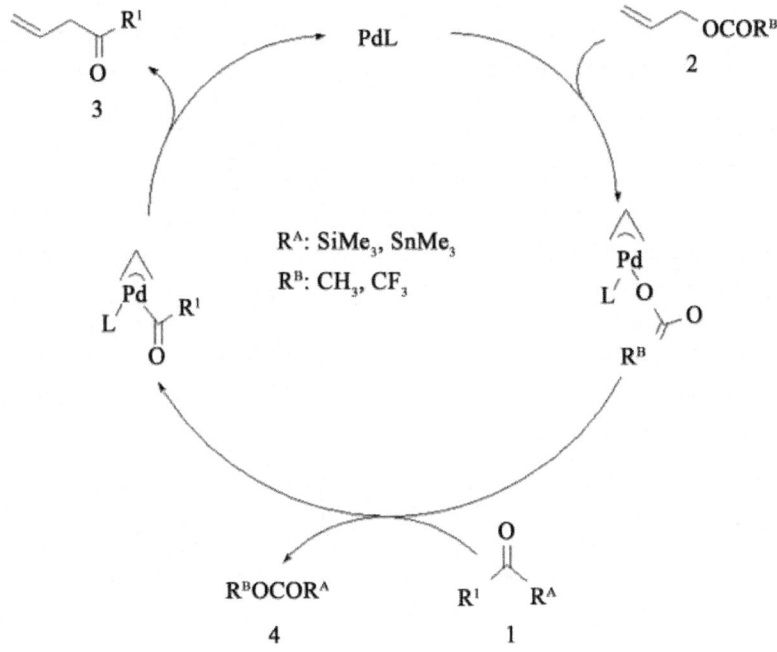

Scheme 3: Catalytic cycle with Pd(II) complex.

COMPUTATIONAL DETAILS

All geometry optimizations were calculated with the DFT method, where the B3PW91 [20] [21] functional was used for the exchange-correlation term. We ascertained that each equilibrium structure exhibited no imaginary frequencies, and each transition state had only one imaginary frequency. In these calculations, the following basis set system was employed. The (541/541/211/1) [22] [23] basis set was used to represent the valence electrons of Pd, where the effective core potentials (ECPs) were employed to replace core electrons (up to 3d) [24] . The LANL2DZ basis set was used for Sn. For C and O atoms, the usual 6-311G(d) basis sets were employed. For Si atom, the 6-311G(2d) basis set was used. For H atom, the 6-31G(d,p) basis set were employed. To investigate the solvent effects of tetrahydrofuran (THF), we performed single point calculations with the SMD [25] method using the optimized structures. The values of free energies were calculated with thermodynamic cycle [26] .

All of these calculations were carried out with the Gaussian 09 program package [27] .

RESULTS AND DISCUSSION

The reaction mechanism proposed by Tsuji and co-workers were shown in Scheme 4. In the first step of this reaction mechanism, Pd dinuclear complex 5 is formed to by catalyst and allylic ester. 5 is decomposed into two molecules, Pd mononuclear complex 6 is formed. 1 is close to 6, and intermediate 7 generates.

Although these molecular structures are not known precisely, Pd complex such as 7 were assumed to be the active species in the catalytic cycle for the acylation with allylic ester. However, the reaction mechanism starting from 7 has not yet been investigated. We proposed the reaction mechanism from 7 as shown in Scheme 5. The O(1) atom of 7 attacks the Sn or Si atom through the transition state, TS_{7-8}, to produce 8. The C(1)-C(2) reductive elimination from 8 occurs through the TS_{8-9} to yield 9.

The Sn-C and Sn-O(1) distances of 7 in Reaction I were calculated to be 2.238 and 2.749Å, respectively (see Figure 1). On the other hand, since the Sn-C and Sn-O(1) distances of 8 in Reaction I were 4.900 and 2.073Å, respectively, the Sn-C bond is broken and new Sn-O(1) bond is formed. In reaction (8 → 9), the Pd-C(1) and Pd-C(2) distances of 8 and 9 in Reaction I were 2.072 and 3.963 Å, respectively. As the reaction progresses, we find out that the Pd-C(2) bond have been dissociated. Moreover, 8 in Reaction I was a π-aryl complex, but 9 in Reaction I was not. Reaction II and III showed these results similar to Reaction I.

The Si-C and the Si-O(1) distances of 7 in Reaction II were 1.939 and 3.030Å, respectively (seeFigure 2). On the other hand, the Si-C and Si-O(1) distances of 8 in Reaction II were calculated to be 4.838 and 1.740Å, respectively. When comparing Reaction I and II, the Si-C and Si-O(1) bonds are stronger than the Sn-C and Sn-O(1) bonds. From these results, it was suggested that dissociations of the Sn-C and Sn-O bonds are easier than those the Si-C and Si-O bonds.

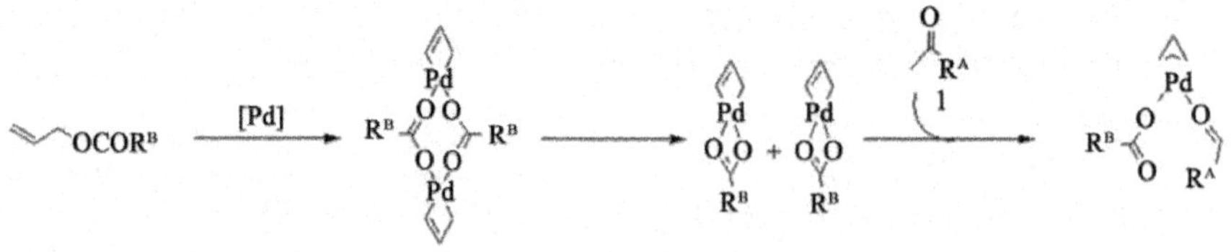

Scheme 4: Reaction mechanism proposed by Tsuji and co-workers.

R^A: $SiMe_3$, $SnMe_3$
R^B: CH_3, CF_3

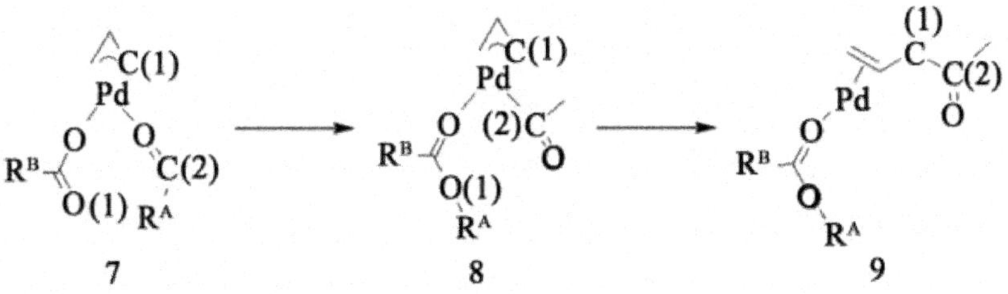

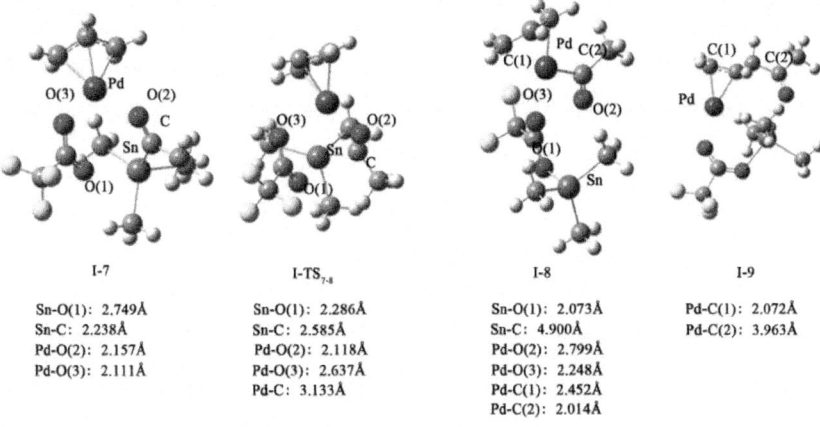

Scheme 5: Proposed reaction mechanism from 7.

Figure 1: Optimized structures of I-7, I-TS$_{7-8}$, I-8 and I-9.

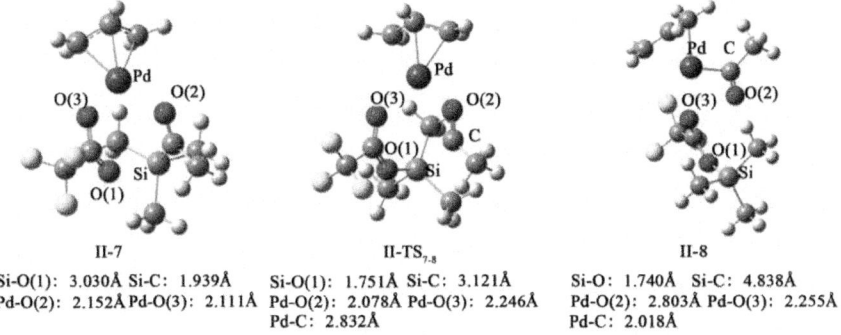

Figure 2: Optimized structures of II-7, II-TS$_{7-8}$ and II-8.

The $\Delta G^{\ddagger}$ values of the reaction (7 → 8) for Reaction I, II and III were calculated to be 27.1, 39.1 and 30.1 kcal/mol, respectively. Additionally, the $\Delta G^{\ddagger}$ values of the step (8 → 9) for Reaction I, II and III were 21.2, 20.4 and 23.1 kcal/mol, respectively. Intermediates 9 for Reaction I, II and III are much more stable than 7 by 21.2, 21.7 and 34.8 kcal/mol, respectively. As a result, the reactions leading to 9 in Reaction I, II and III will occur easily.

In the proposed reaction mechanism, 9 is dissociated to Pd(0), 3 and 4. Figure 3 displays the free energy change between 9 and three product compounds in Reaction III. Product compounds (Pd(0), 3 and 4) in Reaction III are much more unstable than 9 by 29.2 kcal/mol, it can be considered than this reaction will not proceed. Therefore, we assumed a new reaction shown in Scheme 6.

In this new reaction, 2 attacks 9 to form 10. 4 dissociates from 10, and a five coordinated intermediate 11 is generated. We examined the validity of this reaction mechanism.

The Pd-C(1) and Pd-C(2) bonds of 9 + 2 in Reaction I were dissociated, respectively. Those distances of 10 in Reaction I were 2.123 and 2.162 Å, respectively, and new Pd-C(1) and Pd-C(2) bonds were formed. The $\Delta G^{\ddagger}$ values of the reaction (9 → 10) for Reaction I was calculated to be 2.0 kcal/mol. Additionally, 10 in Reaction I was more stable by 8.3 kcal/mol than 9. Reaction II and III showed these results similar to Reaction I. In reactiton from 10 to 11, 4 dissociates from 10, and a five coordinated intermediate 11 is generated. The geometry changes and relative free energies of this reaction in Reaction I, II and III are shown in Figure 4 and Figure 5, respectively. In this reaction, new Pd-O(1) bond was formed. This reaction proceeds with no barrier, and 11 was more stable by 13.8 kcal/mol than 10 in Reaction I. From these results, it is considered that reaction proceeds. The succeeding isomerization (10 → 11) occurs, to yield a π-aryl complex 12. The $\Delta G^{\ddagger}$ values of this reaction for Reaction I, II and III were calculated to be 10.4, 10.4 and 13.5 kcal/mol, respectively. Moreover, 12 in Reaction I, II and III were more stable by 10.5, 10.5 and 3.4 kcal/mol than 11, respectively, and this reaction occurs easily.

Scheme 6: Proposed reaction mechanism from 9.

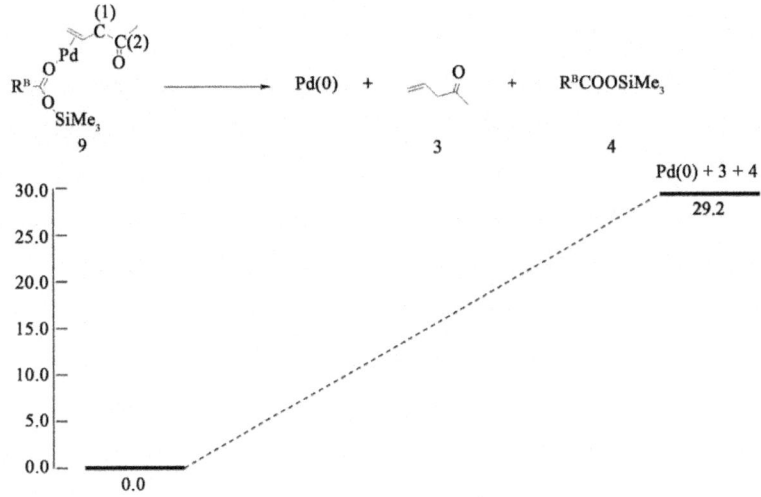

Figure 3: Relative free energies from 9 to Pd(0), 3 and 4 in Reaction III.

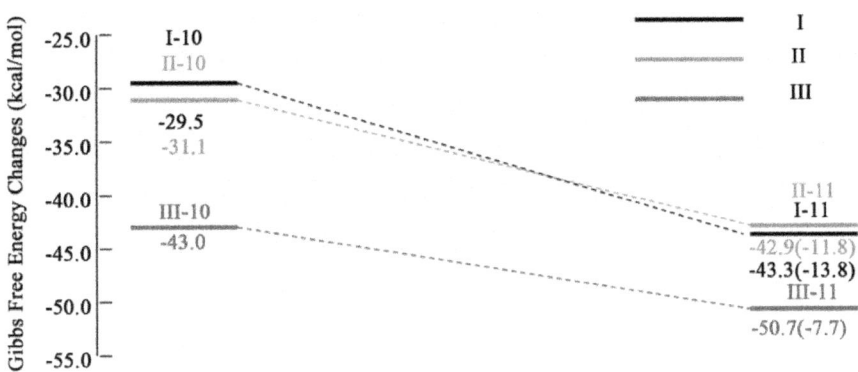

Figure 4: Relative free energies from 10 to 11 in Reaction I, II and III.

It is considered that the dissociation of 12 gives a product 3 and a mononuclear complex 6. The geometry changes and relative free energies of this reaction in Reaction I, II and III are shown in Figure 6 and Figure 7, respectively. The $\Delta G^{\ddagger}$ values of this reaction for Reaction I, II and III were calculated to be 10.5, 10.5 and 9.1 kcal/mol, respectively. Finally, the reaction between mononuclear complex 6 and 1 are formed an active species 7, and the catalytic cycle is completed.

We investigated the catalytic cycle starting from 7, as shown in Scheme 7. Figure 8-10 display the relative free energies from 7 in Reaction I, II and III, respectively. The rate-determining step is the nucleophilic attack of carbonyl oxygen to R^A (7→8), their $\Delta G^{\ddagger}$ values are 27.1, 39.1 and 30.1 kcal/mol, respectively. The $\Delta G^{\ddagger}$ value of the rate-determining step in Reaction I is lower than that in Reaction II by 12.0 kcal/mol. This result showed the same tendency as the experimental one.

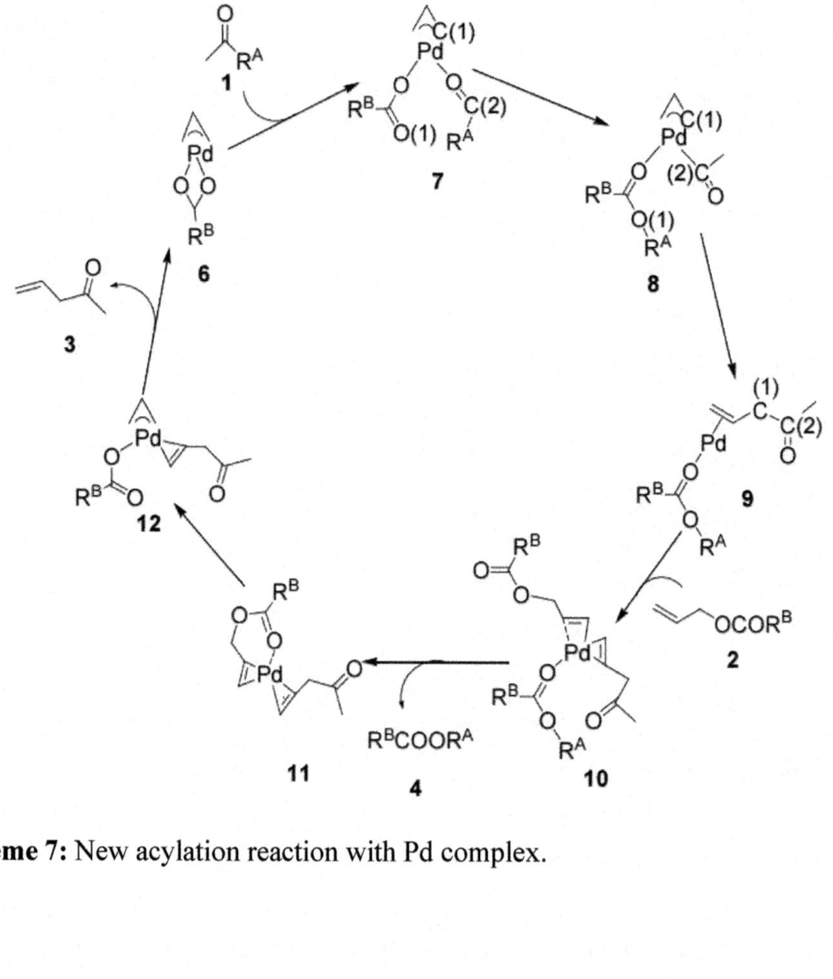

Scheme 7: New acylation reaction with Pd complex.

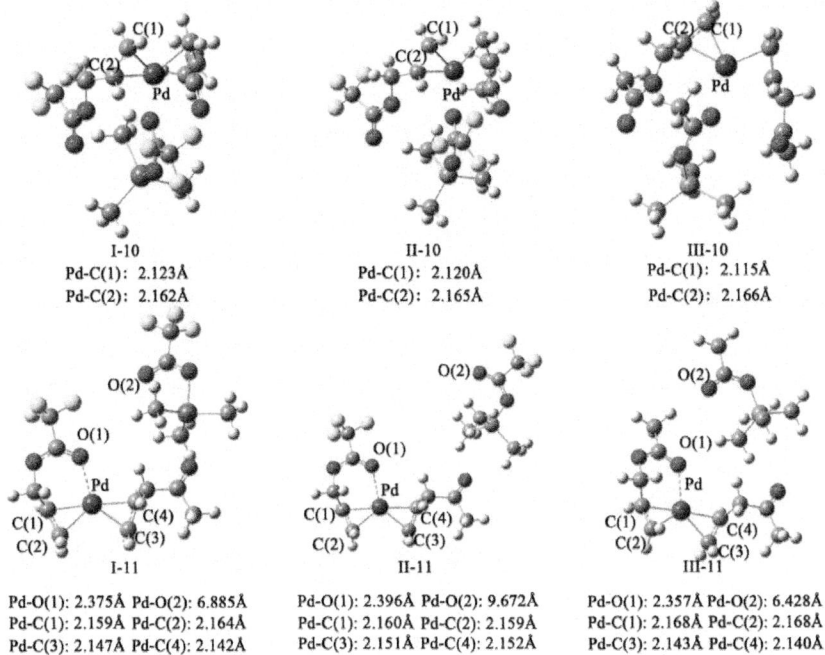

Figure 5: Optimized structures of I-10, II-10, III-10, I-11, II-11 and III-11.

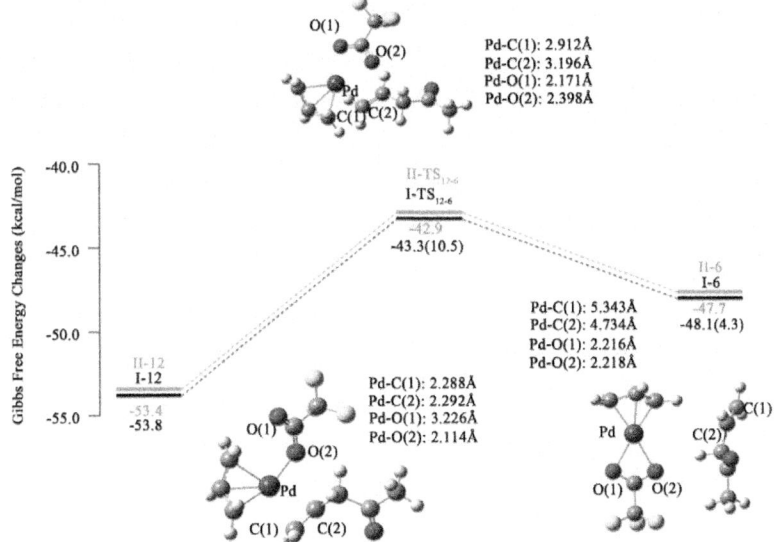

Figure 6: Optimized geometries and relative free energies from 12 to 6 in Reaction I and II.

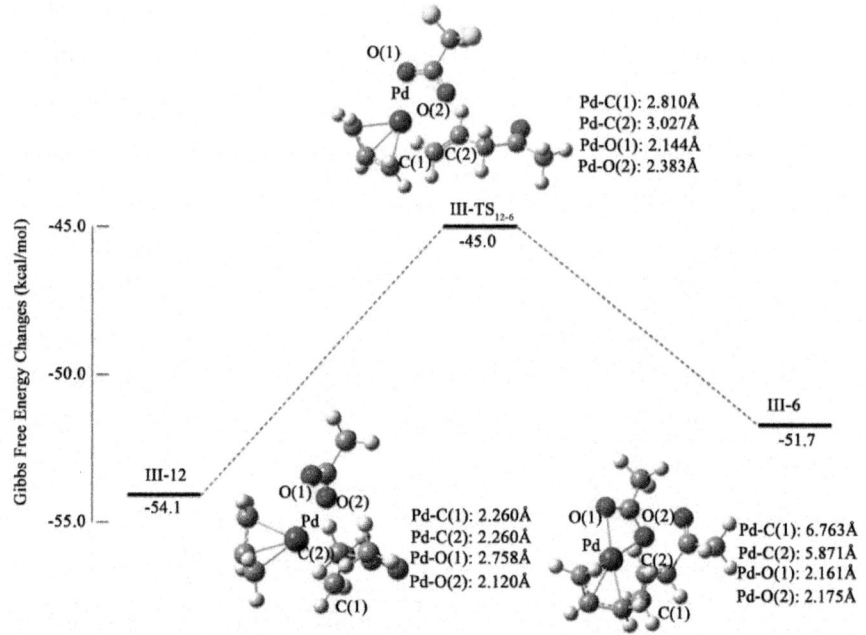

Figure 7. Optimized geometries and relative free energies from 12 to 6 in Reaction III.

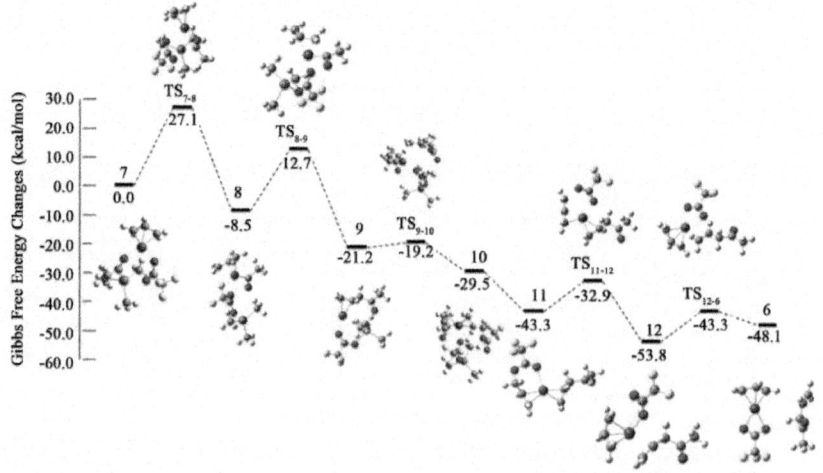

Figure 8: Optimized geometries and relative free energies from 7 in Reaction I.

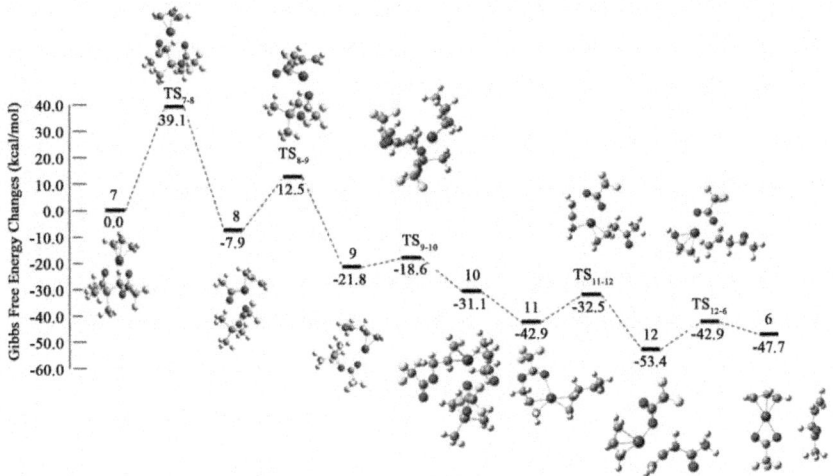

Figure 9: Optimized geometry and relative free energies from 7 in Reaction II.

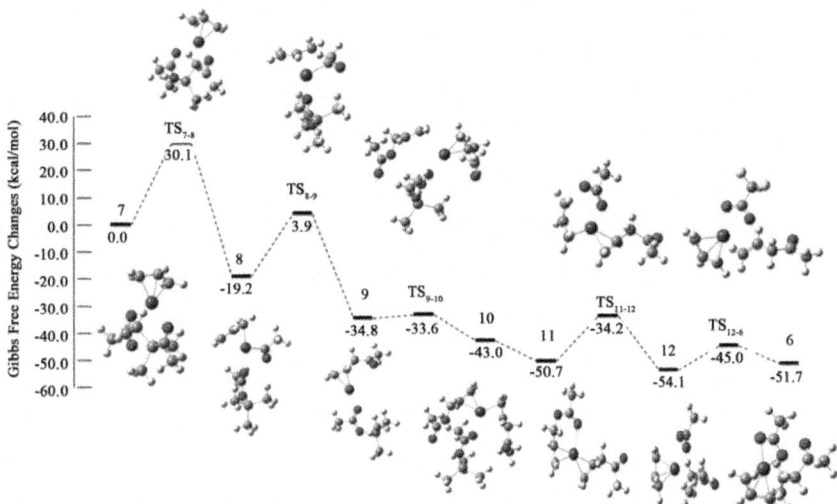

Figure 10: Optimized geometries and relative free energies from 7 in Reaction III.

To investigate in details about these differences, we calculated the C-SiMe$_3$ and C-SnMe$_3$ bond energies. The C-SiMe$_3$ bond energy is stronger than the C-SnMe$_3$ one by 18.6 kcal/mol. In other words, it is easy to break the C-SnMe$_3$ bond than the C-SiMe$_3$ bond. Therefore, it is suggested that Reaction I (R^A = SnMe$_3$) is more favorable than Reaction II.

On the other hand, since the $\Delta G^{\ddagger}$ value in Reaction III is lower than that in Reaction II by 9.0 kcal/mol. Since only reactions of Reaction I and II occur in the experiment, this result obtained by our theoretical calculation is different with experimental one. To elucidate this cause, we focused on a formation reaction of the active species. At the first step of the catalytic cycle, 6 reacts with 1 to form the active species 7. 7 in Reaction I and II are more stable than 6 by 2.5 and 1.6 kcal/mol, respectively. On the other hand, 7 in Reaction III is less stable than 6 by 5.2 kcal/mol. From these results, it is suggested that the active species 7 generate only in Reaction I and III. In other words, since the active species 7 in Reaction II does not form, the catalytic reaction cannot proceed. As a result, even if the $\Delta G^{\ddagger}$ value of the rate-determining step in Reaction II is low, it is considered that the reaction does not occur.

From all results, it is suggested that this catalytic reaction starts from 6 without reaction III and the calculated results well explain the experimental one.

CONCLUSIONS

In the present study, we investigated the acylation of allylic esters with acylsilanes and acylstannanes in the presence of a palladium complex using the DFT method. Firstly, we examined the reaction mechanism starting from 7. The results of reaction mechanisms are summarized as follows:

1) The O(1) atom of 8 attacks the Sn or Si atom in TS_{8-9} to produce 9.

2) The C(1)-C(2) reductive elimination from 9 proceeds through the TS_{9-10} to yield 10.

3) 2 attacks 10 to form 11.

4) 11 released 4 to produce a five coordinated intermediate 12.

5) 12 changes into structure to another π-acyl complex 6.

6) The product 3 was released from 6 and 7 forms again for another catalytic cycle.

The rate-determining step is the nucleophilic attack of carbonyl oxygen to R^A (7→8), their $\Delta G^{\ddagger}$ values are 27.1, 39.1 and 30.1 kcal/mol, respectively.

Therefore, we illustrated three kinds of reactions by the difference in the substituents (R^A and R^B).

CITE THIS PAPER

ShoheiSanada,MichinoriSumimoto,KenjiHori,   (2015)   Theoretical Study on Pd-Catalyzed Acylation of Allylic Esters with Acylsilanes and

Acylstannanes. *International Journal of Organic Chemistry,***05**,246-255. doi: 10.4236/ijoc.2015.54024

REFERENCES

1. Geoffrey, G.L., Sheridan, J.B., Bassner, S.L. and Kelley, C. (1989) Migratory-Insertion of Carbon Monoxide into Metal-Acyl Bonds. Pure and Applied Chemistry, 61, 1723-1729.

2. Blangetti, M., Rosso, H., Prandi, C., Deagostino, A. and Venturello, P. (2013) Suzuki-Miyaura Cross-Coupling in Acylation Reactions, Scope and Recent Developments. Molecules, 18, 1188-1213. http://dx.doi. org/10.3390/molecules18011188

3. Katritzky, A.R. and Pastor, A. (2000) Synthesis of a-Dicarbonyl Compounds Using 1-Acylbenzotriazoles as Regioselective C-Acylating Reagents. The Journal of Organic Chemistry, 65, 3679-3682. http:// dx.doi.org/10.1021/jo991878f

4. Saragoni, V.G. and Contreras, R.R. (1993) Theoretical Study of the C-vs. 0-Acylation of Metal Enolates. Frontier Molecular Orbital Analysis including Solvent Effects. Journal of Quantum Chemistry, 27, 713-721. http://dx.doi.org/10.1002/qua.560480863

5. 5. Mermerian, A.H. and Fu, G.C. (2003) Catalytic Enantioselective Synthesis of Quaternary Stereocenters via Intermolecular C-Acylation of Silyl Ketene Acetals: Dual Activation of the Electrophile and the Nucleophile. Journal of the American Chemical Society, 125, 4050-4051. http://dx.doi.org/10.1021/ja028554k

6. 6. Bairgrie, L.M., Leung-Toung, R. and Tidwell, T.T. (1988) Oxygen and Carbon-Acylation of Enolates by Ketenes. Tetrahedron Letters, 29, 1673-1676. http://dx.doi.org/10.1016/S0040-4039(00)82014-6

7. 7. Yoshida, K. and Yamashita, Y. (1966) The Acylation of Enolate Anion by Acid Halides and Dimethylketene. Tetrahedron Letters, 7, 693-696. http://dx.doi.org/10.1016/S0040-4039(00)90247-8

8. 8. Durman, J. and Warren, S. (1985) Acylation of Extended Rnolatr Ions from A-Phenyltrio(Phs-)Crotonatr Esters. Tetrahedron Letters, 26, 2895-2898. http://dx.doi.org/10.1016/S0040-4039(00)98865-8

9. 9. Seebach, D. (1969) Methods and Possibilities of Nucleophilic Acylation. Angewandte Chemie International Edition, 8, 639-649. http:// dx.doi.org/10.1002/anie.196906391

10. 10. Lin, M.H. and Rajan Babu, T.V. (2000) Metal-Catalyzed Acyl Transfer

Reactions of Enol Esters: Role of Y5(OiPr)13O and (thd)2Y(OiPr) as Transesterification Catalysts. Organic Letters, 2, 997-1000. http://dx.doi. org/10.1021/ol0057131

11.   11. Naik, S., Bhattacharjya, G., Kavala, V.R. and Patel, B.K. (2004) Mild and Eco-Friendly Chemoselective Acylation of Amines in Aqueous Medium. ARKIVOC, 1, 55-63.

12.   12. Naik, S., Bhattacharjya, G., Talukdar, B. and Patel, B.K. (2004) Chemoselective Acylation of Amines in Aqueous Media. European Journal of Organic Chemistry, 2004, 1254-1260. http://dx.doi. org/10.1002/ejoc.200300620

13.   13. Ohshima, T., Iwasaki, T., Maegawa, Y., Yoshiyama, A. and Mashima, K. (2008) Enzyme-Like Chemoselective Acylation of Alcohols in the Presence of Amines Catalyzed by a Tetranuclear Zinc Cluster. Journal of the American Chemical Society, 130, 2944-2945. http://dx.doi. org/10.1021/ja711349r

14.   14. Negishi, E., Matsushita, H., Chatterjee, S. and John, R.A. (1982) Selective Carbon-Carbon Bond Formation via Transition Metal Catalysis. 29. A Highly Regio- and Stereospecific Palladium-Catalyzed Allylation of Enolates Derived from Ketones. The Journal of Organic Chemistry, 47, 3188-3190. http://dx.doi.org/10.1021/jo00137a038

15.   15. Musteata, M., Musteata, V., Dinu, A., Florea, M., Hoang, V., Trong-On, D., Kaliaguine, S. and Parvulescu, V.I. (2007) Acylation of Different Amino Derivatives with Fatty Acids on UL-MFI-Type Catalysts. Pure and Applied Chemistry, 79, 2059-2068. http://dx.doi.org/10.1351/ pac200779112059

16.   16. Tsuji, Y., Kajita, S., Isobe, S. and Funato, M. (1993) Palladium-Catalyzed Silylation of Allylic Acetates with Hexamethyldisilane or (Trimethylsilyl)tributylstannane. The Journal of Organic Chemistry, 58, 3607-3608.http://dx.doi.org/10.1021/jo00066a005

17.   17. Obora, Y., Ogawa, Y., Imai, Y., Kawamura, T. and Tsuji, Y. (2001) Palladium Complex Catalyzed Acylation of Allylic Esters with Acylsilanes. Journal of the American Chemical Society, 123, 10489-104893.http://dx.doi.org/10.1021/ja010674p

18.   18. Obora, Y., Nakanishi, M., Tokunaga, M. and Tsuji, Y. (2002) Palladium Complex Catalyzed Acylation of Allylic Esters with Acylstannanes: Complementary Method to the Acylation with Acylsilanes. The Journal of Organic Chemistry, 67, 5835-5837. http://dx.doi.org/10.1021/jo0202482

19.   19. Becke, A.D. (1993) Density-Functional Thermochemistry. III. The Role of Exact Exchange. The Journal of Chemical Physics, 98, 5648-

5652. http://dx.doi.org/10.1063/1.464913

20. Becke, A.D. (1988) Density-Functional Exchange-Energy Approximation with Correct Asymptotic Behavior. Physical Review A, 38, 3098-3100. http://dx.doi.org/10.1103/PhysRevA.38.3098

21. Perdew, J.P. and Wang, Y. (1992) Accurate and Simple Analytic Representation of the Electron-Gas Correlation Energy. Physical Review B, 45, 13244-13249. http://dx.doi.org/10.1103/PhysRevB.45.13244

22. Hay, P.J. and Wadt, W.R. (1985) Ab Initio Effective Core Potentials for Molecular Calculations. Potentials for K to Au including the Outermost Core Orbitals. The Journal of Chemical Physics, 82, 299-310. http://dx.doi.org/10.1063/1.448975

23. Couty, M. and Hall, M.B. (1996) Basis Sets for Transition Metals: Optimized Outer p Functions. Journal of Computational Chemistry, 17, 1359-1370. http://dx.doi.org/10.1002/(SICI)1096-987X(199608)17:11<1359::AID-JCC9>3.0.CO;2-L

24. Ehlers, A.W., Bohme, M., Dapprich, S., Gobbi, A., Hollwarth, A., Jonas, V., Kohler, K.F., Stegmann, R., Veldkamp, A. and Frenking, G. (1993) A Set of f-Polarization Functions for Pseudo-Potential Basis Sets of the Transition Metals SC-Cu, Y-Ag and La-Au. Chemical Physics Letters, 208, 111-114. http://dx.doi.org/10.1016/0009-2614(93)80086-5

25. Marenich, A.V., Cramer, C.J. and Truhlar, G. (2009) Universal Solvation Model Based on Solute Electron Density and on a Continuum Model of the Solvent Defined by the Bulk Dielectric Constant and Atomic Surface Tensions. The Journal of Physical Chemistry B, 113, 6378-6396. http://dx.doi.org/10.1021/jp810292n

26. Krishnamurthy, V.M., Bohall, B.R., Semetey, V. and Whitesides, G.M. (2006) The Paradoxical Thermodynamic Basis for the Interaction of Ethylene Glycol, Glycine, and Sarcosine Chains with Bovine Carbonic Anhydrase II: An Unexpected Manifestation of Enthalpy/Entropy Compensation. Journal of the American Chemical Society, 128, 5802-5812. http://dx.doi.org/10.1021/ja060070r

27. Frisch, M.J. Trucks, G.W., Schlegel, H.B., Scuseria, G.E., Robb, M.A., Cheeseman, J.R., Scalmani, G., Barone, V., Mennucci, B., Petersson, G.A., Nakatsuji, H., Caricato, M., Li, X., Hratchian, H.P., Izmaylov, A.F., Bloino, J. Zheng, G. Sonnenberg, J.L. Hada, M. Ehara, M. Toyota, K. Fukuda, R. Hasegawa, J. Ishida, M. Nakajima, T. Honda, Y., Kitao, O., Nakai, H., Vreven, T., Montgomery Jr., J.A., Peralta, J.E., Ogliaro, F., Bearpark, M., Heyd, J.J., Brothers, E., Kudin, K.N., Staroverov, V.N., Kobayashi, R., Normand, J., Raghavachari, K., Rendell, A., Burant, J.C.,

Iyengar, S.S., Tomasi, J., Cossi, M., Rega, N., Millam, J.M., Klene, M., Knox, J.E., Cross, J.B., Bakken, V., Adamo, C., Jaramillo, J., Gomperts, R., Stratmann, R.E., Yazyev, O., Austin, A.J., Cammi, R., Pomelli, C., Ochterski, J.W., Martin, R.L., Morokuma, K., Zakrzewski, V.G., Voth, G.A., Salvador, P., Dannenberg, J.J., Dapprich, S., Daniels, A.D., Farkas, O., Foresman, J.B., Ortiz, J.V., Cioslowski, J., Fox, D.J. (2009) Gaussian 09. Gaussian, Inc., Wallingford.

Chapter 5

CHEMICALLY MODIFIED URIDINE MOLECULES INCORPORATING ACYL RESIDUES TO ENHANCE ANTIBACTERIAL AND CYTOTOXIC ACTIVITIES

Sarkar M. A. Kawsar[1], Hamida A. Ara[1], Sheikh Aftab Uddin[2], Mohammed K. Hossain[3], Shagir A. Chowdhury[1], Abul F. M. Sanaullah[1], Mohammad A. Manchur[4], Imtiaj Hasan[5], Yukiko Ogawa[6], Yuki Fujii[6], Yasuhiro Koide[5], Yasuhiro Ozeki[5]

[1]Laboratory of Carbohydrate and Protein Chemistry, Department of Chemistry, Faculty of Science, University of Chittagong, Chittagong, Bangladesh

[2]The Institute of Marine Science, Faculty of Science, University of Chittagong, Chittagong, Bangladesh

[3]Department of Pharmacy, Faculty of Biological Science, University of Chittagong, Chittagong, Bangladesh

[4]Department of Microbiology, Faculty of Biological Science, University of Chittagong, Chittagong, Bangladesh

[5]Laboratory of Glycobiology and Marine Biochemistry, Department of Life and Environmental System Science, Graduate School of NanoBio Sciences, Yokohama City University, Yokohama, Japan

[6]Divisions of Microbiology, Graduate School of Pharmaceutical Science, Nagasaki International University, Nagasaki, Japan

ABSTRACT

A new N-acetylsulfanilylation series of uridine have been synthesized in good yield using direct acylation method and afforded the 5'-O-N-acetylsulfanilyluridine. In order to obtain newer products, the 5'-O-N-acetylsulfanilyluridine derivative was further transformed to a series of 2',3'-di-O- acyl derivatives containing a wide variety of functionalities in a single molecular framework. The chemical structures of the newly synthesized compounds were confirmed on the basis of their FTIR, ^{1}H-NMR spectroscopy, physicochemical properties and elemental analysis. All the synthesized uridine derivatives were tested for their in vitro antibacterial activity against six human pathogenic bacterial strains and for comparison standard antibiotic Ampicillin was also determined. The study revealed that the selectively acylated derivatives 5'-O-N-acetylsulfanilyl-2',3'-di-O-lau- royluridine and 5'-O-N-acetylsulfanilyl-2',3'-di-O-pivaloyluridine showed highest inhibition against Staphylococcus aureus and Bacillus cereus, respectively. We also observed that the introduction of hexanoyl, decanoyl, lauroyl, myristoyl and pivaloyl groups, the antibacterial functionality of the compound uridine increases. Another noteworthy observation was that the uridine derivatives were found comparatively more effective against Gram-positive microorganisms than those of Gram-negative microorganisms. In addition, the test chemicals were also tested for cytotoxicity by brine shrimp lethality bioassay and compounds showed different rate mortality with different concentrations.

INTRODUCTION

Nucleosides are glycosylamines consisting of a nucleobase (i.e., nitrogeneous base) which β-glycosidic bonded to a sugar (ribose or deoxyribose) and the nitrogen-containing compound is either a pyrimidine base (cytosine, thymine or uracil) or a purine base (adenine or guanine). When nucleoside attached with one or more phosphate groups at the C-5' is called a nucleotide. Nucleotide derivatives are necessary for life, as they are building blocks of nucleic acids and have thousands of other roles in cell metabolism and regulation. Nucleosides are required for DNA and RNA synthesis, and the nucleoside adenosine has a function in a variety of signaling processes [1]. In medicine, several nucleoside analogues are used as antiviral or anticancer agents [2].

Uridine (1) (Figure 1) is a molecule (known as nucleoside) that is formed when uracil is attached to a ribose ring via β-N_1-glycosidic bond. Uridine is one of the four basic components of robonucleic acid (RNA). Upon digestion of foods containing RNA, uridine is released from RNA and is absorbed intact in the gut. Uridine is found in sugarcane, tomatos, broccoli, liver, pancreas etc. Uridine has anti depression activity, asthmatic airway inflammation, hepatocyte proliferation [3] [4] .

Infectious diseases worldwide have been known to be a cause of morbidity, disability and mortality. Approximately 15 million people die each year due to infectious diseases-nearly all live in developing countries [5] . Indiscriminate and unconcerned use of antibiotics has led to increased microbial resistance. Consequently, newer agents have been brought in at increased economic costs to the patient but they too have become inefficacious in due course and pose worldwide a great threat to human health. So, noble, emerging and reemerging infectious diseases have become a focus for the development of new cost-effective drug in both developed and developing countries. So, the finding of new drug is very important for the treatment.

Selective acylation is very important in the field of nucleoside chemistry because of its usefulness for the synthesis of biologically active products. During the last few years, many workers have investigated selective acylation and alkylation of hydroxyl groups of the carbohydrate moieties of nucleosides and nucleotides using various methods [6] - [8]. Different methods for acylation of carbohydrates and nucleosides have so far been developed and employed successfully [9] [10]. Of these, direct method has been found to be the most encouraging method for acylation of carbohydrates and nucleosides [11].

From literature survey revealed that a large number of biologically active compounds possess aromatic and heteroaromatic nucleus and acyl substituents [12] [13]. It is also known that, if an active nucleus is linked to another nucleus, the resulting molecule may possess greater potential for biological activity [14].

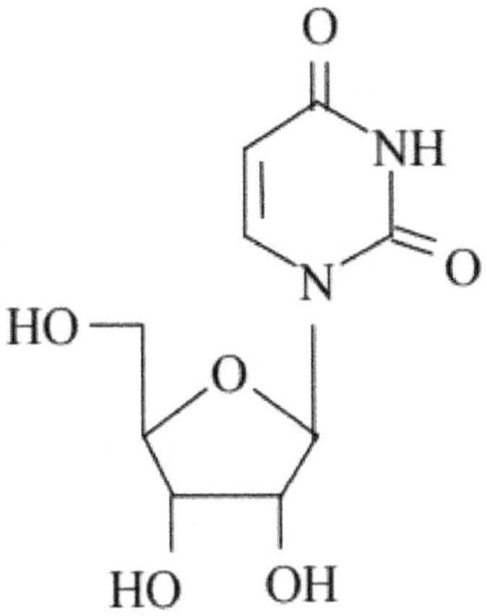

Figure 1: Uridine (1).

The benzene and substituted benzene nuclei play important role as common denominator of various biological activities [15] . Results of an ongoing research work on selective acylation of carbohydrates [16] [17] and nucleosides [18] [19] and also evaluation of antimicrobial activities reveal that in many cases the combination of two or more aromatic or heteroaromatic nuclei [14] . It is also found that nitrogen, sulfur and halogen containing substitution products showed marked antimicrobial activities i.e., enhance the biological activity of the parent compound [20] - [25] .

Encouraged by literature reports and our own findings, we synthesized a series of uridine derivatives (Scheme 1) deliberately incorporating a wide variety of probable biologically active components to the ribose moiety. Antibacterial screening of these compounds was carried out using a variety of bacterial strains and cytotoxicity test was also performed by brine shrimp lethality bioassay and the results are reported here as first time.

RESULTS AND DISCUSSION

Chemistry

The objective of the present research work reported in this paper was to perform regioselective N-acetylsulfani- lylation of uridine (1) using the direct method (Scheme 1). A number of rarely used acylating agents were employed for this purpose (Scheme 2 and Table 1) and the physicochemical properties of the synthesized compounds were presented in the (Table 2). The structure of the acylated products were ascertained by analyzing their FTIR and ^{1}H-NMR spectra [29] - [31] . In continuation of a project going on the carbohydrate and protein chemistry laboratory, we intended to prepare a series of uridine derivatives for use as test compounds for biological evaluation. Keeping this objective in mind, we thus prepared a set of derivatives containing a wide variety of substituents in a single molecular framework. The reaction pathways have been summarized in the Scheme 1 and Scheme 2 and Table 1.

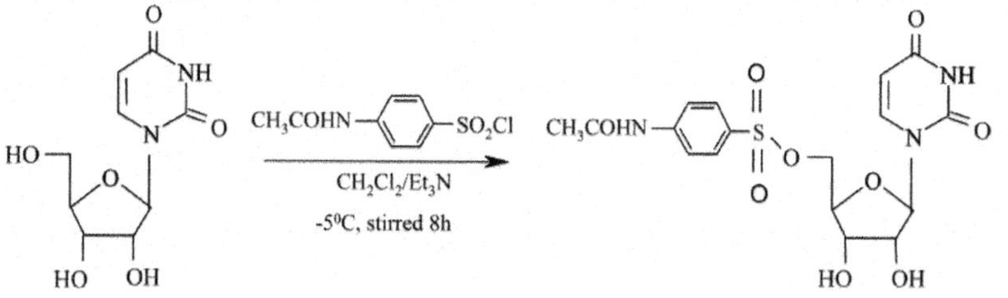

Uridine (1)

5'-O-N-acetylsulfanilyluridine (2)
(61%)

Scheme 1: Synthesis of 5'-O-N-acetylsulfanilyluridine (2).

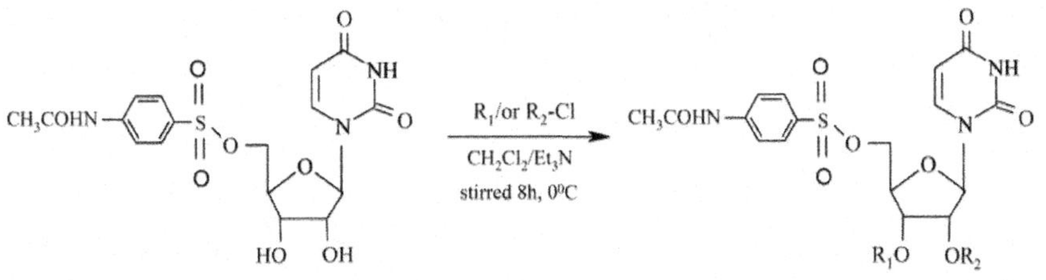

5'-O-N-acetylsulfanilyluridine (2) → 2',3'-di-O-acylated uridine derivatives (3-14)

Scheme 2: Synthesis of 2',3'-di-O-acylated uridine derivatives (3-14).

Table 1: Synthesized of uridine derivatives (2-14).

Compound no	$R_1 = R_2$	Compound no	$R_1 = R_2$
2	H	9	$CH_3(CH_2)_{14}CO-$
3	CH_3CO-	10	$(CH_3)_3CCO-$
4	$CH_3(CH_2)_3CO-$	11	C_6H_5CO-
5	$CH_3(CH_2)_4CO-$	12	$3-Cl.C_6H_4CO-$
6	$CH_3(CH_2)_8CO-$	13	$4-Cl.C_6H_4CO-$
7	$CH_3(CH_2)_{10}CO-$	14	$2,6-di.Cl.C_6H_3CO-$
8	$CH_3(CH_2)_{12}CO-$		

Table 2: Physicochemical properties of the synthesized of uridine derivatives (2-14).

Compound no	RT (h)	R_f	Yield (%)	Physical State
2	8	0.5	61	needles, m.p. 210°C - 212°C
3	7	0.5	71	semi solid
4	6	0.52	68	pasty mass
5	6	0.51	63	thick syrup
6	6	0.5	63	syrup
7	6.5	0.52	60	needles, m.p. 74°C - 75°C
8	7	0.53	58	needles, m.p. 65°C - 70°C
9	6	0.51	65	needles, m.p. 58°C - 60°C
10	6.5	0.5	70	pasty mass
11	6.5	0.51	66	needles, m.p. 138°C - 140°C
12	6	0.52	63	needles, m.p. 123°C - 125°C
13	6.5	0.5	60	needles, m.p. 128°C - 130°C
14	6	0.52	65	thick syrupy

Spectral Characterization

Our initial effort was to react uridine (1) with equivalent amount of N-acetylsulfanilyl chloride and triethylamine in anhydrous dichloromethane at room temperature, followed by removal of solvent and silica gel column chromatographic purification, furnished the N-acetylsulfanilyl derivative (2). The IR spectrum of compound 2 showed absorption bands at: 1740 cm^{-1} (-CO stretching), 3510 cm^{-1} (-OH stretching), 3360 cm^{-1} (-NH stretching) and 1365 cm^{-1} (-SO$_2$ stretching). In its ^{1}H-NMR spectrum displayed two two-proton doublets at d 7.78 (2H, J = 8.8 Hz) and d 7.68 (2H, J = 8.8 Hz) corresponding to the aromatic ring protons, one one-proton singlet at d 7.53 (-NH), one three-proton singlet at d 2.22 (CH$_3$CON-) thereby suggesting the introduction of one N- acetylsulfanilyl (4-acetamidobenzenesulfonyl) group in the molecule. Also, the C-5' proton was deshielded considerably to d 5.77 (1H, dd, J = 2.2 and 12.2 Hz, 5›a) and 5.73 (1H, J = 2.2 and 12.3 Hz, 5›b) from its usual value, suggested the introduction of the N-acetylsulfanilyl group at position 5›. Complete analysis of the IR and ^{1}H-NMR spectra of this compound was in agreement with the structure accorded as 5'-O-N-acetylsulfanilyluri- dine (2).

The structure of compound (2) was further confirmed by preparation of its acetyl derivatives (3). Thus, acetylation of compound 2 with acetic anhydride, followed by removal of solvent and purification by silica gel column chromatography, provided the tri-O-acetyl derevative (3). Complete analysis of the rest of the IR and ^{1}H-NMR spectra was in supported the structure ascertained as 5'-O-N-acetylsulfanilyl-2',3'-di-O-acetyluridine (3). Thus, pentanoylation of compound 2 in CH$_2$Cl$_2$/Et$_3$N using silica gel chromatographic purification afforded the pentanoyl derivative (4). In its ^{1}H-NMR spectrum, the resonance peaks three four-proton multiplets at δ 2.42 {2 × CH$_3$(CH$_2$)$_2$CH$_2$CO-} and δ 1.64 {2 × CH$_3$CH$_2$CH$_2$CH$_2$CO-} and δ 1.42 {2 × CH$_3$CH$_2$(CH$_2$)$_2$CO-} and one six-proton multiplet at δ 0.89 {2 × CH$_3$(CH$_2$)$_3$CO-} showed the presence of two pentanoyl groups in the molecule. Thus, complete analysis of the IR and ^{1}H-NMR spectra established the structure of this compound as 5'-O-N-acetylsulfanilyl-2',3'-di-O-pentanoyluridine (4). Reaction of compound 2 with hexanoyl chloride as usual method, provided the hexanoyl derivative (5). The ^{1}H-NMR spectrum displayed characteristics two four-proton multiplets at δ 2.38 and δ 1.63, an eight-proton multiplet at δ 1.29 and one six-proton multiplet at δ 0.95 showing the attachment of two hexanoyl groups in the compound. The rest of the IR and ^{1}H-NMR spectra enabled us to assign the structure of the hexanoyl derivative as 5'-O-N-acetylsulfanilyl-2',3'-di-O-hexanoyluridine (5).

We then treated compound 2 with decanoyl chloride in anhydrous CH$_2$Cl$_2$/ Et$_3$N, we obtained the 2',3'-di-O- decanoyl derivative (6). Complete analysis of the IR and ^{1}H-NMR spectra was consistent with the structure of the compound assigned as 5'-O-N-acetylsulfanilyl-2',3'-di-O-decanoyluridine (6). The N-acetylsulfanilyl derivative (2) was then subjected to lauroylation using same method and obtained in good yield as needless. It's ¹H-NMR spectrum displayed two four-proton multiplets at 2.37, δ 1.64, a thirty two-proton multiplet at δ 1.24 and a nine-proton multiplet at δ 0.87, therefore, indicating the presence of two lauroyl groups. So, analysis of rest of the IR and ¹H-NMR spectra, the structure of the lauroyl derivative was assigned as 5'-O-N-acetylsulfa- nilyl-2',3'-di-O-lauroyluridine (7). The diol (2) reacted with myristoyl chloride in anhydrous dichloromethane and triethylamine followed by removal of solvent and chromatographic purification, it yielded the myristate (8). As same analysis of the IR and ¹H-NMR spectra of the compound 9 was in agreement with the structure established as 5'-O-N-acetylsulfanilyl-2',3'-di-O-palmitoyluridine (9).

Our next effort was to treat compound 2 with unimolecular amount of pivaloyl chloride in anhydrous CH_2Cl_2/Et_3N by silica gel chromatography furnished the pivaloyl derivative (10). The downfield shift of H-2' proton to δ 5.31 (1H, m) and H-3' proton to δ 5.10 (1H, m) from their precursor diol (2) δ values showed the attachment of the pivaloyl groups at positions 2' and 3'. The rest of the IR and ¹H-NMR spectra was compatible with the structure assigned as 5'-O-N-acetylsulfanilyl-2',3'-di-O-pivaloyluridine (10). Benzoylation of the N-acetylsulfanilyl derivative (2) with benzoyl chloride and further purification procedure provided the benzoyl derivative (11) in good yields. In its ¹H-NMR spectrum, the aromatic proton peaks at δ 8.10 (4H, m), δ 7.66 (2H, m), δ 7.49 (4H, m) corresponded to two benzoyl groups present in the compound. Complete analysis of the IR and ¹H-NMR spectrum suggested that the structure of this compound may be assigned as 5'-O-N-acetylsulfa- nilyl-2',3'-di-O-benzoyluridine (11).

Encouraged by the results obtained so far, we then used 3-chlorobenzoyl chloride as the next acylating agent. Treatment of compound 2 with 3-chlorobenzoyl chloride and silica gel chromatographic purification, furnished the 2',3'-substitution product (12). In its ¹H-NMR spectrum, the peaks at δ 7.92 (2H, J = 7.5 Hz), δ 7.90 (2H, s), δ 7.79 (2H, J = 7.7 Hz) and δ 7.48 (2H, J = 7.6 Hz) corresponded to eight aromatic protons and were due to the presence of two 3-chlorobenzoyl groups. By complete analysis of the IR and ¹H-NMR spectra, the structure of the compound was ascertained as 5'-O-N-acetylsulfanilyl-2',3'-di-O-(3-chlorobenzoyl)uridine (12). After similar solvent removal and chromatographic techniques, the 4-chlorobenzoyl derivative (13) was isolated in good yield. In its ¹H-NMR spectrum showed the deshielding of H-2' and H-3' protons from their precursor compound (2) and the resonance of other protons in their anticipated positions confirmed the structure of this compound as 5'-O-N-acetylsulfanilyl-2',3'-di-O-(4-chlorobenzoyl)uridine (13).

Final confirmation of the structure of the compound 2 was provided by preparation of its 2,6-di-O-benzoyl derivative (14). The ¹H-NMR spectrum displayed a six proton multiplet at δ 7.46 (6H, Ar-H) corresponded to two 2,6-dichlorobenzoyl groups present in the compound. The H-2' and H-3' protons resonated at d 5.49 (1H, J = 5.1 Hz) and d 5.35 (1H, m) shifted downfield from their precursor compound (2), which were indicative of the attachment of the 2,6-dochlorobenzoyl groups at 2' and 3' positions. Complete analysis of the IR and ¹H-NMR spectrum was in accord with the structure of this compound assigned as 5'-O-N-acetylsulfanilyl-2',3'-di-O- (2,6-dichlorobenzoyl)uridine (14).

Thus, selective acylation of uridine (1) with a number of acylating agents by using the direct method was carried out successfully. The study was found to be very promising since in all the cases, a single monosubstitution product was isolated reasonably high yields. Furthermore, the N-acetylsulfanilyl acylation product was converted into a series of acyl derivatives (3-14). These newly synthesized products may be used as important precursors for the modification of the uridine molecule at different positions. All the acylation products of uridine (1) and their derivatives have been employed for determining their antibacterial activity against a number of human pathogens and cytotoxicity activity by brine shrimp lethality bioassay.

Antibacterial Activities

The antibacterial screening results in (Table 3) and (Figure 2) showed that, in case of Gram-positive organism, 5'-O-N-acetylsulfanilyl-2',3'-di-O-lauroyluridine (7) exhibited the highest activity (14.0 mm), whereas 5'-O-N-N.B: * = marked inhibition, ** = standard antibiotic, NF = not found, dw = dry weight. Acetylsulfanilyl-2›,3›-di-O-pivaloyluridine (10) showed lowest inhibition (4.0 mm) against the same organism, Staphylococcus aureus.

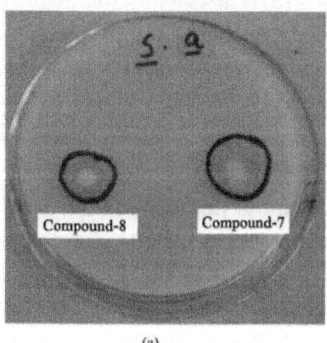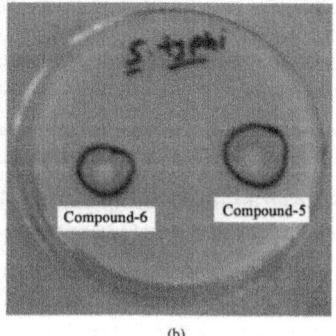

(a)   (b)

Figure 2: (a) Growth of inhibition zone observed against Gram-positive S. aureus by two test chemicals 7 and 8 and (b) against Gram-negative S. typhi by test chemicals 5 and 6.

The inhibition by compound 8 in case of Bacillus cereus (10.0 mm), Staphylococcus aureus (12.0 mm), by 9 in case of Bacillus magaterium (11.0 mm), and by 10 in case of Bacillus cereus (12.0 mm), Bacillus magaterium (10.0 mm) were very significant. But compound 2, 3, 11 and 13 were unable to show any inhibition against all tested Gram-positive microorganisms. The results in Table 2 revealed that that the test chemical 5 was highly active towards the growth of all the Gram-negative bacteria. The inhibitions of growth of bacteria were very remarkable in many cases which was in conformity with our previous work [32] - [35] . Of the test chemicals 6, 8, 9, and 12 were very effective towards the inhibition of growth of maximum Gram-negative bacterial strains used. In general, it has been observed that antibacterial results of the selectively acylated uridine derivatives obtained by using various acylating agents follow the order for Gram-positive organisms: 7 > 10 > 9 > 8 > 4 > 5 and Gram-negative bacteria follow the order: 5 > 6 > 8 > 9 > 10 > 12.

Table 3: Antibacterial activity of synthesized uridine derivatives with standard antibiotic.

Compound No.	Zone of inhibition (mm) at 200 μg dw/disc					
	Gram + Ve bacteria			Gram − Ve bacteria		
	B. cereus	B. megaterium	S. aureus	E. coli	S. typhi	S. paratyphi
2	NF	NF	NF	NF	NF	NF
3	NF	NF	NF	NF	NF	NF
4	5.4	NF	7	NF	8	7
5	4.5	6	6	*10	*13	*12
6	NF	5	6	NF	*10	9
7	NF	NF	*14	NF	NF	NF
8	*10	5.5	*12	8	6	7
9	NF	*11	5	6	NF	8
10	*12	*10	4	NF	5.5	NF
11	NF	NF	NF	NF	NF	NF
12	8	NF	NF	NF	4	5
13	NF	NF	NF	NF	NF	NF
14	7	NF	NF	NF	NF	NF
**Ampicillin	*14	*12	*19	*16	*20	*18

N.B: * = marked inhibition, ** = standard antibiotic, NF = not found, dw = dry weight

From the results we observed that the introduction of some specific functionalities in the test chemicals improved their antibacterial activities. In this series the presence of hexanoyl, lauroyl, myristoyl, palmitoyl, pivaloyl and 3-chlorobenzoyl groups might be responsible for the enhancement of the antibacterial capacity of the test chemicals. The results reported in Table 3 revealed that the hydrophobicity is the primary contributor to antibacterial activity. Here the hydrophobicity of the molecules increased gradually from compound 4 to 9. The hydrophobicity of materials is an important parameter with respect to such bioactivity as toxicity or alteration of membrane integrity, because it is directly related to membrane permeation [36] . Hunt [37] also proposed that the potency of aliphatic alcohols is directly related to their lipid solubility through the hydrophobic interaction between alkyl chains from alcohols and lipid regions in the membrane. We believe that a similar hydrophobic interaction might occur between the acyl chains of uridine accumulated in the lipid like nature of the bacteria membranes. As a consequence of their hydrophobic interaction, bacteria lose their membrane permeability, ultimately causing death of the bacteria [36] - [38] .

Cytotoxicity Test by Brine Shrimp Lethality Assay

The brine shrimp lethality assay is considered a useful tool for assessment of toxicity. It is based on the ability to kill laboratory-cultured Artemia nauplii brine shrimp. The assay is considered a useful tool for preliminary assessment of toxicity and it has been used for the detection of fungal toxins, plant extract toxicity, cyanobacteria toxins and cytotoxicity testing of dental materials. The cytotoxic activity of the acylated derivatives of uridine (Scheme 1 and Scheme 2) in the brine shrimp lethality bioassay is presented in (Figure 3) and shows the per- centage of mortality of shrimps at 24 hrs. The number of survived nauplii in each vial was counted for every concentration of each compound and the results were noted. However, some compounds showed a significant cytotoxicity activity in the brine shrimp lethality bioassay indicating that these compounds are biologically active. The compounds showed different rate mortality with different concentrations.

It is evident from the results of brine shrimp lethality testing that the test compounds 4, 5, 8, 13 and 14 showed highest levels of toxicity indicating its higher mortality. The mortality of brine shrimp was found to increase with the increase of concentrations of compounds. The 24 hours LD_{50} value of compound 2 showed 51.3369, and 43.9389, 62.8891, 63.0692, 50.3787, 51.3369, 43.9389, 50.1332, 51.3369, 55.7344, 54.3787, 68.7442, 69.0568 for compounds 3, 4, 5, 6, 7, 8, 9, 10, 11, 12, 13 and 14 respectively. To our knowledge, this is the first report on cytotoxic study of the newly synthesized uridine derivatives by brine shrimp lethality.

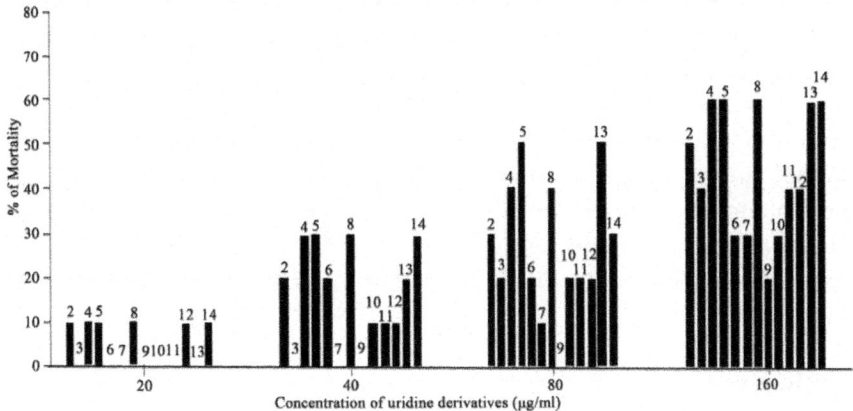

Figure 3: Graph of concentration (g/ml) versus percentage of brine shrimp mortality of the tested compounds.

CONCLUSION

In the present investigation, we synthesized and characterized a novel series of uridine derivatives and their antibacterial & cytotoxic activities assayed successfully. Simplicity and feasible reaction condition, good yield and safety are the key characteristics of this procedure. From the antibacterial results, we observed that com- pound 8, 9 and 10 were very sensitive towards all of both Gram-positive and Gram-negative bacterial organisms. Antibacterial activity of the new synthesized compounds bearing benzene and various acyl moieties revealed that some tested compounds showed good to moderate activities against selected human pathogenic strains. It has been suggested that the presence of some particular functional groups/atoms in the test compounds enhanced their sensitivities towards the growth of bacteria and also cytotoxic activities.

EXPERIMENTAL

General

All reagents used were commercially available (Sigma-Aldrich) and were used as received, unless otherwise specified. FTIR spectra were recorded by $CHCl_3$ techniques at the Chemistry Department, University of Chittagong, Bangladesh, with an IR Affinity Fourier Transform Infrared Spectrophotometer (SHIMADZU). ^{1}H- NMR (400 MHz) spectra were recorded for solutions in

CDCl$_3$using TMS as internal standard with a Bruker DPX-400 spectrometer at the Bangladesh Council of Scientific and Industrial Research (BCSIR) Laboratories, Dhaka, Bangladesh. Evaporations were carried out under reduced pressure using VV-1 type vacuum rotary evaporator (Germany) with a bath temperature below 40°C. Melting points were determined on an electro-thermal melting point apparatus (England) and are uncorrected. Column chromatography was performed with silica gel G$_{60}$. Thin layer chromatography (t.l.c) was performed on Kieselgel GF$_{254}$ and spots were detected by spraying the plates with 1% H$_2$SO$_4$ and heating at 150°C - 200°C until coloration took place.

Synthesis

Synthesis of 5'-O-N-Acetylsulfanilyluridine Derivative (2)

A suspension of uridine (1) (500 mg, 2.048 mmol) mixed with anhydrous dichloromethane (3 ml) and triethylamine (0.15 ml) which was treated with N-acetyl sulfanilyl chloride (0.703 mg). The reaction mixture was stirred at room temperature for 8 hrs. The progress of the reaction was monitored by T.l.c (methanol-chloroform, 1:12) which showed complete conversion of the starting material into faster moving product. Purification by passage through a silica gel column chromatography with methanol-chloroform (1:12) furnished the titled compound (2) as a white crystalline solid. Recrystallization from chloroform-hexane gave the 5'-O-N-acetylsulfanilyluridine (2).

5'-O-N-Acetylsulfanilyluridine (2)

FTIR (v/cm^{-1}): 1740 (C=O), 3510 (-OH), 3360 (-NH), 1365 (-SO$_2$). ^{1}H-NMR (CDCl$_3$) δ: 9.10 (1H, s, -NH), 7.78 (2H, d, J = 8.8 Hz, Ar-H), 7.76 (1H, d, J = 7.8 Hz, H-6), 7.68 (2H, d, J = 8.8 Hz, Ar-H), 7.53 (1H, s, -NH), 5.89 (1H, d, J = 5.5 Hz, H-1'), 5.81 (1H, s, 2'-OH), 5.77 (1H, dd, J = 2.2 and 12.2 Hz, H-5'a), 5.73 (1H, dd, J = 2.2 and 12.3 Hz, H-5'b), 5.66 (1H, d, J = 8.1 Hz, H-5), 5.42 (1H, s, 3'-OH), 4.42 (1H, dd, J = 2.2 and 5.4 Hz, H-4'), 4.20 (1H, d, J = 5.6 Hz, H-2'), 4.13 (1H, dd, J = 7.4 and 5.4 Hz, H-3'), 2.22 (3H, s, CH$_3$CON-). Anal. Calcd: C, 46.26; H, 4.31% for C$_{17}$H$_{19}$O$_9$N$_3$S (425). Found: C, 46.36; H, 4.35%.

General Procedure for the Synthesis of 5'-O-N-Acetylsulfanilylu-ridine Derivatives (3 - 14)

A stirred and cooled (O°C) solution of the 5'-O-N-acetylsulfanilyluridine (2) (100 mg, 0.226 mmol) in anhydrous dichloromethane (3 ml) and triethylamine

(0.15 ml) were separately treated with acetic anhydride (Ac$_2$O) (0.084 ml), pentanoyl chloride (0.13 ml), hexanoyl chloride (0.15 ml), decanoyl chloride (0.23 ml), lauroyl chloride (0.25 ml), myristoyl chloride (0.31 ml), palmitoyl chloride (0.34 ml), pivaloyl chloride (0.14 ml), benzoyl chloride (0.08 ml), 3-chlorobenzoyl chloride (0.13 ml), 4-chlorobenzoyl chloride (0.15 ml), and 2,6-di-chloro- benzoyl chloride (0.14 ml) respectively. The reaction mixture was stirred at room temperature for 8 hrs. The progress of the reaction was monitored by T.l.c (methanol-chloroform, 1:16) which showed complete conversion of the starting material into faster moving product (R$_f$ = 0.50). Excess reagent was destroyed by the addition of a few pieces of ice and the reaction mixture was extracted with chloroform (CHCl$_3$) (3 × 10 ml). The combined organic extract was washed successively with dilute hydrochloric acid (HCl), saturated aqueous sodium hydrogen carbonate (NaHCO$_3$) solution and water (H$_2$O). The organic layer was dried magnesium sulfate (MgSO$_4$), filtered, and the filtrate was evaporated off. The resulting syrupy residue was passed through silica gel column chromatography and eluted with methanol-chloroform (1:16) to afford compounds 2',3'-di-O-acetyl derivative 3, 4, 5, 6, 7, 8, 9, 10, 11, 12, 13 and 14, respectively.

5'-O-N-Acetylsulfanilyl-2,3'-di-O-Acetyluridine (3)

FTIR (v/cm^{-1}): 1768, 1760, 1684 (C = O), 3322 (-NH), 1362 (-SO$_2$). ^{1}H-NMR (CDCl$_3$) δ: 9.08 (1H, s, -NH), 7.70 (2H, d, J = 8.7 Hz, Ar-H), 7.68 (1H, d, J = 7.8 Hz, H-6), 7.62 (2H, d, J = 8.8 Hz, Ar-H), 7.51 (1H, s, -NH), 5.79 (1H, d, J = 5.5 Hz, H-1'), 5.69 (1H, dd, J = 2.2 and 12.2 Hz, H-5'a), 5.65 (1H, d, J = 2.1 and 12.3Hz, H-5'b), 5.61 (1H, d, J = 8.0 Hz, H-5), 5.13 (1H, d, J = 5.5 Hz, H-2'), 5.05 (1H, dd, J = 7.3 and 5.2 Hz, H-3'), 4.44 (1H, m, H-4'), 2.22 (3H, s, CH$_3$CON-), 2.11, 2.03 (2 × 3H, 2 × s, 2 × CH$_3$CO-). Anal. Calcd: C, 48.0; H, 4.38% for C$_{21}$H$_{23}$O$_{11}$N$_3$S (509). Found: C, 48.06; H, 4.42%.

5'-O-N-Acetylsulfanilyl-2',3'-di-O-Pentanoyluridine (4)

FTIR (v/cm^{-1}): 1738 (C = O), 3360 (-NH), 1358 (-SO$_2$). ^{1}H-NMR (CDCl$_3$) δ: 8.49 (1H, s, -NH), 7.92 (2H, d, J = 8.7 Hz, Ar-H), 7.49 (1H, d, J = 7.9 Hz, H-6), 7.47 (2H, d, J = 8.7 Hz, Ar-H), 7.35 (1H, s, -NH), 6.18 (1H, d, J = 5.6 Hz, H-1'), 5.72 (1H, dd, J = 2.2 and 12.2 Hz, H-5'a), 5.63 (1H, dd, J = 2.1 and 12.2 Hz, H-5'b), 5.48 (1H, d, J = 7.8 Hz, H-5), 4.52 (1H, d, J = 5.2 Hz, H-2'), 4.48 (1H, dd, J = 7.7 and 5.6 Hz H-3'), 4.43 (1H, m, H-4'), 2.42 {4H, m, 2 × CH$_3$(CH$_2$)$_2$CH$_2$CO-}, 2.26 (3H, s, CH$_3$CON-), 1.64 (4H,m, 2 × CH$_3$CH$_2$CH$_2$CH$_2$CO-), 1.42 {4H, m, 2 × CH$_3$CH$_2$(CH$_2$)$_2$CO-}, 0.88 {6H, m, 2 × CH$_3$(CH$_2$)$_3$CO-}. Anal. Calcd: C, 53.20; H, 5.75% for C$_{27}$H$_{35}$O$_{11}$N$_3$S (593). Found: C, 53.26; H, 5.77%.

5′-O-N-Acetylsulfanilyl-2′,3′-di-O-Hexanoyluridine (5)

FTIR (v/cm⁻¹): 1778, 1710 (C = O), 3320 (-NH), 1365 (-SO₂). ¹H-NMR (CDCl₃) δ: 8.88 (1H, s, -NH), 7.97 (2H, d, J = 8.8 Hz, Ar-H), 7.50 (1H, d, J = 7.8 Hz, H-6), 7.48 (2H, d, J = 8.8 Hz, Ar-H), 7.36 (1H, s, -NH), 6.16 (1H, d, J = 5.7 Hz, H-1'), 5.70 (1H, dd, J = 2.2 and 12.2 Hz, H-5'a), 5.66 (1H, dd, J = 2.1 and 12.0 Hz, H-5'b), 5.49 (1H, d, J = 7.8 Hz, H-5), 4.53 (1H, d, J = 5.2 Hz, H-2'), 4.49 (1H, dd, J = 7.8 and 5.6 Hz H-3'), 4.46 (1H, m, H-4'), 2.38 {4H, m, 2 × CH₃(CH₂)₃CH₂CO-}, 2.31 (3H, s, CH₃CON-), 1.63 {4H, m, 2 × CH₃(CH₂)₂CH₂CH₂CO-}, 1.29 {8H, m, 2 × CH₃(CH₂)₂CH₂CH₂CO-}, 0.95 {6H, m, 2 × CH₃(CH₂)₄CO-}. Anal. Calcd: C, 54.63; H, 6.12% for C₂₉H₃₉O₁₁N₃S (621). Found: C, 54.66; H, 6.17%.

5′-O-N-Acetylsulfanilyl-2′, 3′-di-O-Decanoyluridine (6)

FTIR (v/cm⁻¹): 1720, 1685 (C=O), 3368 (-NH), 1351 (-SO₂). ¹H-NMR (CDCl₃) δ: 8.98 (2H, s, -NH), 7.76 (2H, d, J=8.7 Hz, Ar-H), 7.73 (1H, d, J=7.7 Hz, H-6), 7.64 (2H, d, J = 8.8 Hz, Ar-H), 7.54 (1H, s,-NH), 5.79 (1H, m, H-1'), 5.68 (2H, m, H-5a' and H-5'b), 5.61 (1H, d, J = 8.2 Hz, H-5), 5.10 (1H, m, H-2'), 4.93 (1H, m, H-3'), 4.63 (1H, m, H-4'), 2.34 {4H, m, 2 × CH₃(CH₂)₇CH₂CO-}, 2.20 (3H, s, CH₃CON-), 1.64 {4H, m, 2 × CH₃(CH₂)₆CH₂CH₂CO-}, 1.25 {24H, m, 2 × CH₃(CH₂)₆CH₂CH₂CO-}, 0.89 {6H, m, 2 × CH₃(CH₂)₈CO-}. Anal. Calcd: C, 59.28; H, 7.34% for C₃₇H₅₅O₁₁N₃S (733). Found: C, 59.37; H, 7.38%.

5′-O-N-Acetylsulfanilyl-2′,3′-di-O-Lauroyluridine (7)

FTIR (v/cm⁻¹): 1708 (C = O), 3355 (-NH), 1360 (-SO₂). ¹H-NMR (CDCl₃) δ: 8.92 (1H, s, -NH), 7.78 (2H, m, Ar-H), 7.71 (1H, d, J = 7.8 Hz, H-6), 7.68 (2H, m, Ar-H), 7.55 (1H, s, -NH), 5.72 (1H, m, H-1'), 5.53 (1H, m, H-5'a), 5.50 (1H, m, H-5'b), 5.48 (1H, d, J = 8.1 Hz, H-5), 5.13 (1H, d, J = 5.6 Hz, H-2'), 4.95 (1H, m, H-3'), 4.72 (1H, m, H-4'), 2.37 {4H, m, 2 × CH₃(CH₂)₉CH₂CO-}, 2.22 (3H, s, CH₃CON-), 1.64 {4H, m, 2 × CH₃(CH₂)₈CH₂CH₂CO-}, 1.24 {32H, m, 2 × CH₃(CH₂)₈CH₂CH₂CO-}, 0.87 {6H, m, 2 × CH₃(CH₂)₁₀CO-}. Anal. Calcd: C, 61.11; H, 7.83% for C₄₁H₆₃O₁₁N₃S (789). Found: C, 61.16; H, 7.91%.

5′-O-N-Acetylsulfanilyl-2′,3′-di-O-Myristoyluridine (8)

FTIR (v/cm⁻¹): 1720 C = O), 3370 (-NH), 1368 (-SO₂). ¹H-NMR (CDCl₃) δ: 8.01 (2H, d, J = 8.7 Hz, Ar-H), 7.52 (1H, d, J = 7.8 Hz, H-6), 7.47 (2H, m, Ar-H), 7.32 (2H, s, -NH), 6.18 (1H, d, J = 5.8 Hz, H-1'), 6.15 (1H, m, H-5'a), 5.80 (1H, m, H-5'b), 5.51 (1H, d, J = 7.7 Hz, H-5), 5.48 (1H, d, J = 5.3 Hz, H-2'), 5.38 (1H, m, H-3'), 4.47 (1H, dd, J = 2.2 and 5.4 Hz, H-4'), 2.34 {4H, m, 2 × CH₃(CH₂)₁₁CH₂CO-}, 2.30 (3H, s, CH₃CON-), 1.29 {44H, m, 2

× CH$_3$(CH$_2$)$_{11}$CH$_2$CO-}, 0.87 {6H, m, 2 × CH$_3$(CH$_2$)$_{12}$CO-}. Anal. Calcd: C, 62.72; H, 8.25% for C$_{45}$H$_{71}$O$_{11}$N$_3$S (845). Found: C, 62.79; H, 8.29%.

5'-O-N-Acetylsulfanilyl-2',3'-di-O-Palmitoyluridine (9)

FTIR (v/cm^{-1}): 1718, 1680 (C = O), 3366 (-NH), 1358 (-SO$_2$). ^{1}H-NMR (CDCl$_3$) δ: 8.78 (1H, s, -NH), 7.75 (2H, d, J = 8.8 Hz, Ar-H), 7.68 (1H, m,H-6), 7.61 (2H, m, Ar-H), 7.53 (1H, s, -NH), 5.73 (1H, m, H-1'), 5.51 (2H, m, H-5'a and H-5'b), 5.47 (1H, d, J = 8.1 Hz, H-5), 5.22 (1H, d, J = 5.5 Hz, H-2'), 5.01 (1H, m, H-3'), 4.90 (1H, m, H-4'), 2.30 {4H, m, 2 × CH$_3$(CH$_2$)$_{13}$CH$_2$CO-}, 2.23 (3H, s, CH$_3$CON-), 1.22{52H, m, 2 × CH$_3$(CH$_2$)$_{13}$CH$_2$CO-}, 0.89 {9H, m, 3 × CH$_3$(CH$_2$)$_{14}$CO-}. Anal. Calcd: C, 64.12; H, 8.62% for C$_{49}$H$_{79}$O$_{11}$N$_3$S (901). Found: C, 64.19; H, 8.71%.

5'-O-N-Acetylsulfanilyl-2',3'-di-O-Pivaloyluridine (10)

FTIR (v/cm^{-1}): 1740 (C = O), 3350 (-NH), 1366 (-SO$_2$). ^{1}H-NMR (CDCl$_3$) δ: 9.01 (1H, s, -NH), 7.77 (2H, d, J = 8.8 Hz, Ar-H), 7.73 (1H, d, J = 7.8 Hz, H-6), 7.63 (2H, d, J = 8.8 Hz, Ar-H), 7.54 (1H, s, -NH), 5.80 (1H, d, J = 5.4 Hz, H-1'), 5.68 (1H, m, H-5'a), 5.59 (1H, m, H-5'b), 5.49 (1H, d, J = 8.2 Hz, H-5), 5.31 (1H, m, H-2'), 5.10 (1H, m, H-3'), 4.88 (1H,, m, H-4'), 2.20 (3H, s, CH$_3$CON-), 1.21 {18H, s, 2 × (CH$_3$)$_3$CCO-}. Anal. Calcd: C, 53.20; H, 5.75% for C$_{27}$H$_{35}$O$_{11}$N$_3$S (593). Found: C, 53.26; H, 5.81%.

5'-O-N-Acetylsulfanilyl-2',3'-di-O-Benzoyluridine (11)

FTIR (v/cm^{-1}): 1685 (C = O), 3369 (-NH), 1349 (-SO$_2$). ^{1}H-NMR (CDCl$_3$) δ: 8.9 (1H, s, -NH), 8.10 (4H, m, Ar-H), 7.88 (2H, d, J = 8.7 Hz, Ar-H), 7.71 (1H, d, J = 7.8 Hz, H-6), 7.66 (2H, m, Ar-H), 7.52 (2H, m, Ar-H), 7.49 (4H, m, Ar-H), 7.42 (1H, s, -NH), 5.78 (1H, d, J = 5.5 Hz, H-1'), 5.57 (2H, m, H-5'a and H-5'b), 5.47 (1H, d, J = 7.8 Hz, H-5), 5.13 (1H, d, J = 5.6 Hz, H-2'), 4.88 (1H, m, H-3'), 4.56 (1H, m, H-4'), 2.22 (3H, s, CH$_3$CON-). Anal. Calcd: C, 57.32; H, 4.16% for C$_{31}$H$_{27}$O$_{11}$N$_3$S (633). Found: C, 57.37; H, 4.18%.

5'-O-N-Acetylsulfanilyl-2',3'-di-O-(3-Chlorobenzoyl)Uridine (12)

FTIR (v/cm^{-1}): 1738 (C = O), 3348 (-NH), 1360 (-SO$_2$). ^{1}H-NMR (CDCl$_3$) δ: 8.10 (1H, s, -NH), 7.92 (2H, d, J = 7.7 Hz, Ar-H), 7.90 (2H, s, Ar-H), 7.83 (1H, d, J = 7.6 Hz, H-6), 7.79 (2H, d, J = 7.7 Hz, Ar-H), 7.58 (2H, d, J = 8.8 Hz, Ar -H), 7.49 (2H, d, J = 8.8 Hz, Ar -H), 7.48 (2H, t, J = 7.6 Hz, Ar -H), 7.42 (1H, s, -NH), 6.19 (1H, d, J = 6.5 Hz, H-1'), 5.58 (1H, m, H-5'a), 5.83 (1H, m, H-5'b), 5.79 (1H, d, J = 7.8 Hz, H-5), 4.78 (1H, d, J = 5.2 Hz, H-2'), 4.72 (1H, dd, J = 7.6 and 5.5 Hz, H-3'), 4.36 (1H, m, H-4'), 2.12 (3H, s, CH$_3$CON-).

Anal. Calcd: C, 51.81; H, 3.48% for $C_{31}H_{25}O_{11}N_3SCl_2$ (702). Found: C, 51.88; H, 3.53%.

5'-O-N-Acetylsulfanilyl-2',3'-di-O-(4-Chlorobenzoyl)Uridine (13)

FTIR (v/cm^{-1}): 1680 (C = O), 3357 (-NH), 1362 (-SO$_2$). ^{1}H-NMR (CDCl$_3$) δ: 8.87 (1H, s, -NH), 8.03 (4H, m, Ar-H), 7.96 (2H, d, J = 8.8 Hz, Ar-H), 7.86 (1H, d, J = 7.7 Hz, H-6), 7.47 (2H, d, J = 8.6 Hz, Ar -H), 7.46 (4H, m, Ar-H), 7.39 (1H, s, -NH), 6.23 (1H, d, J = 5.5 Hz, H-1'), 5.81 (1H, m, H-5'a), 5.62 (1H, dd, J = 2.1 and 12.2 Hz, H-5'b), 5.37 (1H, m, H-5), 4.78 (1H, m, H-2'), 4.72 (1H, m, H-3'), 4.62 (1H, m, H-4'), 2.17 (3H, s, CH$_3$CON-). Anal. Calcd: C, 51.81; H, 3.48% for $C_{31}H_{25}O_{11}N_3SCl_2$ (702). Found: C, 51.89; H, 3.55%.

5'-O-N-Acetylsulfanilyl-2',3'-di-O-(2,6-Dichlorobenzoyl)Uridine (14)

FTIR (v/cm^{-1}): 1735 (C = O), 3335 (-NH), 1349 (-SO$_2$). ^{1}H-NMR (CDCl$_3$) δ: 8.0 (2H, d, J = 8.8 Hz, Ar-H), 7.50 (1H, d, J = 7.8 Hz, H-6), 7.46 (6H, m, Ar -H), 7.42 (2H, m, Ar-H), 7.33 (1H, s, -NH), 6.29 (1H, d, J = 5.2 Hz, H-1'), 6.13 (1H, m, H-5'a), 5.82 (1H, m, H-5'b), 5.53 (1H, d, J = 7.6 Hz, H-5), 5.49 (1H, d, J = 5.1 Hz, H-2'), 5.35 (1H, m, H-3'), 4.42 (1H, m, H-4'), 2.12 (3H, s, CH$_3$CON-). Anal. Calcd: C, 47.27; H, 2.92% for $C_{31}H_{23}O_{11}N_3SCl_4$ (771). Found: C, 47.33; H, 2.97%.

Antibacterial Screening Studies

Test Microorganisms

Test tube cultures of bacterial pathogens were obtained from the Microbiology Laboratory, Department of Microbiology, University of Chittagong. The synthesized test compounds (Scheme 1 and Scheme 2) were subjected to antibacterial screening against three Gram-positive and three Gram-negative bacterial strains (Table 4).

Nutrient Agar (NA) media was used throughout the study (Table 5). In all cases, a 2% solution (in CHCl$_3$) of the chemicals was used.

Antibacterial Activity Assay

The in vitro antibacterial spectrum of the synthesized chemicals were done by disc diffusion method [26] with little modification [27] . Sterilized paper discs of 4 mm in diameter and Petri dishes of 150 mm in diameter were used throughout the experiment. The autoclaved Mueller-Hinton agar medium, cooled to 45°C, was poured into sterilized Petri dishes to a depth of 3 to 4 mm and after solidification of the agar medium the plates were transferred to an incubator at 37°C for 15 to 20 minutes to dry off the moisture that developed on the agar surface. The plates were inoculated with the standard bacterial suspensions (as McFarland 0.5 standard) followed by spread plate method and allowed to dry for three to five minutes. Dried and sterilized filter paper discs were treated separately with 50 µg dry weight/disc from 2% solution (in $CHCl_3$) of each test chemical using a micropipette, dried in air under aseptic condition and were placed at equidistance in a circle on the seeded plate. A control plate was also maintained in each case without any test chemical. These plates were kept for 4 - 6 hours at low temperature (4°C - 6°C) and the test chemicals diffused from disc to the surrounding medium. The plates were then incubated at 35°C ± 2°C for 24 hours to allow maximum growth of the microorganisms. The antibacterial activity of the test agent was determined by measuring the mean diameter of zone of inhibitions (in millimeter). Each experiment was repeated thrice. All the results were compared with the standard antibacterial antibiotic Ampicillin (20 µg/disc, BEXIMCO Pharm. Bangladesh Ltd.).

Cytotoxic Activity by Brine Shrimp Lethality Bioassay

Brine shrimp lethality assay of the acylated derivatives of uridine was performed according to McLaughlin et al. method [28] . The tested compounds were dissolved in DMSO and prepared 20, 40, 80 and 160 ml by adding

Table 4: List of used bacteria.

Types of bacteria	Name of tested bacteria	Strain no.
	Bacillus cereus	BTCC 19
Gram + Ve	*Bacillus megaterium*	BTCC 18
	Staphylococcus aureus	ATCC 6538
	Escherichia coli	ATCC 25922
Gram − Ve	*Salmonella typhi*	AE 14612
	Salmonella paratyphi	AE 146313

Table 5: The composition of Nutrient Agar (NA) media.

Sl no.	Name of ingredients	Amounts (g/or ml)
1	Beef extract	0.5 g
2	Peptone	1.0 g
3	Agar	1.5 g
4	NaCl	1.0 g
5	Distilled water	1000 mL
	Final pH 7.0 with 5N NaOH	

NaCl solution to each vial up to 1 ml volume which were designated as type-A, B, C and D, respectively. Three sets of experiment were done for each concentration and 10 brine shrimps nauplii were placed in each vial. A control experiment was performed in a vial containing 10 nauplii in 1 ml seawater. After 24 hrs of incubation at RT, the vials were observed using a magnifying glass and the number of survivors in each vial was counted and noted. No deaths were found in the controls. From the data, the average percentage of mortality of nauplii was calculated for each concentration.

ACKNOWLEDGEMENTS

The authors are grateful to the Ministry of Education (MOE), Government of the People's Republic of Bangladesh for the financial support (Ref.: 37-01.0000.078.02.018.13-50, 2014-2015) to carrying out this research work. We are also thankful to the Chairman, Bangladesh Council of Scientific and Industrial Research (BCSIR) Laboratories, Dhaka, for providing the ^{1}H-NMR spectra.

CITE THIS PAPER

Sarkar M. A.Kawsar, Hamida A.Ara,Sheikh AftabUddin, Mohammed K.Hossain,Shagir A.Chowdhury,Abul F. M.Sanaullah,Mohammad A.Manchur, ImtiajHasan, YukikoOgawa, YukiFujii, YasuhiroKoide,YasuhiroOzeki, (2015) Chemically Modified Uridine Molecules Incorporating Acyl Residues to Enhance Antibacterial and Cytotoxic Activities. *International Journal of Organic Chemistry*,**05**,232-245. doi: 10.4236/ijoc.2015.54023

REFERENCES

1.   Wen, J. and Xia, Y. (2012) Adenosine Signaling: Good or Bad in Erectile Function. Arteriosclerosis, Thrombosis, and Vascular Biology, 32, 845-850. http://dx.doi.org/10.1161/ATVBAHA.111.226803

2.   Jordheim, L.P., Durantel, D., Zoulim, F. and Dumontet, C. (2013)

Advances in the Development of Nucleoside and Nucleotide Analogues for Cancer and Viral Diseases. Nature Reviews Drug Discovery, 12, 447-464. http://dx.doi.org/10.1038/nrd4010

3.   Carlezon Jr., W.A., Mague, S.D., Parow, A.M., Stoll, A.L., Cohen, B.M. and Renshaw, P.F. (2005) Antidepressant- Like Effects of Uridine and Omega-3 Fatty Acids Are Potentiated by Combined Treatment in Rats. Biological Psychiatry, 57, 343-350. http://dx.doi.org/10.1016/j.biopsych.2004.11.038

4.   Jonas, D.A., Elmadfa, I., Engel, K.H., Heller, K.J., Kozianowski, G., Konig, A., Muller, D., Narbonne, J.F., Wackemagel, W. and Kleiner, J. (2001) Safety Considerations of DNA in Food. Annals of Nutrition & Metabolism, 45, 235-254. http://dx.doi.org/10.1159/000046734

5.   WHO (2004) The Global Burden of Disease Update. World Health Organization.   www.who.int/healthinfo/global_burden_disease/2004_report_update/en/index.html

6.   Williams, J.M. and Richardson, A.C. (1967) Selective Acylation of Pyranosides-I. Benzoylation of Methyl α-D-Gly- copyranosides of Mannose, Glucose and Galactose. Tetrahedron, 23, 1369-1378. http://dx.doi.org/10.1016/0040-4020(67)85091-9

7.   Hiroyuki, H., Horoshi, A., Hiromichi, T. and Tadashi, M. (1990) Introduction of an Alkyl Group into the Sugar Portion of Uracilnucleosides by the Use of Gilman Reagents. Chemical & Pharmaceutical Bulletin, 38, 355-360. http://dx.doi.org/10.1248/cpb.38.355

8.   Willard, J.J., Brimacombe, J.S. and Brueton, R.P. (1964) The Synthesis of 3-O-[(benzylthio)carbonyl]-β-D-Glucopy- ranose and Methyl 2,4,6-tri-O-α and β-D-Glucopyranosides. Canadian Journal of Chemistry, 42, 2560-2567. http://dx.doi.org/10.1139/v64-374

9.   Wagner, D., Verheyden, J.P.H. and Moffatt, J.G. (1974) Preparation and Synthetic Utility of Some Organotin Derivatives of Nucleosides. Journal of Organic Chemistry, 39, 24-30.http://dx.doi.org/10.1021/jo00915a005

10.   Kim, S., Chang, H. and Kim, W. (1985) Regioselective Acylation of Some Glycopyranoside Derivatives. Journal of Organic Chemistry, 50, 1751-1752. http://dx.doi.org/10.1021/jo00210a036

11.   Kabir, A.K.M.S., Dutta, P. and Anwar, M.N. (2005) Synthesis of Some New Derivatives of D-Mannose. Chittagong University Journal of Science, 29, 1-8.

12.   Ichinari, M., Nakayama, K. and Hayase, Y. (1988) Synthesis of 2,4-Dioxoimidazolidines from 2-Arylimino-1,3-thiaz- ines and Their Antifungal Activity. Heterocycles, 27, 2635-2641. http://dx.doi.

org/10.3987/COM-88-4699

13. Gawande, N.G. and Shingare, M.S. (1987) Synthesis of Some Thiazolylthiosemicarbazides, Triazoles, Oxazoles, Thiadiazoles & Their Microbial Activity. Indian Journal of Chemistry, 26, 387-394.

14. Gupta, R., Paul, S., Gupta, A.K., Kachroo, P.L. and Bani, S. (1997) Synthesis and Biological Activities of Some 2-Substituted Phenyl-3-(3-alkyl/aryl-5,6-dihydro-s-triazolo[3,4-b][1,3,4]thiazolo-6-yl)-indoles. Indian Journal of Chemistry, 36, 707-710.

15. Singh, H., Shukla, K.N., Dwivedi, R. and Yadav, L.D.S. (1990) Cycloaddition of 4-Amino-3-mercapto-1,2,4-triazole to Heterocumulenes and Antifungal Activity of the Resulting 1,2,4-Triazolo[3,4-c]-1,2-dithia-4,5-diazines. Journal of Agricultural & Food Chemistry, 38, 1483-1486. http://dx.doi.org/10.1021/jf00097a011

16. Kawsar, S.M.A., Islam, M.M., Chowdhury, S.A., Hasan, T., Hossain, M.K., Manchur, M.A. and Ozeki, Y. (2013) Design and Newly Synthesis of Some 1,2-O-isopropylidene-α-D-glucofuranose Derivatives: Characterization and Antibacterial Screening Studies. Hacettepe Journal of Biology and Chemistry, 41, 195-206.

17. Kawsar, S.M.A., Faruk, M.O., Rahman, M.S., Fujii, Y. and Ozeki, Y. (2014) Regioselective Synthesis, Characterization and Antimicrobial Activities of Some New Monosaccharide Derivatives. Scientia Pharmaceutica, 82, 1-20. http://dx.doi.org/10.3797/scipharm.1308-03

18. Kabir, A.K.M.S., Matin, M.M. and Kawsar, S.M.A. (1998) Synthesis and Antibacterial Activities of Some Uridine Derivatives. Chittagong University Journal of Science, 22, 13-18.

19. Kabir, A.K.M.S., Kawsar, S.M.A., Bhuiyan, M.M.R. and Hossain, S. (2003) Synthesis and Characterization of Some Uridine Derivatives. Journal of Bangladesh Chemical Society, 16, 6-14.

20. Kawsar, S.M.A., Ferdous, J., Mostafa, G. and Manchur, M.A. (2014) A Synthetic Approach of D-Glucose Derivatives: Spectral Characterization and Antimicrobial Studies. Chemistry & Chemical Technology, 8, 19-27.

21. Kawsar, S.M.A., Kabir, A.K.M.S., Bhuiyan, M.M.R., Ferdous, J. and Rahman, M.S. (2013) Synthesis, Characterization and Microbial Screening of Some New Methyl 4,6-O-(4-methoxybenzylidene)-α-D-glucopyranoside Derivatives. Journal of Bangladesh Academy of Sciences, 37, 145-158.

22. Kawsar, S.M.A., Kabir, A.K.M.S., Manik, M.M., Hossain, M.K. and Anwar, M.N. (2012) Antibacterial and Mycelial Growth Inhibition of Some Acylated Derivatives of D-Glucopyranoside. International Journal

of Bioscience, 2, 66-73.

23. Kabir, A.K.M.S., Kawsar, S.M.A., Bhuiyan, M.M.R., Rahman, M.S. and Chowdhury, M.E. (2009) Antimicrobial Screening Studies of Some Derivatives of Methyl α-D-Glucopyranoside. Pakistan Journal of Scientific and Industrial Research, 52, 138-142.

24. Kabir, A.K.M.S., Kawsar, S.M.A., Bhuiyan, M.M.R. and Banu, B. (2008) Synthesis of Some New Derivatives of Methyl 4,6-O-cyclohexylidene-α-D-glucopyranoside. Journal of Bangladesh Chemical Society, 21, 72-80.

25. Kabir, A.K.M.S., Kawsar, S.M.A., Bhuiyan, M.M.R., Islam, M.R. and Rahman, M.S. (2004) Biological Evaluation of Some Mannopyranoside Derivatives. Bulletin of Pure & Applied Sciences, 23, 83-91.

26. Bauer, A.W., Kirby, W.M.M., Sherris, J.C. and Turck, M. (1966) Antibiotic Susceptibility Testing by a Standardized Single Disc Method. American Journal of Clinical Pathology, 45, 493-496.

27. Miah, M.A.T., Ahmed, H.U., Sharma, N.R., Ali, A. and Miah, S.A. (1990) Antifungal Activity of Some Plant Extracts. Bangladesh Journal of Botany, 19, 5-10.

28. McLaughlin, J.L. (1991) In: Hostettmann, K., Ed., Methods in Plant Biochemistry: Assays for Bioactivity, Vol. 1, Academic Press, London, 1-31.

29. Loss, A. and Lutteke, T. (2015) Using NMR Data on GLYCOSCIENCES. de. In: Lütteke, T. and Frank, M., Eds., Glycoinformatics, Methods in Molecular Biology, Vol. 1273, Springer, New York, 87-95. http://dx.doi.org/10.1007/978-1-4939-2343-4_6

30. Brauer, B., Pincu, M., Buch, V., Bar, I., Simons, J.P. and Gerber, R.B. (2011) Vibrational Spectra of α-Glucose, β-Glucose, and Sucrose: Anharmonic Calculations and Experiment. The Journal of Physical Chemistry A, 115, 5859-5872. http://dx.doi.org/10.1021/jp110043k

31. Kawsar, S.M.A., Khaleda, M., Asma, R., Manchur, M.A., Koide, Y. and Ozeki, Y. (2015) Infrared, 1H-NMR Spectral Studies of some Methyl 6-O-myristoyl-α-D-glucopyranoside Derivatives: Assessment of Antimicrobial Effects. International Letters of Chemistry, Physics and Astronomy, 58, 122-136. http://dx.doi.org/10.18052/www.scipress.com/ILCPA.58.122

32. Kawsar, S.M.A., Faruk, M.O., Mostafa, G. and Rahman, M.S. (2014) Synthesis and Spectroscopic Characterization of Some Novel Acylated Carbohydrate Derivatives and Evaluation of Their Antimicrobial Activities. Chemistry & Biology Interface, 4, 37-47.

33. Kawsar, S.M.A., Hasan, T., Chowdhury, S.A., Islam, M.M., Hossain, M.K. and Manchur, M.A. (2013) Synthesis, Spectroscopic Characterization and in Vitro Antibacterial Screening of Some D-Glucose Derivatives. International Journal of Pure and Applied Chemistry, 8, 125-135.

34. Kawsar, S.M.A., Kabir, A.K.M.S., Bhuiyan, M.M.R., Siddiqa, A. and Anwar, M.N. (2012) Synthesis, Spectral and Antimicrobial Screening Studies of Some Acylated D-Glucose Derivatives. Rajiv Gandhi University of Health Sciences (RGUHS) Journal of Pharmaceutical Sciences, 2, 107-115.

35. Kabir, A.K.M.S., Kawsar, S.M.A., Bhuiyan, M.M.R., Rahman, M.S. and Banu, B. (2008) Biological Evaluation of Some Octanoyl Derivatives of Methyl 4,6-O-cyclohexylidene-α-D-glucopyranoside. The Chittagong University Journal of Biological Science, 3, 53-64.

36. Kim, Y.M., Farrah, S. and Baney, R.H. (2007) Structure-Antimicrobial Activity Relationship for Silanols, a New Class of Disinfectants, Compared with Alcohols and Phenols. International Journal of Antimicrobial Agents, 29, 217-222. http://dx.doi.org/10.1016/j.ijantimicag.2006.08.036

37. Hunt, W.A. (1975) The Effects of Aliphatic Alcohols on the Biophysical and Biochemical Correlates of Membrane Function. Advances in Experimental Medicine and Biology, 56, 195-210. http://dx.doi.org/10.1007/978-1-4684-7529-6_9

38. Judge, V., Narasimhan, B., Ahuja, M., Sriram, D., Yogeeswari, P., Clercq, E.D., Pannecouque, C. and Balzarini, J. (2013) Synthesis, Antimycobacterial, Antiviral, Antimicrobial Activity and QSAR Studies of N2-Acyl Isonicotinic Acid Hydrazide Derivatives. Medicinal Chemistry, 9, 53-76. http://dx.doi.org/10.2174/157340613804488404

Chapter 6

A NOVEL ONE-POT AND EFFICIENT PROCEDURE FOR SYNTHESIS OF NEW FUSED URACIL DERIVATIVES FOR DNA BINDING

Bothaina A. Mousa[1], Ashraf H. Bayoumi[2*], Makarem M. Korraa[1], Mohamed G. Assy[3], Samar A. El-Kalyoubi[4]

[1]Department of Organic Chemistry, Faculty of Pharmacy, Cairo University, Giza, Egypt

[2]Department of Organic Chemistry, Faculty of Pharmacy (Boys), Al-Azhar University, Cairo, Egypt

[3]Department of Organic Chemistry, Faculty of Science, Zagazig University, Zagazig, Egypt

[4]Department of Organic Chemistry, Faculty of Pharmacy (Girls), Al-Azhar University, Cairo, Egypt

ABSTRACT

Hydrazinolysis of 6-chloro-1-methyluracil followed by condensation of the product with different aromatic aldehyde gives the respective hydrazones which undergoes oxidative cyclization using thionyl chloride to obtain pyrazolo[3,4-d]pyrimidines in good yields. On the other hand, nitrosation of 6-aminouracils followed by the reaction with different arylidineanilines gives new xanthine derivatives. Finally, indenopyrrolopyrimidine and indenopteridine are obtained in good yields via the reaction of 6-aminouracils and 5,6-diaminouracil with ninhydrin respectively. The newly synthesized compounds show binding, chelation and fragmentation of the nucleic acid DNA.

INTRODUCTION

The importance of fused pyrimidines, common source for the development of new potential therapeutic agents [1] [2] , is well known.

Fused pyrimidines continue to attract considerable attention because of their great practical usefulness, primarily due to very wide spectrum of biological activities. This is evident especially from publications of regular reviews on the chemistry of systems where the pyrimidine ring is fused to various heterocycles such as purines, quinazolines, pyridopyrimidines, triazolopyrimidines, pyrazolopyrimidines, pyrimidoazepines, furopyrimidines and pyralopyrimidines.

5-Fluorouracil [3] -[5] and methotrexate (MTX) [6] -[8] are the oldest antifolate anticancer drugs [9] , which are widely used as chemotherapeutic drugs. They compete with the normal substrates, folic acid and dihydrofolate, for the active site on the enzyme dihydrofolate reductase (DHFR) [10] -[12] .

Pyrido[2,3-d]pyrimidines possess dihydrofolate reductase inhibiting and antitumour activity [13] . Similarly, in recent years, considerable attention has been focused on the development of new methodology to synthesize many kinds of pyrazolopyrimidine ring [14] . Indeed, pyrazolopyrimidines [15] [16] and purines [17] represent an important class of heterocyclic compounds having wide range of pharmaceutical and biological activities. Therefore, versatile and widely applicable methods for the synthesis pyrazolopyrimidines and purines are of considerable interest. The existing methods for the preparation of pyrazolopyrimidines are based on heterocyclic hydrazones or hydrazine precursors. Pyrimidines and their derivatives are considered to be important for drugs. A large number of pyrimidine derivatives are reported to exhibit antimycobacterial [18] , antitumor [19] , antiviral [20] , anticancer [21] [22] activities. In the present study, a series of new pyrimidine fused ring analogs have been synthesized and their biological effects are determined.

MATERIAL AND METHODS

Chemistry

All melting points were determined with an Electrothermal Mel.-Temp. II apparatus and were uncorrected. Element analyses were performed at the Micro Analytical Unit, Chemistry Department, Mansoura University. The infrared (IR) spectra were recorded using potassium bromide disc technique on Nikolet IR 200 FT IR at Pharmaceutical Analytical Unit, Faculty of Pharmacy, Al-Azhar University. The proton nuclear magnetic resonance (^{1}H-NMR) spectra were recorded on Varian Gemini 300 MHz Spectrometer using DMSO-d_6 as a solvent and tetramethylsilane (TMS) as an internal standard (Chemical shift in δ, ppm), Faculty of Science, Chemistry Department, Cairo University. Mass spectra were recorded on DI-50 unit of Shimadzu GC/MS-QP 5050A at the Regional Center for Mycology and Biotechnology at Al-Azhar University. All reactions were monitored by TLC using precoted plastic sheets silica gel (Merck 60 F_{254}) and spots were visualized by irradiation with UV light (254 nm). The used solvent system was chloroform: methanol (9:1) & ethyl acetate: toluene (1:1).

6-(2-Arylidenehydrazin-1-yl)-1-methyluracils (4a-f) [23]

A mixture of 6-hydrazinyl-1-methyluracil (3) (0.4 g, 2.5 mmol) and the appropriate aromatic aldehyde (2.5 mmol) in ethanol (25 ml) was stirred at room temperature for 1.5 - 2 hours. The formed precipitate was filtered, washed with ethanol and crystallized from DMF/ethanol (2:1) into yellow crystals.

Benzaldehyde(3-methyl-2,6-dioxo-1,2,3,6-tetrahydropyrimidin-4-yl) hydrazine 4a [23] : Yield: 81%, m.p. = 276°C - 277°C [23] .

4-Methoxybenzaldehyde(3-methyl-2,6-dioxo-1,2,3,6-tetrahydropyrimidin-4-yl)hydrazine 4b [23] : Yield: 94%, m.p. = 266°C - 268°C [23] .

4-Hydroxybenzaldehyde(3-methyl-2,6-dioxo-1,2,3,6-tetrahydropyrimidin-4-yl)hydrazine 4c: Yield: 79%, m.p. = 254°C - 256°C. IR= 3300-3136 overlapped (OH &NH), 3010 (CH-arom.), 2840 (CH-aliph.), 1706 (2 C = O), 1644 (C = N), 832 (p-substituted phenyl). Anal. Calcd for $C_{12}H_{12}N_4O_3$(260.25), Calcd.: C, 55.38, H, 4.65, N, 21.35, Found: C, 55.42, H, 4.70, N, 21.65.

3-Chlorobenzaldehyde(3-methyl-2,6-dioxo-1,2,3,6-tetrahydropyrimidin-4-yl)hydrazine 4d: Yield: 91%, m.p. = 261°C - 263°C. IR= 3252 (NH), 3102 (CH-arom.), 2900 (CH aliph.), 1728 (2 C = O), 1636 (C = N), 700 & 786 (m-substituted phenyl). Anal. Calcd for $C_{12}H_{11}ClN_4O_2$ (278.69), Calcd.: C, 51.72, H, 3.98, N, 20.10, Found: C, 51.41, H, 4.45, N, 20.28.

4-Chlorobenzaldehyde(3-methyl-2,6-dioxo-1,2,3,6-tetrahydropyrimidin-4-yl)hydrazine 4e [23] : Yield: 92%, m.p. = 273°C - 275°C [23] .

4-Hydroxy-3-methoxybenzaldehyde(3-methyl-2,6-dioxo-1,2,3,6-tetrahydropyrimidin-4-yl) hydrazine 4f: Yield: 82%, m.p. = 250°C - 252°C. IR= 3494 (OH), 3180 (NH), 3036 (CH-arom.), 2920 (CH aliph.), 1708 (2 C = O), 1644 (C = N), 820, 760 (substituted phenyl). Anal. Calcd for $C_{13}H_{14}N_4O_4$ (290.27), Calcd.: C, 53.79, H, 4.86, N, 19.30, Found: C, 53.60, H, 4.53, N, 19.10.

3-Aryl-7-methyl-1H-pyrazolo[3,4-d]pyrimidine-4,6(5H,7H)-diones (5a-f)

A mixture of the appropriate 6-(2-arylidenehydrazin-1-yl)-1-methyluracil (4a-f) (1.2 mmol) and excess of thionyl chloride (2 ml) was heated under reflux for 5 - 7 minutes. The excess thionyl chloride was evaporated under reduced pressure. An adequate amount of aqueous ammonia solution was added to the residue. The formed precipitate was filtered, washed with ethanol and crystallized from DMF/ethanol (3:1).

7-Methyl-3-phenyl-1H-pyrazolo[3,4-d]pyrimidine-4,6(5H,7H)-dione 5a: Yield: 69%, m.p. = 212°C - 214°C. IR = 3172 (NH), 3050 (CH arom.), 2940, 2868 (CH aliph.), 1716, 1672 (2 C = O), 1563 (C = N) & (C = C). ^{1}H-NMR (DMSO-d$_6$) δ ppm: 11.80 (bs, 1H, NH), 10.86 (s, 1H, NH), 8.06 (s, 2H, arom.), 7.48 (s, 3H, arom.), 3.52 (s, 3H, NCH$_3$). Anal. Calcd for C$_{12}$H$_{10}$N$_4$O$_2$ (242.23), Calcd.: C, 59.50, H, 4.16, N, 23.13, Found: C, 58.96, H, 4.35, N, 23.01.

3-(4-Methoxyphenyl)-7-methyl-1H-pyrazolo[3,4-d]pyrimidine-4,6(5H,7H)-dione 5b: Yield: 54%, m.p. = 215°C - 217°C. IR = 3173 (NH), 3052 (CH arom.), 2943, 2815 (CH aliph.), 1683 (br, 2 C = O), 1563 (C = N) & (C = C), 844 (p-substituted phenyl). MS: m/z (%) = 272 (M$^+$, 2.52), 85 (100). Anal. Calcd for C$_{13}$H$_{12}$N$_4$O$_3$ (272.25), Calcd.: C, 57.35, H, 4.44, N, 20.58, Found: C, 57.01, H, 4.44, N, 20.32.

3-(4-Hydroxyphenyl)-7-methyl-1H-pyrazolo[3,4-d]pyrimidine-4,6(5H,7H)-dione 5c: Yield: 51%, m.p. = 210°C - 212°C. IR = 3430 (OH), 3174 (NH), 3053 (CH arom.), 2939, 2817 (CH aliph.), 1697 (2 C = O), 1583 (C = N) & (C = C), 840 (p-substituted phenyl). Anal. Calcd for C$_{12}$H$_{10}$N$_4$O$_3$(258.23), Calcd.: C, 55.81, H, 3.90, N, 21.70, Found: C, 55.56, H, 3.79, N, 21.41.

3-(3-Chlorophenyl)-7-methyl-1H-pyrazolo[3,4-d]pyrimidine-4,6(5H,7H)-dione 5d: Yield: 67%, m.p. = 239°C - 241°C. IR = 3171 (NH), 3051 (CH arom.), 2920, 2853 (CH aliph.), 1716, 1676 (2 C = O), 1564 (C = N) & (C = C), 756, 668 (m-substituted phenyl). Anal. Calcd for C$_{12}$H$_9$ClN$_4$O$_2$(276.67), Calcd.: C, 52.09, H, 3.28, N, 20.25, Found: C, 51.65, H, 3.68, N, 20.40.

3-(4-Chlorophenyl)-7-methyl-1H-pyrazolo[3,4-d]pyrimidine-4,6(5H,7H)-dione 5e: Yield: 72%, m.p. = 195°C - 197°C. IR = 3171 (NH), 3050 (CH arom.), 2936, 2854 (CH aliph.), 1713, 1670 (2 C = O), 1565 (C = N) & (C = C), 824 (p-substituted phenyl). Anal. Calcd for C$_{12}$H$_9$ClN$_4$O$_2$(276.67), Calcd.: C, 52.09, H, 3.28, N, 20.25, Found: C, 52.09, H, 3.18, N, 20.08.

3-(4-Hydroxy-3-methoxyphenyl)-7-methyl-1H-pyrazolo[3,4-d] pyrimidine-4,6(5H,7H)-dione 5f: Yield: 63%, m.p. = 192°C - 194°C. IR = 3426 (OH), 3172 (NH), 3051 (CH arom.), 2930, 2879 (CH aliph.), 1721, 1671 (2 C = O), 1561 (C = C), 843, 665 (substituted phenyl). Anal. Calcd for C$_{13}$H$_{12}$N$_4$O$_4$ (288.25), Calcd.: C, 54.17, H, 4.20, N, 19.44, Found: C, 54.52, H, 4.02, N, 19.31.

6-Amino-1-[(2-chlorophenyl)methyl]-5-nitrosouracil (7)

A mixture of 6-amino-1-[(2-chlorophenyl)methyl]uracil (6) (2.0 g, 7.9 mmol) was suspended in water (90 ml) in the presence of glacial acetic acid (0.4 ml) and sodium nitrite (0.54 g, 7.9 mmol) in water (5 ml) was stirred at room temperature for 1/2 hr. The formed cherry red precipitate was filtered, washed with ethanol and crystallized from ethanol into violet crystals 7, Yield:

95%, m.p. = 235°C - 237°C. IR = 3479 (N-OH), 3338, 3251 (NH$_2$ & NH), 3077 (CH arom.), 2979, 2804 (CH aliph.), 1690, 1638 (2 C = O), 751 (o-substituted phenyl). Anal. Calcd for C$_{11}$H$_9$ClN$_4$O$_3$ (280.66), Calcd.: C, 47.07, H, 3.23, N, 19.96, Found: C, 47.03, H, 3.20, N, 19.72.

8-Aryl-3-[(2-chlorophenyl)methyl]-7-hydroxyxanthines (8a-d)

A mixture of 6-amino-1-[(2-chlorophenyl)methyl]-5-nitrosouracil (7) (0.3 g, 1.06 mmol) and the appropriate N-arylidene aniline (1.06 mmol) in glacial acetic acid (3 ml) was heated under reflux for 8 - 10 hours. After cooling, the formed precipitate was filtered, washed with ethanol and crystallized from DMF/ethanol (2:1) into colourless crystals.

3-(2-Chlorobenzyl)-8-(4-chlorophenyl)-7-hydroxyxanthine (8a): Yield: 86%, m.p. = >330°C. IR = 3300 - 2900 (br, OH), 3143 (NH), 3042 (CH arom.), 2823 (CH aliph.), 1695 (C = O), 1548 (C = C), 838 (p-substituted phenyl), 747 (o-substituted phenyl). ^{1}H-NMR (DMSO-d$_6$) δ 14.01 (s, 1H, OH, exchangeable), 11.33 (s, 1H, NH, exchangeable), 8.06 - 8.04 (d, 2H, arom.), 7.56 - 7.52 (d, 2H, arom.), 7.49 (d, 1H, arom.), 7.33 - 7.22 (m, 2H, arom.), 7.08 - 7.05 (d, 1H, arom.), 5.24 (s, 2H, NCH$_2$). Anal. Calcd for C$_{18}$H$_{12}$Cl$_2$N$_4$O$_3$ (403.21), Calcd.: C, 53.62, H, 3.00, N, 13.89, Found: C, 53.81, H, 3.15, N, 13.80.

8-(4-Bromophenyl)-3-(2-chlorobenzyl)-7-hydroxyxanthine (8b): Yield: 81%, m.p. = >330°C. IR = 3300 - 2900 (br, OH), 3150 (NH), 3024 (CH arom.), 2940, 2819 (CH aliph.), 1697 (C = O), 1552 (C = C), 835 (p-substituted phenyl), 749 (o-substituted phenyl). Anal. Calcd for C$_{18}$H$_{12}$BrClN$_4$O$_3$(447.66), Calcd.: C, 48.29, H, 2.70 , N, 12.52, Found: C, 48.49, H, 3.20, N, 12.86.

3-(2-Chlorobenzyl)-7-hydroxy-8-(4-nitrophenyl)xanthine 8c: Yield: 89%, m.p. = >330°C. IR = 3300 - 3000 (br, OH), 3147 (NH), 3025 (CH arom.), 2925, 2850 (CH aliph.), 1693 (C = O), 1559 (C = C), 1520, 1343 (NO$_2$), 859 (p-substituted phenyl), 752 (o-substituted phenyl). Anal. Calcd for C$_{18}$H$_{12}$ClN$_5$O$_5$ (413.77), Calcd.: C, 52.25, H, 2.92, N, 16.93, Found: C, 52.40, H, 2.90, N, 17.22.

3-(2-Chlorobenzyl)-8-(4-fluorophenyl)-7-hydroxyxanthine 8d: Yield: 63%, m.p. = >330°C. IR = 3300 - 3000 (br, OH), 3155 (NH), 3025 (CH arom.), 2923, 2849 (CH aliph.), 1695 (C = O), 1562 (C = C), 843 (p-substituted phenyl), 747 (o-substituted phenyl). ^{1}H-NMR (DMSO-d$_6$) δ 13.98 (s, 1H, OH), 11.28 (s, 1H, NH), 8.08 - 8.04 (m, 2H, arom.), 7.50 - 7.47 (d, 1H, arom.), 7.33 - 7.24 (m, 4H, arom.), 7.04 - 7.02 (d, 1H, arom.), 5.23 (s, 2H, NCH$_2$). MS: m/z (%) = 388 (M$^+$+2, 1.8), 386 (M$^+$, 4.53), 125 (100). Anal. Calcd for C$_{18}$H$_{12}$ClFN$_4$O$_3$(386.76), Calcd.: C, 55.90, H, 3.13, N, 14.49, Found: C, 56.06, H, 3.50, N, 14.14. 1-Benzyl[or(2-chlorophenyl)methyl]-4b,9b-dihydroxy-9b,10-dihydroindeno[2',1':4,5] pyrrolo[2,3- d]pyrimidine-2,4,5(1H,3H,4bH)-triones

(9a,b) A mixture of the appropriate 6-amino-1-benzyl-[or (2-chlorophenyl) methyl]uracil (6a,b) (1.2 mmol) and ninhydrin (0.2 g, 1.2 mmol) in ethanol (20 ml) was heated under reflux for 1 hour. The formed precipitate on hot was filtered, washed with ethanol and crystallized from ethanol.

1-Benzyl-4b,9b-dihydroxy-9b,10-dihydroindeno[2',1':4,5]pyrrolo[2,3-d] pyrimidine-2,4,5 (1H,3H,4bH)-trione 9a: Yield: 68%, m.p. = 270°C - 272°C. IR = 3544 - 3000 (br, OH), 3286, 3182 (NH), 3025 (CH arom.), 2922, 2845 (CH aliph.), 1709, 1656 (C = O), 1553 (C = C), 769, 702 (monosubstituted phenyl). ^{1}H-NMR (DMSO-d_6) δ 10.39 (s, 1H, NH), 9.37 (s, 1H, NH), 7.85 - 7.80 (m, 2H, arom.), 7.70 - 7.68 (d, 1H, arom.), 7.59 - 7.57 (d, 1H, arom.), 7.29 - 7.27 (m, 3H, arom.), 7.18 - 7.16 (m, 2H, arom.), 6.81 (s, 1H, OH), 5.98 (s, 1H, OH), 4.94 - 4.90 (d, 1H, NCH$_2$), 4.80 - 4.67 (d, 1H, NCH$_2$). Anal. Calcd for $C_{20}H_{15}N_3O_5$ (377.35) Calcd.: C, 63.66, H, 4.01, N, 11.14, Found: C, 63.46, H, 4.10, N, 10.82.

1-(2-Chlorophenyl)methyl-4b,9b-dihydroxy-9b,10-dihydroindeno[2',1':4,5]pyrrolo[2,3-d]pyrimidine- 2,4,5(1H,3H,4bH)-trione 9b: Yield: 71%, m.p. = 272°C - 273°C. IR = 3600 - 2900 (br, OH), 3286, 3182 (NH), 3025 (CH arom.), 2844 (CH aliph.), 1712, 1661 (C = O), 1560 (C = C), 762 (o-substituted phenyl). ^{1}H-NMR (DMSO-d_6) δ 10.48 (s, 1H, NH, exchangeable), 9.41 (s, 1H, NH, exchangeable), 7.79 - 7.77 (d, 2H, arom.), 7.72 - 7.69 (d, 1H, arom.), 7.60 - 7.49 (m, 2H, arom.), 7.35 - 7.21 (m, 2H, arom.), 6.83 - 6.80 (d, 1H, arom.), 6.78 (s, 1H, OH, exchangeable), 5.97 (s, 1H, OH, exchangeable), 5.04 - 4.98 (d, 1H, NCH$_2$), 4.81 - 4.75 (d, 1H, NCH$_2$). MS: m/z (%) = 414 (M$^+$+2, 0.2), 412 (M$^+$, 0.47), [395 (1.2), 393 (3.25, M$^+$−H$_2$O)], 44 (100). Anal. Calcd for $C_{20}H_{14}ClN_3O_5$ (411.79) Calcd.: C, 58.33, H, 3.43, N, 10.20, Found: C, 58.31, H, 3.42, N, 9.87.

1,3-Dimethyl-2H-indeno[2,1-g]pteridine-2,4,6-(1H,3H)-trione (11)

Two methods were applied for the synthesis of 11:

A) A mixture of 5,6-diamino-1,3-dimethyluracil hydrochloride (10) (0.2 g, 1.00 mmol) and ninhydrin (0.18 g, 1.00 mmol) in ethanol (10 ml) and drops of TEA was added to adjust pH = 8. The reaction mixture was stirred at room temperature for 30 minutes. The formed precipitate was filtered, washed with ethanol and crystallized from ethanol into yellow crystals.

B) A mixture of 5,6-diamino-1,3-dimethyluracil hydrochloride (10) (0.2 g, 1.00 mmol) and ninhydrin (0.18 g, 1.00 mmol) in water (15 ml) and few drops of ammonium hydroxide solution was added to adjust pH = 8 was stirred at room temperature for 1 hour. The formed precipitate was filtered, washed with ethanol and crystallized from ethanol. Yield: A 60.7%, B 59%, m.p. = >320°C. IR = 3066 (CH arom.), 2934, 2870 (CH aliph.), 1723, 1672 (C = O), 1567 (C =

N), 1508 (C = C). ^{1}H-NMR (DMSO-d$_6$) δ 7.97 - 7.95 (d, 1H, arom.), 7.85 - 7.81 (m, 2H, arom.), 7.74 - 7.69 (d, 1H, arom.), 3.66 (s, 3H, NCH$_3$), 3.41 (s, 3H, NCH$_3$). MS: m/z (%) = 294 (M$^+$, 100). Anal. Calcd for C$_{15}$H$_{10}$N$_4$O$_3$ (294.26) Calcd.: C, 61.22, H, 3.43, N, 19.04, Found: C, 61.03, H, 3.06, N, 18.60.

Biological Evaluation

Nucleic Acids Preparation

For extraction of genomic DNA, yeast cells were washed with cold phosphate borate sodium chloride (PBS) buffer and lysed in a buffer containing 50 mM Tris-HCl (pH 8.0), 1 mM EDTA, 0.2% Triton X-100 for 20 min at 4˚C. After centrifugation at 14,000 rpm for 15 min, the supernatant was treated with proteinase K (0.5 mg/ml) and 1% SDS for 1 h at 50˚C. DNA was extracted twice with buffered phenol/chloroform and precipitated with 140 mM NaCl and 2 volumes of ethanol at −20˚C overnight. DNA precipitates were washed twice with 70% ethanol, air-dried and dissolved in TE buffer, and treated for 1 h at 37˚C with RNase A according to reported method [24] . Finally, DNA preparations were electrophoresed in 1% agarose gels.

Agrose Gel Preparation and Visualization of DNA

1% agarose gel was prepared by adding 1 gm ultra agarose to 100 ml Tris-Acetate-EDTA (TAE) buffer and heated in a microwave oven then cooled to ~60˚C before pouring in gel tray.

Examination of the gel was carried out using ultraviolet illuminated box. Ethidium bromide (0.1 mg/ml) solution was used to stain the nucleic acid (DNA bands) in the gel as it intercalates between DNA bases and give florescence. The gel was photographed using polarized camera.

Nucleic Acid Affinity, Binding and Fragmentation Assay

The test compounds were dissolved in DMSO at 20 µg/µl concentrations, mixed with 2 µg/µl DNA and incubated at room temperature for 2 hrs. The mixtures were mixed with the gel loading buffer and then electrophoresed in the agarose gel (1% w/v) at 80 V for 1.5 hrs. As positive control for affinity, binding and fragmentation, methotrexate (20 µg/µl) was mixed with DNA , and as negative control DMSO was mixed with equal amount of DNA . After running, agarose gels were stained with ethidium bromide and visualized using polarized camera.

RESULTS AND DISCUSSION

Chemistry

6-Chlorouracils (1) were prepared by the alkaline hydrolysis of 2,4,6-trichloropyrimidines [25] [26] . Methylation of 6-chlorouracil (1) was carried out with methyl iodide in the presence of potassium carbonate applying a reported procedure [27] . 6-Hydrazinyl-1-methyluracil (3) [28] was prepared in a good yield by the reaction of 6-chloro-1-methyluracil (2) with alcoholic hydrazine hydrate at room temperature following a reported method [23] . In this investigation the title compounds were furnished through the hydrazinolysis of 6-chloro-1-methy- luracil (2). Condensation of the hydrazinylpyrimidine 3 with aromatic aldehydes gave the respective hydrazones 4a-f. Oxidative cyclization of 4 using thionyl chloride produced pyrazolopyrimidines 5a-f in good yields.

Thus, refluxing of compounds 4a-f with thionyl chloride resulted in intramolecular cyclization affording pyrazolopyrimidines 5a-f presumably via the formation of the 5-chlorosulfinyl derivatives A which loses (SO) group and HCl to form Xa-f. The structure of target compounds was confirmed by element analysis in addition to IR, ^{1}H-NMR spectral data.

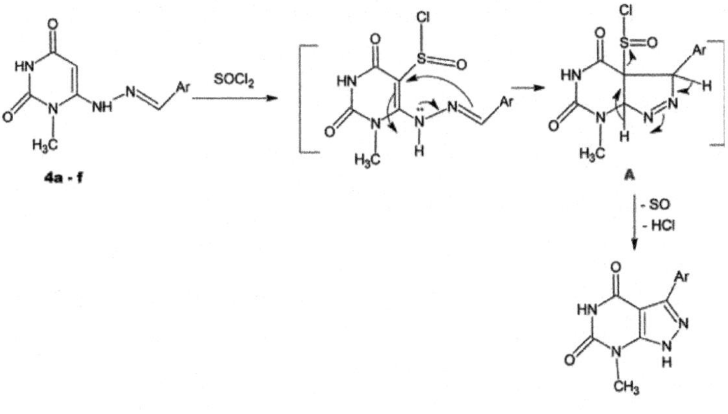

4a - f

5a - f

Compounds 6a,b were prepared in good yields by the condensation of ethyl cyano-acetate with N-benzyluea [29] or N-[(2-chlorophenyl)methyl]urea [30] in sodium ethoxide or methoxide. In this work, it was in need to prepare first the unavailable starting material, 6-amino-1-[(2-chlorophenyl)methyl]-5-nitrosouracil (7). Reaction of 6-amino-1-[(2-chlorophenyl)methyl]uracil (6b) with aqueous sodium nitrite in the presence of acetic acid afforded a high yield of the coloured nitroso derivative 7 [31] . Thus, reaction of 7 with different arylidene aniline in acetic acid took place by the elimination of aniline to give 8a-d.

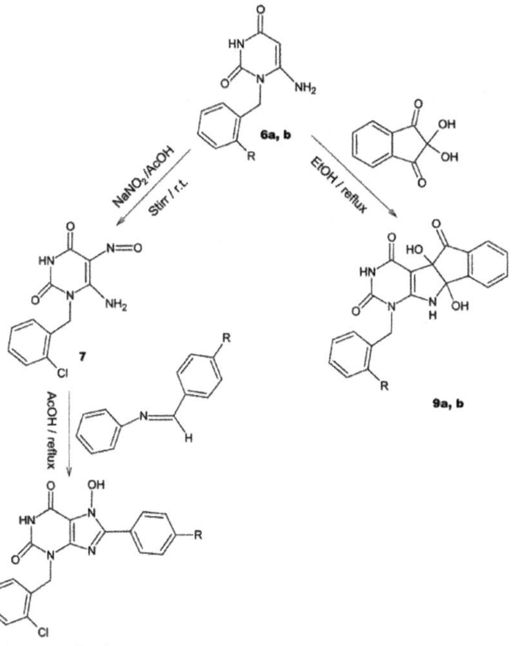

An expected mechanism might be as follows:

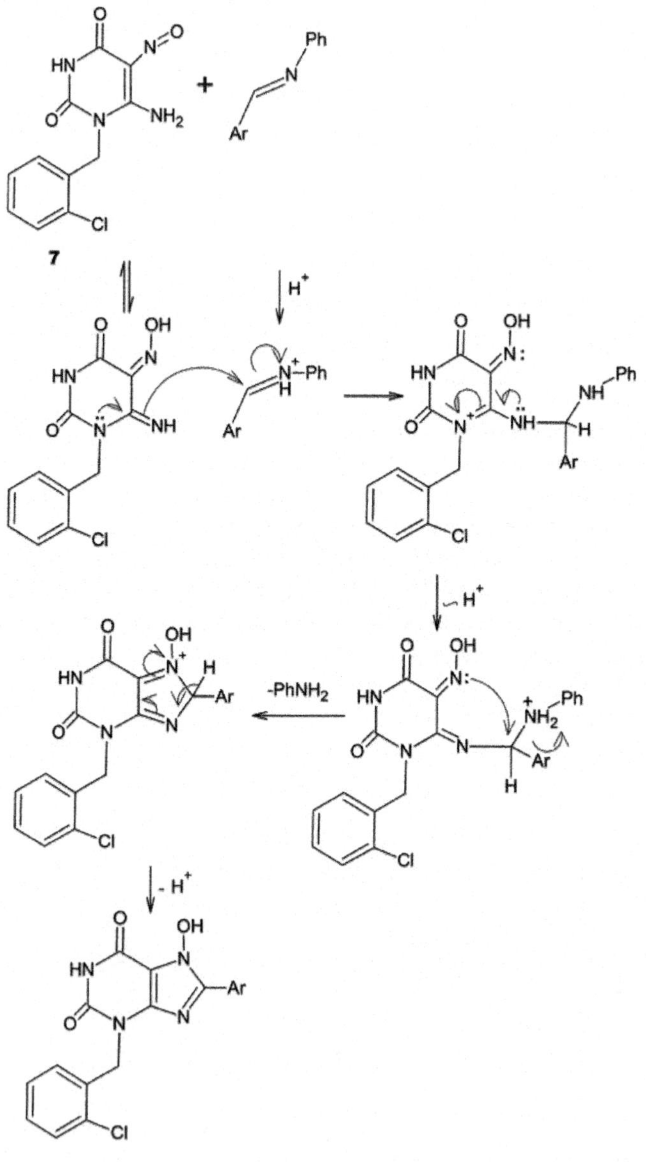

8a - d

On the other hand, the reaction of aminouracils 6a,b by refluxing with ninhydrin in DMF resulted in the formation of indenopyrrolopyrimidines 9a,b in a moderate yields. It was reported that the 2-position of the ninhydrin is more reactive towards nitrogen [32] , oxygen [32] [33] and carbon based nucleophiles [32] -[34] . The cyclization affording 9a,b presumably occurred via the formation of nonisolable acyclic intermediate. The latter might be formed via the attack of the more nucleophilic carbon at 5-position of uracil to the more reactive center at 2-position of ninhydrin. Cyclization could be affected via the addition of the amino group to the carbonyl at 1-position of ninhydrin moiety affording the final product 9a,b. ^{1}H-NMR showed the two benzylic hydrogens as two doublets at $\delta = 4.94 - 4.67$ ppm which indicated that they were not magnetically equivalent. This observation may be attributed to the presence of stereoisomers resulted from the two asymmetric carbons at 4b and 9b positions.

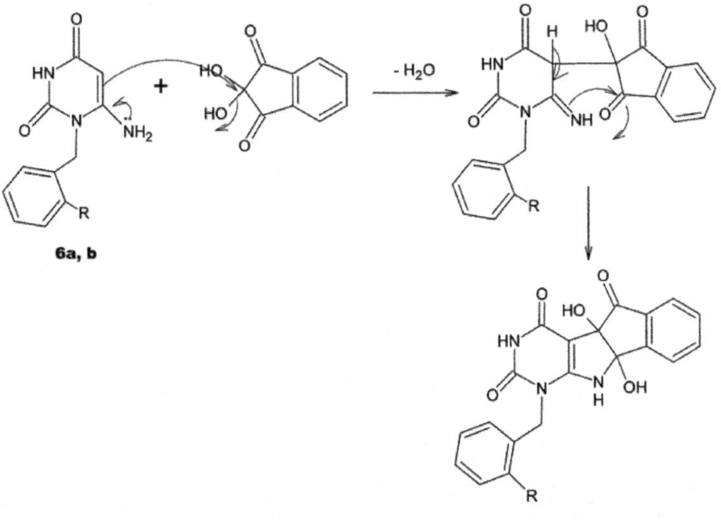

6a, b

9a. b

The project now directed towards the possible utility of diaminouracils for the synthesis of the title compound 11. Thus, the reaction of 5,6-diamino-1,3-dimethyluracil hydrochloride (10) [29] [35] -[37] with ninhydrin in the presence of triethylamine or ammonium hydroxide afforded the title compound 11.

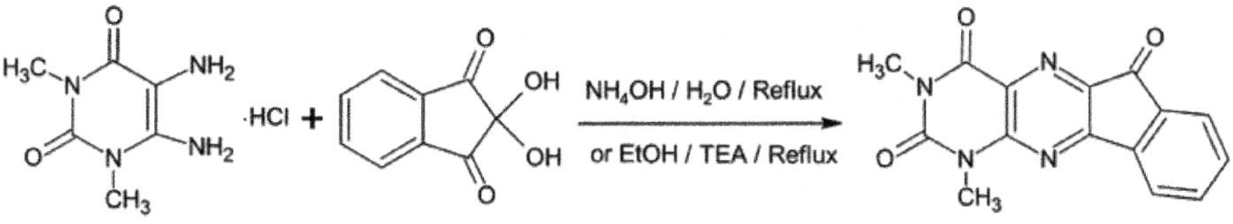

10

11

The formation of 11 from the aminouracil 10 and ninhydrin may be proceed through first condensation between the more reactive NH_2 at 5-position with the more electrophilic center at C-2 of ninhydrin. Attack of the less reactive

NH_2 group at 6-position to one of the C = O groups of the reagent afforded the cyclized tautomer B which was stabilized by loss of H_2O to give 11.

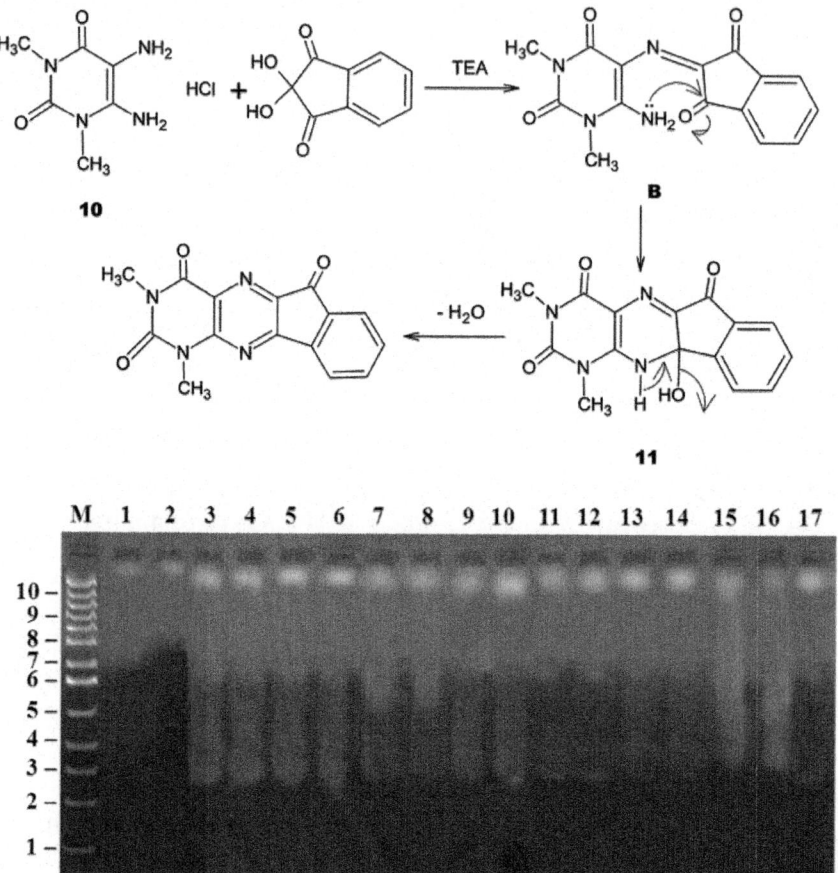

Figure 1: Gel electrophoresis 1% w/v agrose of untreated and treated DNA. Lane M: Molecular weight marker (left side); Lane 1: Untreated nucleic acid; Lane 2: DMSO treated nucleic acid (negative control); Lane 3: Methotrexate treated nucleic acid (positive control); Lanes 4-17: Compounds (8a-d, 7, 9a, b, 11, 5a-f) treated nucleic acid.

Biological Evaluation

The newly synthesized compounds were subjected to nucleic acid binding assay using agarose gel electrophoresis method.

Nucleic Acids Binding Assay

Different synthetic drugs induced DNA damage was evaluated by measuring the level of genomic DNA fragmentation and detecting DNA ladders on agarose gel electrophoresis (Figure 1). Compared with the vehicle control group (lane 2 negative control and lane 3 positive control), there was no significant change in genomic DNA fragmentation in some treated groups. There were major differences in the response of extracted DNA (from Lanes 4-17 in Figure 1). It is possible that drugs exert its effect solely by indirect mechanisms. This contrast may have been due to different enzyme(s) being with differing susceptibilities to drugs.

CONCLUSION

Our results describe a simple and efficient method for the synthesis of different novel fused uracils. Heteroannulation on the C-5 of uracil usually requires forcing conditions and complex synthetic pathways. Our synthetic compounds concern with the reactions of uracils with different benzylideneaniline, araldehydes and ninhydrin which have a biological screen.

ACKNOWLEDGEMENTS

The authors wish to thank Dr. Yassin El-Ayouty, professor of Microbiology, Botany and Microbiology Department, Faculty of Science, Zagazig University, Zagazig, Egypt, for carrying out the biological activities of the new synthesized compounds.

REFERENCES

1. Brown, D.J. (1984) Pyrimidines and Their Benzo Derivatives. Comprehensive Heterocyclic Chemistry, 3, 57-155. http://dx.doi.org/10.1016/B978-008096519-2.00035-7

2. Wamhoff, H., Dzenis, J. and Hirota, K. (1992) Uracils: Versatile Starting Materials in Heterocyclic Synthesis. Advances in Heterocyclic Chemistry, 55, 129-259.http://dx.doi.org/10.1016/S0065-2725(08)60222-6

3. González-Vallinas, M., Molina, S., Vicente, G., de la Cueva, A., Vargas, T., Santoyo, S., García-Risco, M.R., Fornari, T., Reglero, G. and de Molina, A.R. (2013) Antitumor Effect of 5-Fluorouracil Is Enhanced by Rosemary Extract in Both Drug Sensitive and Resistant Colon Cancer Cells. Pharmacological Research, 72, 61-68.http://dx.doi.org/10.1016/j.phrs.2013.03.010

4. Innominato, P.F., Lévi, F.A. and Bjarnason, G.A. (2010) Chronotherapy and the Molecular Clock: Clinical Implications in Oncology. Advanced Drug Delivery Reviews, 62, 979-1001.http://dx.doi.org/10.1016/j.addr.2010.06.002

5.   Isanbor, C. and O'Hagan, D. (2006) Fluorine in Medicinal Chemistry: A Review of Anti-Cancer Agents. Journal of Fluorine Chemistry, 127, 303-319.http://dx.doi.org/10.1016/j.jfluchem.2006.01.011

6.   Muzzalupo, R., Tavano, L. and La Mesa, C. (2013) Alkyl Glucopyranoside-Based Niosomes Containing Methotrexate for Pharmaceutical Applications: Evaluation of Physico-Chemical and Biological Properties. International Journal of Pharmaceutics, 458, 224-229.http://dx.doi.org/10.1016/j.ijpharm.2013.09.011

7.   Wu, Z.Q., Shah, A., Patel, N. and Yuan, X.D. (2010) Development of Methotrexate Proline Prodrug to Overcome Resistance by MDA-MB-231 Cells. Bioorganic Medicinal Chemistry Letters, 20, 5108-5112. http://dx.doi.org/10.1016/j.bmcl.2010.07.024

8.   Pectasides, D., Pectasides, E., Papaxoinis, G., Xiros, N., Kamposioras, K., Tountas, N. and Economopoulos, T. (2010) Methotrexate, Paclitaxel, Ifosfamide, and Cisplatin in Poor-Risk Nonseminomatous Germ Cell Tumors. Urologic Oncology: Seminars and Original Investigations, 28, 617-623. http://dx.doi.org/10.1016/j.urolonc.2008.10.013

9.   Banerjee, D., Mayer-Kuckuk, P., Capiaux, G., Budak-Alpdogan, T., Gorlick, R. and Bertino, J.R. (2002) Novel Aspects of Resistance to Drugs Targeted to Dihydrofolate Reductase and Thymidylate Synthase. Biochimica et Biophysica Acta (BBA)—Molecular Basis of Disease, 1587, 164-173. http://dx.doi.org/10.1016/S0925-4439(02)00079-0

10.   Mackey, J.R., Baldwin, S.A., Young, J.D. and Cass, C.E. (1998) Nucleoside Transport and Its Significance for Anticancer Drug Resistance. Drug Resistance Updates, 1, 310-324.http://dx.doi.org/10.1016/S1368-7646(98)80047-2

11.   Marques, S.M., Enyedy, E.A., Supuran, C.T., Krupenko, N.I., Krupenko, S.A. and Santos, M.A. (2010) Pteridine-Sul- fonamide Conjugates as Dual Inhibitors of Carbonic Anhydrases and Dihydrofolate Reductase with Potential Antitumor Activity. Bioorganic & Medicinal Chemistry, 18, 5081-5089. http://dx.doi.org/10.1016/j.bmc.2010.05.072

12.   Mauritz, R., Peters, J., Priest, D.G., Assaraf, Y.G., Drori, S., Kathmann, I., Noordhuis, P., Bunni, M.A., Rosowsky, A., Schornagel, J.H., Pinedo, H.M. and Jansen, G. (2002) Multiple Mechanisms of Resistance to Methotrexate and Novel Antifolates in Human CCRF-CEM Leukemia Cells and Their Implications for Folate Homeostasis. Biochemical Pharmacology, 63, 105-115. http://dx.doi.org/10.1016/S0006-2952(01)00824-3

13.   Gangjee, A., Adair, O. and Queener, S.F. (1999) Pneumocystis carinii and

Toxoplasma gondii Dihydrofolate Reductase Inhibitors and Antitumor Agents: Synthesis and Biological Activities of 2,4-Diamino-5-methyl-6-[(monosubstituted anilino)methyl]pyrido[2,3-d]pyrimidines. Journal of Medicinal Chemistry, 42, 2447-2455.http://dx.doi.org/10.1021/jm990079m

14. Elnagdi, M.H., Al-Awadi, N. and Erian, A.N. (1996) In Compensative Heterocyclic Chemistry II. In: Katritzky, A.R., Rees, C.W. and Scriven, E.F.V., Eds., Pergamon Press, Oxford, 431-488.

15. Rashad, A.E., Mahmoud, A.E. and Ali, M.M. (2011) Synthesis and Anticancer Effects of Some Novel Pyrazolo[3,4-d]- pyrimidine Derivatives by Generating Reactive Oxygen Species in Human Breast Adenocarcinoma Cells. European Journal of Medicinal Chemistry, 46, 1019-1026. http://dx.doi.org/10.1016/j.ejmech.2011.01.013

16. Elnagdi, M.H., Elmoghayar, M.R.H. and Elgemeie, G.F. (1987) Chemistry of Pyrazolo-Pyrimidines. Advances in Heterocyclic Chemistry, 41, 319-376.http://dx.doi.org/10.1016/S0065-2725(08)60164-6

17. Zatloukal, M., Jorda, R., Gucký, T., Řezníčková, E., Voller, J., Pospíšil, T., Malínková, V., Adamcová, V., Kryštof, V. and Strnad, M. (2013) Synthesis and in Vitro Biological Evaluation of 2,6,9-Trisubstituted Purines Targeting Multiple Cyclin- Dependent Kinases. European Journal of Medicinal Chemistry, 61, 61-72.http://dx.doi.org/10.1016/j.ejmech.2012.06.036

18. Kumar, A., Sinha, S. and Chauhan, P.M. (2012) Synthesis of Novel Antimycobacterial Combinatorial Libraries of Structurally Diverse Substituted Pyrimidines by Three Component Solid Phase Reactions. Bioorganic Medicinal Chemistry Letters, 12, 667-669.http://dx.doi.org/10.1016/S0960-894X(01)00829-0

19. Baraldi, P.G., Pavani, M.G., Nunez, M., Brigid, P., Vitali, B., Gambari, R. and Romagnoli, R. (2002) Antimicrobial and Antitumor Activity of n-Heteroimmine-1,2,3-dithiazoles and Their Transformation in Triazolo-, Imidazo-, and Pyrazolopirimidines. Bioorganic Medicinal Chemistry, 10, 449-456. http://dx.doi.org/10.1016/S0968-0896(01)00294-2

20. Nasr, M.N. and Gineinah, M.M. (2002) Pyrido[2, 3-d]pyrimidines and Pyrimido[5',4':5, 6]pyrido[2, 3-d]pyrimidines as New Antiviral Agents: Synthesis and Biological Activity. Archiv der Pharmazie, 335, 289-295. http://dx.doi.org/10.1002/1521-4184(200208)335:6<289::AID-ARDP289>3.0.CO;2-Z

21. Nagarapu, L., Vanaparthi, S., Bantu, V. and Kumar, C.G. (2013) Synthesis of Novel Benzo[4,5]thiazolo[1,2- a]pyrimi- dine-3-carboxylate

Derivatives and Biological Evaluation as Potential Anticancer Agents. European Journal of Medicinal Chemistry, 69, 817-822.http://dx.doi.org/10.1016/j.ejmech.2013.08.024

22. Sondhi, S.M., Johar, M., Rajvanshi, S., Dastidar, S.G., Shukla, R., Raghubir, R., et al. (2001) Anticancer, Antiinflammatory and Analgesic Activity Evaluation of Heterocyclic Compounds Synthesized by the Reaction of 4-Isothiocyanato- 4-methylpentan-2-one with Substituted o-Phenylenediamines, o-Diaminopyridine and (Un)Substituted o. Australian Journal of Chemistry, 54, 69-74. http://dx.doi.org/10.1071/CH00141

23. Youssif, S. and Assy, M. (1996) Fervenulin, 4-Deazafervenulin and 5-Deazaalloxazines Analogue: Synthesis and Antimicrobial Activity. Journal of Chemical Research, 442, 2546.

24. Herrmann, M., Lorenz, H.M., Voll, R., Grünke, M., Woith, W. and Kalden, J.R. (1994) A Rapid and Simple Method for the Isolation of Apoptotic DNA Fragments. Nucleic Acids Research, 22, 5506-5507. http://dx.doi.org/10.1093/nar/22.24.5506

25. Ishikawa, I., Itoh, T., Melik-Ohanjanian, R.G., Takayangi, H., Mizunc, Y. and Ogura, H. (1990) Synthesis and X-Ray Analysis of 1-Benzyl-6-chlorouracil. Heterocycles, 31, 1641-1646. http://dx.doi.org/10.3987/COM-90-5472

26. Cresswell, R.M. and Wood, H.C.S. (1960) The Biosynthesis of Pteridines. Part I. The Synthesis of Riboflavin. Journal of the Chemical Society, 4768-4775.

27. Youssif, S. and Pfleiderer, W. (1998) Purines XIV.. Reactivity of 8-Bromo-3,9-dimethylxanthine towards Some Nucleophilic Reagents. Journal of Heterocyclic Chemistry, 35, 949-954. http://dx.doi.org/10.1002/jhet.5570350428

28. Youssif, S. (1997) DMF Acetals as Alkylating and Cyclizing Agents: A Facile Route to Substituted Pyrazolo[3,4-d] pyrimidine-4,6(5H,7H)-diones. Chemical Monthly, 128, 493-501. http://dx.doi.org/10.1007/BF00806857

29. Hutzenlaub, W. and Pfleiderer, W. (1979) Purines, XIII. Simplified Syntheses of 7-Methy- and 1,7-Dimethylxanthines and Uric Acids. Liebigs Annalen der Chemie, 1847-1854.

30. Youssif, S. (2004) 6-Aminouracil as Precursors for the Synthesis of Fused Di- and Tricyclic Pyrimidines. Journal of Chemical Research, 341-343.

31. Youssif, S. and Ageli, F. (2008) One-Pot Synthesis of Fused 2-Thiouracils: Pyrimidopyrimidines, Pyridopyrimidines and Imidazolopyrimidines.

Zeitschrift für Naturforschung, 63b, 860-864.

32. Peet, N.P., Huber, E.W. and Huffman, J.C. (1995) Reaction of Ninhydrin with β-Dicarbonyl Compounds. Journal of Heterocyclic Chemistry, 32, 33-41.http://dx.doi.org/10.1002/jhet.5570320106

33. Prabhakar, K.R., Veerapur, V.P., Bansal, P., Vipan, K.P., Reddy, K.M., Barik, A., Reddy, B.K.D., Reddanna, P., Priyadarsini, K.I. and Unnikrishnan, M.K. (2006) Identification and Evaluation of Antioxidant, Analgesic/Anti-In- flammatory Activity of the Most Active Ninhydrin-Phenol Adducts Synthesized. Bioorganic & Medicinal Chemistry, 14, 7113-7120.http://dx.doi.org/10.1016/j.bmc.2006.06.068

34. Klumpp, D.A., Fredrick, S., Lau, S., Jin, K.K., Bau, R., Prakash, G.K.S. and Olah, G.A. (1999) Acid-Catalyzed Condensations of Ninhydrin with Aromatic Compounds. Preparation of 2,2-Diaryl-1,3-indanediones and 3-(Diarylmethylene)- isobenzofuranones. The Journal of Organic Chemistry, 64, 5152-5155. http://dx.doi.org/10.1021/jo990197h

35. Ruttink, J. (1946) Investigations in the Purine Series. II. Synthesis of Some Purine Derivatives. Recueil des Travaux Chimiques, 65, 751-767. http://dx.doi.org/10.1002/recl.19460651007

36. Blicke, F.F. and Godt, H.C. (1954) Reactions of 1,3-Dimethyl-5,6-diaminouracil. Journal of the American Chemical Society, 76, 2798-2800. http://dx.doi.org/10.1021/ja01639a058

37. Bredereck, H. and Edenhofer, A. (1955) Synthesen in der Purinreihe, VI. Mitteil.[1]: Synthesen mit 4- und 5-Amino- uracil. Chemische Berichte, 88, 1306-1312.http://dx.doi.org/10.1002/cber.19550880825

Chapter 7

SYNTHESIS, SPECTROSCOPIC CHARACTERIZATION AND ANTIMICROBIAL ACTIVITY OF SOME NEW 2-SUBSTITUTED IMIDAZOLE DERIVATIVES

Asmaa S. Salman, Anhar Abdel-Aziem, Marwa J.S. Alkubbat

Department of Chemistry, Faculty of Science, Al-Azhar University, Girls' Branch, Nasr City, Cairo, Egypt

ABSTRACT

The reaction of imidazole-2-thione derivative 1 with 2-chloro-N-p-tolylacetamide afforded the corresponding 2-($1HI$-imidazol-2-ylthio)-N-p-tolylacetamide 2.Reaction compound 2 with different reagents such as p-chlorobenzaldehyde and p-chlorophenyldiazonium chloride afforded the corresponding arylidene derivative 3 and hydrazone derivative 6. Reactions of 2 with carbon disulfide in dimethylformamide (DMF) in one equivalent potassium hydroxide afforded intermediate potassium sulphide salt 8, which treatment with dilute hydrochloric acid and phenacyl bromide afforded the corresponding 2-[p-tolylcarbamoyl]ethanedithioic acid 9 and 3-[benzo-ylmethylth- io]-N-p-tolyl-3-thioxo-propaneamide 10. While the reaction 2 with carbon disulphide in the presence of two equivalent potassium hydroxide in DMF gave non-isolated potassium salt 11, which was allowed to react with halogenated compounds namely ethyl chloroacetate and methyl iodide afforded the corresponding 3, 3-bis[(ethoxycarbonyl)methylthio]-N-p-tolylacrylamide 12 and 3,3-bis-(methylthio)-N-p-tolylacrylamide 13 respectively. Reaction 2 with phenyl isothiocyanate in ba- sic DMF yielded the intermediate potassium sulphide salt 18. Acidification 18 with dilute hydro- chloric acid afforded the corresponding thiocarbamoyl derivative 19.Treatment of intermediate 18 with methyl iodide, phenacyl bromide and ethylchloroacetate afforded the 3-anilino-3-(methyl- thio)-N-p-tolylacrylamide 20, 2-(1,3-thiazol-2(3H)-ylidene)-N-p-tolylacetamide 21 and 2-(4-oxo-3-phenyl-1,3-thiazolidin-

2-ylidene)-N-p-tolylacetamide 22 respectively.The structure of the newly synthesized compounds has been confirmed by elemental analysis and spectra data.Synthesized compounds 2, 3, 6, 13, 15a, 15b, 17, 20, 21, 22 and 23 were screened for their antibacterial activities in vitro against Gram-positive (Staphylococcus aureus and Bacillus subtilis), Gram-negative(Pseudomonas aeuroginosa and Escherichia coli)and antifungal activities against (Aspergillus fumigates, Syncephalastrumracemosum, Geotrichumcandidum and Candida albicans).

INTRODUCTION

Aromatic heterocycles are valuable synthetic templates for the preparation of new compounds with specific biological, pharmaceutical and material properties. The pursuit of these properties requires efficient synthetic routes that allow rapid construction of diverse aromatic heterocycles with defined substitution patterns. Therefore,α-oxoketendithioacetals and thiocarbamyl as the organic synthetic intermediates have been widely used in the for- mation of alicyclic, aromatic and heterocyclic compounds[1] -[4] .In view of the above and in continuation of our studies on the synthesis of heterocyclic compounds exhibiting biological activity, we report here the synthesis of some novel heterocycles compounds incorporating imidazole moiety fromα-oxoketendithioacetals and thiocarbamyl.

MATERIAL AND METHODS

Experiment

All melting points were determined in open glass capillaries on a Gallenkamp apparatus and are uncorrected. IR spectra (cm^{-1}) were recorded on a Pye-Unicam spectrophotometer type 1200 using KBr discs. 1H-NMR spectra were recorded on a Varian EM-390 (90MHz) spectrometer using TMS as an internal standard and dimethyl sulphoxide (DMSO-d6) as a solvent. Chemical shifts were expressed inδ (ppm) values and mass spectra were determined on FinniganIncos 500 (70 ev). Elemental analyses were determined using a Parkin-Elmer 240CMicroanalyzer.The microanalyses were performed at the Microanalytical Unit, Faculty of Science, Cairo University.

2-[1-(4-Chlorophenyl)-4,5-diphenyl-1H-imidazol-2-ylthio]-N-p-tolylacetamide 2

A mixture of 1 (0.01 mol) and2-chloro-N-p-tolylacetamide (0.01 mol) in DMF (25 ml) contain few drop oftriethylamine was heated under reflux for 6 h.The reaction mixture was left to cool and then poured to ice cooledwater (100 ml). The solid product that formed was filtered off, dried well and recrystallized from ethanol to give 2 as pall yellow crystals. Yield: 70%.M.p.:156°C -158°C;IR (K Br) cm^{-1}:3246(NH), 1683 (C=O), 3041, 2931(CH), 1600(C=N); 1H-NMR (DMSO-d6) δ ppm:4.11(s, 2H, CH$_2$), 2.24 (s, 3H, CH$_3$), 10.36(s, 1H, NH), 7.09-7.47 (m, 18H, Ar-H); MSm/z (%):510 (M$^+$, 26.25), 404(17.60), 375(27.06), 362(8.612), 193 (100), 147 (10.5); Anal.Calcd.for C$_{30}$ H$_{24}$ ClN$_3$OS (510.05): C, 70.64; H, 4.74; Cl, 6.95; N, 8.24; S, 6.29.Found: C, 70.44; H, 4.54; Cl, 6.65; N, 8.04; S, 6.09.

2-[1-(4-Chlorophenyl)-4,5-diphenyl-1H-imidazol-2-ylthio]-3-(4-chlorophenyl)-N-p- tolylacrylamide3

A mixture of 2(0.01 mol) and p-chlorobenzaldehyde (0.01 mol) in ethanol (30 ml) containing few drop ofpiperidine (0.5 ml) was refluxed for 3h.The reaction mixture was left to cool then poured onto ice water contain few drops of HCl and the obtained solid was recrystallized from ethanol to give 3 as white crystals.Yield:60%. M.p.:170°C -171°C;IR (KBr) cm^{-1}: 3247(NH), 1687 (C=O), 3046, 2951 (CH); 1H-NMR (DMSO-d6) δ ppm: 2.24 (s, 3H, CH$_3$), 10.31(s, 1H, NH), 7.09-7.47(m, 23H, Ar-H and =CH); MSm/z (%):632(M$^+$, 11.24), 548(23.22), 520 (398.33), 430 (23.22), 386(27.34), 309 (24.34), 80 (100); Anal.Calcd.for C$_{37}$ H$_{27}$ Cl$_2$ N$_3$OS (632.60): C, 70.25; H, 4.30; Cl, 11.21; N, 6.64; S, 5.07. Found:C, 70.00; H, 4.00; Cl, 11.00; N, 6.44; S, 5.00.

5-(4-Chlorophenyl)-4-[1-(4-chlorophenyl)-4,5-diphenyl-1H-imi-dazol-2-ylthio]-N-p-tolyl- 1H-pyrazole-3-amine 4

A mixture of 3 (0.01 mol) and hydrazine hydrate (0.01 mol) in 30 ml absolute ethanol was added few drops of glacial acetic and refluxed for 8-10 h. After completion of the reaction, excess of solvent was distilled off; the separated solid was filtered, washed with water, and recrystallized from methanol to give 4 as yellow crystals.Yield:55%. M.p.:138°C-140°C;IR (KBr) cm^{-1}: 2922, 3054(CH), 3437 (NH); 1H-NMR (DMSO-d6) δ ppm:1.74(s, 3H, CH$_3$), 7.17-7.49 (m, 23H, Ar-H and NH), 9.40(s, 1H, NH); MSm/z (%):644(M$^+$, 11.02), 630(12.44), 558 (9.13), 537 (5.98), 511(8.03), 483(6.61), 464(12.76), 405(17.64), 333 (9.29), 388 (9.49), 233(100); Anal. Calcd.for C$_{37}$ H$_{27}$ Cl$_2$N$_5$S (644.61): C, 68.94; H, 4.22; Cl, 11.00; N, 10.86; S, 4.97.Found:C, 68.64; H, 4.00; Cl, 10.99; N, 10.66; S, 4.67.

5-(4-Chlorophenyl)-4-[1-(4-chlorophenyl)-4,5-diphenyl-1H-imi-dazol-2-ythio]-1-phenyl-N- p-tolyl-1H-pyrazol-3-amine 5

A mixture of 3 (0.01 mol), phenyl hydrazine (0.01 mol), glacial acetic acid (20 ml)and few drop of HCl wasrefluxed for 8-10 h.The reaction mixture was then left to cool at room temperature, and poured onto ice cold water (100 ml).The solid product was collected by filtration and recrystallized from acetic acid to give 5 as grey crystals. Yield: 60%.M.p.:290°C-292°C; IR (KBr)cm^{-1}:3426 (NH), 3048, 2926 (CH), 1594(C=N);1H-NMR (DMSO- d6) δ ppm:1.90(s, 3H, CH$_3$), 7.13-7.48 (m, 27H, Ar-H), 13.03 (s, 1H, NH); MSm/z (%): 720 (M$^+$, 8.26), 628(7.05), 554 (6.68), 538 (7.90), 615(8.51), 512 (8.26), 423 (6.93), 266(11.42); Anal. Calcd.for C$_{43}$ H$_{31}$Cl$_2$ N$_5$ S (720.71):C, 71.66; H, 4.34; Cl, 9.84; N, 9.72; S, 4.45.Found:C, 71.36; H, 4.04; Cl, 9.54; N, 9.42; S, 4.25.

2-[2-(4-Chlorophenyl)hydrazono]-2-[1-(4-chlorophenyl)-4,5-di-phenyl-1H-imidazol-2- ylthio]-N-p-tolylacetamide 6

To a cold solution of 2 (0.01 mol) in pyridine (30 ml) was added with continuous stirring 4-chloro phenyl-diazo- nium salt (0.01 mol) [prepared by adding sodium nitrite(0.02 mol) in water (8 ml) to a cold solution ofthe p-chloroanilinein the appropriate amount of hydrochloric acid].The reaction mixture was stirred at room temperature for 2 h, and the solid products, so formed, were collected by filtration and recrystallized from benzene to give 6 as red crystals. Yield: 67%.M.p.:118°C-120°C; IR (KBr, cm^{-1}): 3259, 3136 (NH), 1604(C=N), 1686 (C=O); 1H-NMR ((DMSO-d6)δ ppm:2.24(s, 3H, CH$_3$), 7.09-7.47 (m, 22H, Ar-H), 8.58(s, 1H, NH), 10.32(s, 1H, NH); MSm/z (%):648 (M$^+$,40.12), 543(43.83), 471(45.68), 442 (33.33), 362(33.95), 367 (33.95), 287(55.02), 211(35.80), 196(54.94), 77(100); Anal.Calcd.for: C$_{36}$H$_{27}$Cl$_2$N$_5$OS(648.60):C, 66.66; H, 4.20; Cl, 10.93; N, 10.80; S, 4.94. Found:C, 66.36; H, 4.0 0; Cl, 10.73; N 10.50; S, 4.74.

2-[1-(4-Chlorophenyl)-4,5-diphenyl-1H-imidazol-2-ylthio]-2-[p-tolylcarbamoyl] ethanedithioicacid 9

To suspension of finely powdered potassium hydroxide (0.01mol) in dry DMF (20 ml) at 0°C, the acetamide derivative 2 (0.01mol) was added, the resulted mixture was cooled at 10°C in an ice bath, then carbon disulfide (0.01 mol) was added slowly over the course of 10 min. After complete addition, stirring of the reaction mixture was continued for 6 h. Then hydrochloric acid (2 M, 20ml) was added drop wise and stirring continued for additional 1h.Then, the reaction mixture was poured into ice water.The solid product that formed was filtered off, dried, and recrystallized from the ethanol to give 9 as yellow crystals.Yield:75%.M.p.:222°C-224°C; IR (KBr, cm^{-1}: 3231 (NH), 1682(CO),

3043, 2855(CH), 1271(C=S); MS m/z (%): 587(M$^+$ +1, 4.08), 586 (M$^+$, 4.53), 553 (3.18), 520(5.47), 480(3.35), 415(3.46), 233(100), 221(3.80), 77 (96.65); Anal.Calcd.for C$_{31}$H$_{24}$ClN$_3$OS$_3$ (586.19):C, 63.52; H, 4.13; Cl, 6.05; N, 7.17; S, 16.41.Found:C, 63.32; H, 4.00; Cl, 6.00; N, 7.00; S, 16.21.

3-[Benzoylmethylthio]-2-[1-(4-chlorophenyl)-4,5-diphenyl-1H-imidazol-2-ylthio]-N-p- tolyl-3-thioxo-propaneamide 10

To suspension of finely powdered potassium hydroxide (0.01mol) in dry DMF (20 ml) at 0°C, the acetamide derivative 2 (0.01mol) was added, the resulted mixture was cooled at 10°C in an ice bath, then carbon disulfide (0.01 mol) was added slowly over the course of 10 min. After complete addition, stirring of the reaction mixture was continued for 6 h. Then cooled again to 0°C, phenacyl bromide (0.01 mol) was added slowly over the course of 10 min. After complete addition, stirring of the reaction mixture was continued for 6 h.Then poured into crushed ice, the resulting precipitate was filtrated off, dried and recrystallized from ethanol to give 10 as yellow crystals.Yield:80%.M.p.:146°C-148°C; IR (KBr, cm^{-1}):3416 (NH), 1762, 1686 (CO), 1280(C=S), 1600 (C=N), 3052, 2967, 2913(CH); 1H-NMR (DMSO-d6) δ ppm: 2.25 (s, 3H, CH$_3$), 4.39 (s, 2H, CH$_2$), 4.57(s, H, CH), 7.10- 8.02 (m, 23H, Ar-H), 10.31 (s, 1H, NH); Anal. Calcd. For C$_{39}$H$_{30}$ClN$_3$O$_2$S$_3$ (704.32): C, 66.51; H, 4.29; Cl, 5.03; N, 5.97; S, 13.66.Found: C, 66.31; H, 4.09; Cl, 5.00; N, 5.77; S, 13.46.

3,3-Bis[(ethoxycarbonyl)methylthio]-2-[1-(4-chlorophenyl)-4,5-diphenyl-1H-imidazol-2- ylthio]-N-p-tolylacrylamide12

To suspension of finely powdered of potassium hydroxide (0.02mol) in dry DMF(20 ml) at 0°C the acetamide derivative 2 (0.01mol) was added, the resulted mixture was cooled at 10°C in an ice bath, then carbon disulfide (0.01 mol) was added slowly over the course of 10 min. After complete addition, stirring of the reaction mixture was continued for 6 h. Then cooled again to 0°C, ethyl chloroacetate(0.02 mol) was added slowly over the course of 10 min. After complete addition, stirring of the reaction mixture was continued for 6 h.Then poured into crushed ice, the resulting precipitate was filtrated off, dried and recrystallized from benzene to give 12 as yellow crystals. Yield: 65%.M.p.:120°C -122°C; IR (KBr, cm^{-1}): 3252 (NH), 1736, 1685(CO), 3049, 2980, 2925(CH); 1H-NMR (DMSO-d6)δ ppm: 1.13-1.23(m, 6H, 2-CH$_2$?CH$_3$), 4.33(s, 2H, CH$_2$), 4.37(s, 2H, CH$_2$), 4.04- 4.153 (m, 4H, 2CH$_2$-CH$_3$), 2.24(s, 3H, CH$_3$), 7.09-7.51 (m, 18H, Ar-H), 10.32 (s, 1H, NH); Anal. Calcd.for C$_{39}$H$_{36}$ClN$_3$O$_5$S$_3$ (758.37): C, 61.77; H, 4.78; Cl, 4.67; N, 5.54; S, 12.68.Found:C, 61.67; H, 4.58; Cl, 4.47; N, 5.34; S, 12.38.

3,3-Bis(methylthio)-2-[1-(4-chlorophenyl)-4,5-diphenyl-1H-imi-dazol-2-ylthio]-N-p- tolylacrylamide13

Compound 13 was synthesized as mentioned above in synthesis of 12 but using methyl iodide (0.02 mol) ins- tead of ethyl chloroacetate, the resulting product was recrystallized from ethanol to give 13 as yellow crystals. Yield:60%.M.p.:166˚C-168˚C; IR (KBr, cm^{-1}):3438(NH), 1674(CO), 3050, 2919 (CH); 1H-NMR (DMSO-d6)δ ppm2.21(s, 3H, CH$_3$), 7.18-7.68 (m, 19H, Ar-H and NH), 2.28 (s, 6H, 2SCH$_3$); MS m/z (%):614 (M$^+$, 20.64), 611(16.71), 568(13.154), 506(19.41), 415 (13.27), 315 (14.74), 252 (15.48), 206(17.44), 237(15.97), 75(100); Anal. Calcd.for C$_{33}$H$_{28}$ClN$_3$OS$_3$,(614.24):C, 64.53; H, 4.59; Cl, 5.77; N, 6.84; S, 15.66. Found:C, 64.23; H, 4.39; Cl, 5.57; N, 6.64; S, 15.36.

2-[1-(4-Chlorophenyl)-4,5-diphenyl-1H-imidazol-2-ylthio]-3,3-dihydrazino-N-p- tolylacrylamide 14

A mixture of compound 13 (0.01mol) and hydrazine hydrate (80%, 0.02 mol) was heated under reflux for 4 h, then left to cool. The obtained solid product was triturated with ethanol (10 ml), filtered off, washed with ethanol, dried and recrystallized from butanol afford compound 14 as white crystals. Yield:60%.M.p.:210˚C-213˚C; IR(KBr, cm^{-1}):3297, 3269, 3196, 3126 (NH$_2$, NH), 1682(CO), 3061, 2959 (CH); 1H-NMR(DMSO-d6) δ ppm: 2.24 (s, 3H, CH$_3$), 7.09-7.56(m, 19H, Ar-H and NH), 10.33(s, 1H, NH), 10.51 (s.1H, NH), 4.84(s, 2H, NH$_2$), 4.89 (s, 2H, NH$_2$); MS m/z (%):582(M$^+$, 66.04), 517(13.21), 470 (50.94), 444(61.32), 397(100), 381 (58.49), 321 (24.53); Anal.Calcd.for C$_{31}$H$_{28}$ClN$_7$OS(582.11): C, 63.96; H, 4.85; Cl, 6.09; N, 16.84; S, 5.51. Found: C, 63.66; H, 4.65; Cl, 6.00; N, 16.64; S, 5.41.

Reaction of 3,3-Bis(methylthio)-2-[1-(4-chlorophenyl)-4,5-diphe-nyl-1H-imidazol-2- ylthio]-N-p-tolylacrylamide 13 with Amines

A mixture of 13 (0.01 mol) suitable amine such as o-phenylenediamine, o-aminophenol and/or p-chloroaniline(0.01 mol) in DMF(25 ml) was heated under reflux for 6h.The reaction mixture was left to cool and then poured to ice cooled water (100 ml). The solid product that formed was filtered off, dried well and recrystallized from appropriate solvent.

2-(1H-Benzimidazol-2-yl)-2-[1-(4-chlorophenyl)-4,5-diphenyl-1H-imidazol-2-ylthio]-N-p-tolylacetamide 15a

White powder.Yield: 65%.M.p.:238°C-240°C (ethanol and DMF); IR(KBr, cm⁻¹):3245, 3188 (NH); 3055, 2924, 2865(CH), 1659(CO), 1607(C=N); 1H-NMR (DMSO-d6)δ ppm:2.25(s, 3H, CH$_3$), 7.09-7.53(m, 22H, Ar-H); 7.95 (s, 1H, NH), 10.30 (s, 1H, NH), 4.61(s, 1H, CH); MS m/z (%):626 (M⁺, 29.10), 491 (212.31), 424 (29.10), 391 (31.97), 379(35.25), 328(26.64), 75(100); Anal. Calcd.for C$_{37}$H$_{28}$ClN$_5$OS(626.17):C, 70.97; H, 4.51; Cl, 5.66; N, 11.18; S, 5.12.Found:C, 70.67; H, 4.21; Cl, 5.46; N, 11.08; S, 5.02.

2-(1,3-Benzoxazol-2-yl)-2-[1-(4-chlorophenyl)-4,5-diphenyl-1H-imidazol-2-ylthio]-N-p-tolylacetamide15b

Pale grey crystals.Yield: 55%.M.p.:176°C-178°C (ethanol); IR (KBr, cm⁻¹) 3246 (NH); 1687 (CO), 1600 (C=N), 3048, 2922, 2864(CH); 1H-NMR(DMSO-d6) δ ppm: 2.24(s, 3H, CH$_3$), 7.09-7.47(m, 22H, Ar-H); 10.43(s, 1H, NH), 4.16(s, 1H, CH); MSm/z (%): 627 (M⁺, 11.96), 611(11.52), 516 (10.49), 492(8.57), 405(8.86), 330 (9.16), 264 (10.34), 173(10.49), 77(100); Anal. Calcd.for C$_{37}$H$_{27}$Cl N$_4$O$_2$S(627.15):C, 70.86; H, 4.34; Cl, 5.65; N, 8.93; S, 5.11. Found: C, 70.66; H, 4.04; Cl, 5.45; N, 8.73; S, 5.00.

2-[1-(4-Chlorophenyl)-4,5-diphenyl-1H-imidazol-2-ylthio]-3-(4-chlorophenylamino]-3-(methylthio)- N-p-tolylacrylamide 16

White powder.Yield: 65%.M.p.:190°C-192°C(ethanol); IR (KBr, cm⁻¹) 3246, 3191 (NH); 1687(CO), 1599 (C=N); 1H-NMR (DMSO-d6)δ ppm:2.25(s, 3H, CH$_3$), 2.39(s, 3H, SCH$_3$), 7.09-7.62 (m, 23H, Ar-H and NH); 10.29 (brs, 1H, NH); MSm/z (%):693(M⁺, 53.29), 631(43.13)622(40.72), 607(40.72), 471(35.93), 330(53.89), 234(35.93), 257(51.51), 272 (34.13), 77(100); Anal. Calcd.for C$_{38}$H$_{30}$Cl$_2$N$_4$OS$_2$(693.70):C, 65.79; H, 4.36; Cl, 10.22; N, 8.08; S, 9.24.Found:C, 65.59; H, 4.26; Cl, 10.02; N, 8.00; S, 9.04.

4-[1-(4-Chlorophenyl)-4,5-diphenyl-1H-imidazol-2-ylthio]-3-(p-tolylamino)-5- (methylthio)-1H-pyrrole-2-carboxylic Acid 17

A mixture of compound 13 (0.01mol) and glycine (0.01 mol) in ethanol (30 ml) containing triethylamine (5drops) was heated under reflux for 8 h. The formed solid product was filtered off, dried and recrystallized from ethanol to give 17 as a yellow powder. Yield: 62%.M.p.:180°C-182°C; IR(KBr, cm⁻¹):3436(OH), 3247, 3184 (NH); 1773(CO), 1607(C=N); 1HNMR((DMSO-d6)δ ppm:2.25(s, 3H, CH$_3$), 2.29(s, 3H, SCH$_3$), 7.09-7.47(m, 20H, Ar-H and 2NH); 13.29(s, 1H, OH); MSm/z(%):623(M⁺, 7.89), 579(13.92), 575(21.87), 516(10.93), 480(19.28), 470(14.12), 390 (12.33), 293 (15.52), 261(13.12), 214(24.25), 77(100); Anal.Calcd.for C$_{34}$H$_{27}$ClN$_4$O$_2$S$_2$(623.18):C, 65.53; H, 4.37; Cl, 5.69; N, 8.99; S, 10.29.Found:C, 65.23; H, 4.17; Cl, 5.49; N, 8.79; S, 10.09.

3-Anilino-2-[1-(4-chlorophenyl)-4,5-diphenyl-1H-imidazol-2-ylthio]-N-p-tolyl-3- thioxopropan 19

To a stirred solution of powdered potassium hydroxide (0.02mol) in DMF (20 ml), compound 2 (0.02 mol)was added. After stirring for 30 min, phenyl isothiocyanate(0.02 mol) was added to the resulting mixture, stirring was continued for 6 h, and then poured over crushed ice containing hydrochloric acid.The solid product that formed was filtered off, washed with water, dried and recrystallized from ethanol to give19 as yellow crystals. Yield: 64%.M.p.:186°C-188°C; IR(KBr, cm^{-1}):3246, 3191(NH), 3084, 2925(CH), 1656(C=O), 1606(C=N), 1238 (C=S); 1H-NMR(DMSO-d6)δppm:2.24(s, 3H, CH$_3$), 10.40(s, 1H, NH), 10.20(s, 1H, NH), 7.05-7.59(m, 23H, Ar-H), 4.12(s, 1H, CH); MS m/z(%):645 (M$^+$, 23), 629(22.03), 612(23), 554 (18.32), 521(20.91), 512(23), 526(18.12), 316 (22.65), 300 (187.12), 224(29.21), 192(20.21), 80(100); Anal. Calcd.for C$_{37}$H$_{29}$ClN$_4$OS$_2$(645.23): C, 68.87; H, 4.53; Cl, 5.49; N, 8.68; S, 9.94. Found:C, 68.67; H, 4.23; Cl, 5.29 N, 8.48; S, 9.64.

3-Anilino-2-[1-(4-chlorophenyl)-4,5-diphenyl-1H-imidazol-2-ylthio]-3-(methylthio)-N-p- tolylacrylamide 20

To a stirred solution of potassium hydroxide (0.01 mol) in DMF (20 ml) was added compound 2(0.01mol).After the mixture was stirred for 30 min, phenyl isothiocyanate(0.01 mol) was added to the resulting mixture. Stirring was continued for 6 h, and then methyl iodide (0.01 mol) was added and stirring was continued for 6 h. The reaction mixture was poured onto ice-cold water. The solid product that formed was collected by filtration, dried and recrystallized from ethanol to give compound 20 as yellow crystals.Yield: 62%.M.p.:234°C-236°C; IR (KBr, cm^{-1}): 3246, 3191 (NH), 1656 (C=O), 1604 (C=N), 3086, 2874(CH); 1H-NMR (DMSO-d6) δ ppm:2.25(s, 3H, CH$_3$), 2.54 (s, 3H, SCH$_3$), 10.30 (s, 1H, NH), 10.38(s, 1H, NH), 7.03-7.70 (m, 23H, Ar-H); MS m/z (%): 659 (M$^+$, 44.26), 597 (63.93), 476 (27.87), 433(56.84), 366(51.64), 328(63.93), 296(42.62), 73(100); Anal. Calcd.for C$_{38}$H$_{31}$ClN$_4$OS$_2$(659.26):C, 69.23; H, 4.74; Cl, 5.38; N, 8.50; S, 9.73.Found:C, 69.03; H, 4.54; Cl, 5.08; N, 8.30; S, 9.53.

2-[1-(4-Chlorophenyl)-4,5-diphenyl-1H-imidazol-2-ylthio]-2-(3,4-diphenyl-thiazol- 2(3H)-ylidene)-N-p-tolylacetamide 21 and 2-[1-(4-Chlorophenyl)-4,5-diphenyl-1H- imidazol-2-ylthio]-2-(4-oxo-3-phenyl-1,3-thiazolidin-2-ylidene)-N-p-tolylacetamide 22

To a cold suspension of powdered potassium hydroxide (0.01mol) in DMF (20 ml) was added compound 2(0.01 mol) and phenyl isothiocyanate (0.01 mol). The reaction mixture was stirred at room temperature for 6 h, and then treated with phenacyl bromide and/or ethyl chloroacetate(0.01 mol) and the stirring was continued atroom temperature for further 10 h. The reaction mixture was poured into 50 ml of cold water. The result solid products were collected by filtration and recrystallized from a mixture of ethanol/DMF (1:1) to give compounds 21 and 22.

21:Yellow powder.Yield: 65%.M.p.:240˚C -242˚C; IR (KBr, cm^{-1}): 3248(NH), 1659(C=O), 3056, 2937, 2871 (CH), 1604(C=N); 1H-NMR(DMSO-d6) δ ppm:2.24(s, 3H, CH$_3$), 10.40(s, 1H, NH), 7.05-7.59 (m, 29 H, Ar-H andH-5 thiazoline); MS m/z (%):745(M$^+$, 43.29), 695(40.24), 577(41.40), 450 (39.63), 421 (36.59), 313 (43.29), 217 (50.00), 172(46.34), 111(100).Anal.Calcd.for C$_{45}$H$_{33}$Cl N$_4$OS$_2$ (745.35):C, 72.51; H, 4.46; Cl, 4.76; N, 7.52; S, 8.60.Found: C, 72.31; H, 4.26; Cl, 4.46; N, 7.32; S, 8.50.

22:Pale grey crystals.Yield: 62%.M.p.:228˚C-230˚C; IR (KBr, cm^{-1}):3245(NH), 3084, 2929, 2873 (CH), 1726, 1657(C=O); 1H-NMR (DMSO-d6) δ ppm:2.25(s, 3H, CH$_3$), 10.41(s, 1H, NH), 7.06 - 7.60 (m, 23H, Ar-H), 4.13 (s, 2H, CH$_2$); MS m/z (%):670(M$^+$-CH$_3$, 12.66), 654(10.97), 594(19.83), 503(14.77), 475 (12.24), 348 (10.97), 139 (14.98), 125(18.35), 55(100).Anal. Calcd.for C$_{39}$H$_{29}$ClN$_4$O$_2$S$_2$(685.25):C, 68.36; H, 4.27; Cl, 5.17; N, 8.18; S, 9.36. Found: C, 68.06; H, 4.07; Cl, 5.00; N, 8.00; S, 9.16.

2-[5-Benzylidene-4-oxo-3-phenyl-1,3-thiazolidin-2-ylidene]-2-[1-(4-chlorophenyl)-4,5- diphenyl-1H-imidazol-2-ylthio]-N-p-toly-lacetamide 23

To a well-stirred solution of compound 22 (0.01mol) in DMF (20 ml), piperidine(0.2 ml) and benzaldehyde (0.01mol) were added. The reaction mixture was stirred at 80˚C for 3 h.The separated crystals was filtered, dried and recrystallized from ethanol to give 23 as white crystals. Yield: 60%.M.p.:252˚C-254˚C; IR (KBr, cm^{-1}): 3249 (NH), 3090, 2997(CH), 1657, 1734(C=O); 1H-NMR(DMSO-d6)δ ppm:2.25(s, 3H, CH$_3$), 10.53(s, 1H, NH), 7.03-7.70 (m, 29H, Ar-H and =CH); MS m/z (%):773 (M$^+$, 6.72), 723 (77.3), 707(57.14), 570(50.42), 479(54), 388(100), 372 (58.82)Anal.Calcd. forC$_{46}$H$_{33}$ClN$_4$O$_2$S$_2$(773.36):C, 71.44; H, 4.30; Cl, 4.58; N, 7.24; S, 8.29. Found:C, 71.24; H, 4.00; Cl, 4.28; N, 7.04; S, 8.09.

Antimicrobial Assays

Synthesized compounds 2, 3, 6, 13, 15a, 15b, 17, 20, 21, 22 and 23 were screened for their antimicrobial activities in vitro against two species of

Gram-positive bacteria, namely Staphylococcus aureus(RCMB 0100010) and Bacillus subtilis (RCMB 010067), two Gram-negative bacteria, namely Pseudomonas aeuroginosa(RCMB010043) and Escherichia coli (RCMB 010052) and against four species of fungi, namely Aspergillusfumigatus (RCMB 02568), Syncephalastrumracemosum(RCMB 05922), Geotrichumcandidum(RCMB05097) and Candida albicans (RCMB 05036). The antibacterial and antifungal activities were determined by means of inhibition% ± standard deviation at a concentration of 100 µg/ ml of tested samples [5] - [7] . Optical densities of antimicrobial were measured after 24 hours at 37°C to bacteria and measured after 48 hours at 28°C to fungal using a multidetection microplate reader at the Regional Center for Mycology and Biotechnology (Sun Rise-Tecan, USA at 600 nm) Al-Azhar University. Ampicillin, gentamicin and amphotericin B were used as references to evaluate the potency of the tested compounds under the same conditions.

RESULTS AND DISCUSSION

Chemistry

The synthetic procedures adopted to obtain the target compounds are depicted inSchemes 1-4.S-Alkylation of 1-(4-chlorophenyl)-4,5-diphenyl-1H-imiazole-2-thione 1 with 2-chloro-N-p-tolylacetamide afforded the corre- sponding 2-(1H-imidazol-2-ylthio)-N-p-tolylacetamide derivative 2. The assignment of structure 2 was based on both elemental analysis and spectral data.1H-NMR spectrum of 2 in (DMSO-d6) revealed signals at 2.24 ppm corresponding to CH_3 group and a single at 4.11 ppm for CH_2 group. Moreover, mass spectrum showed a molecular ion peak at m/z 510 corresponding to a molecular formula $C_{30}H_{24}ClN_3OS$.Further evidence for the structure of compound 2 was obtained through studying their chemical reactivity via some chemical reactions. Thus, interaction of compound 2 with p-chlorobenzaldehyde yielded the arylidene derivative 3(Scheme 1).1H-NMR spectrum in (DMSO-d6) of 3 show the disappearance of CH_2 protons observed with the respective starting precursors 2 at δ 4.11 ppm, and the appearance multiple signals in the region at δ 6.41-7.27 ppm corresponding to the aromatic protons together with 1Hbenzylidene C=CH proton. Mass spectrum of 3 showed a molecular ion peak at m/z 632 corresponding to the molecular formula $C_{37}H_{27}Cl_2N_3OS$ (see experimental section).

Reaction of arylidene derivatives 3 withhydrazine hydrate in ethanol [8] and few drop of acetic acid gave the corresponding 1H-pyrazol-3-amine derivative 4.While, reaction 3 with phenyl hydrazine in the presence of acetic acid and few drops of hydrochloric acid [9] afforded the corresponding

1-phenyl-1H-pyrazol-3-amine 5.The structure of the pyrazole derivatives 4 and 5 were established on the basis of analytical and spectral data. The IR spectrum of 4 showed the disappearance of absorption band of C=O group and appearance of new absorption band of NH at 3238 cm^{-1}.Mass spectrum of 4 showed a molecular ion peak at m/z 644 corresponding to the molecular formula $C_{37}H_{27}Cl_2N_5S$.

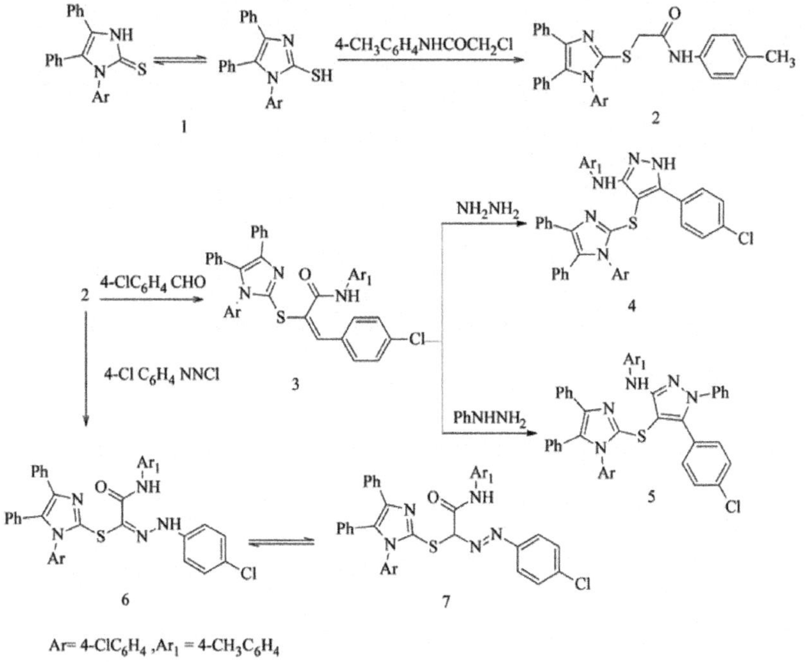
4-CH3C6H4NHCOCH2Cl
4-ClC6H4 CHO
4-Cl C6H4 NNCl
NH2NH2
PhNHNH2
Ar= 4-ClC6H4 ,Ar1 = 4-CH3C6H4

Scheme 1: Formation of compounds 2-6.

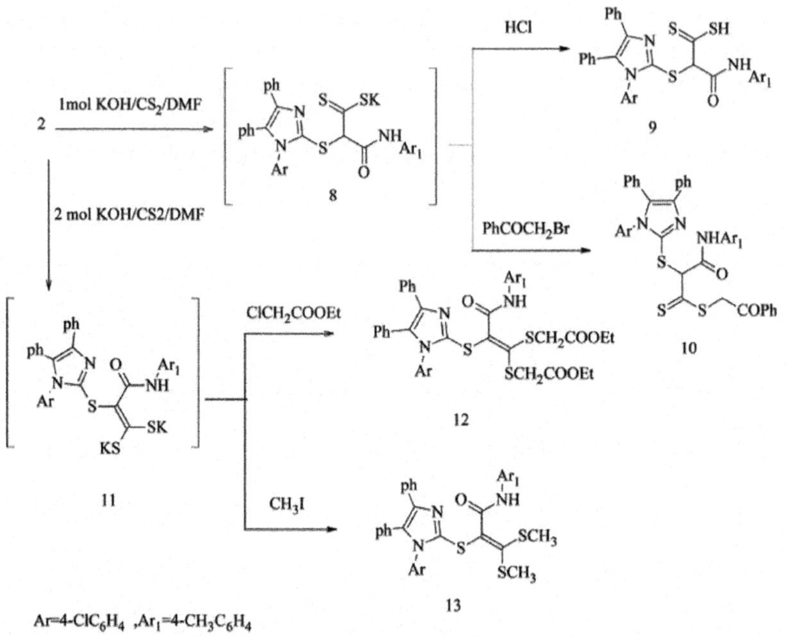

Ar=4-ClC$_6$H$_4$,Ar$_1$=4-CH$_3$C$_6$H$_4$

Scheme 2: Formation of compounds 8-13.

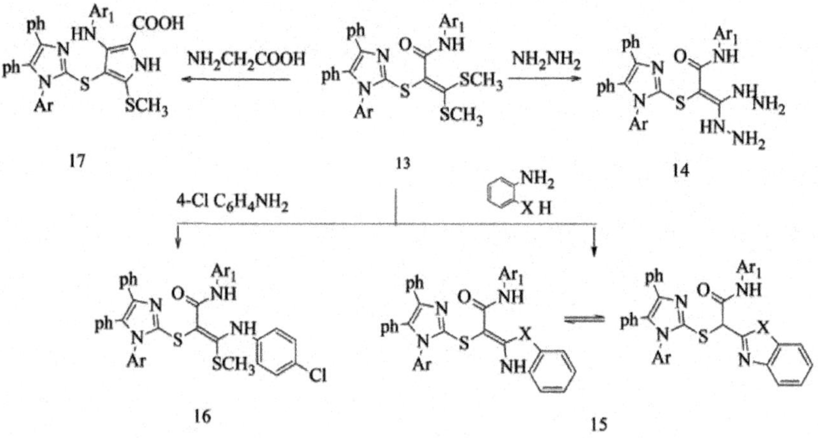

a,X=NH
b,X=O

Ar=p-ClC$_6$H$_4$,Ar$_1$=p-CH$_3$C$_6$H$_4$

Scheme 3: Formation of compounds 14-17.

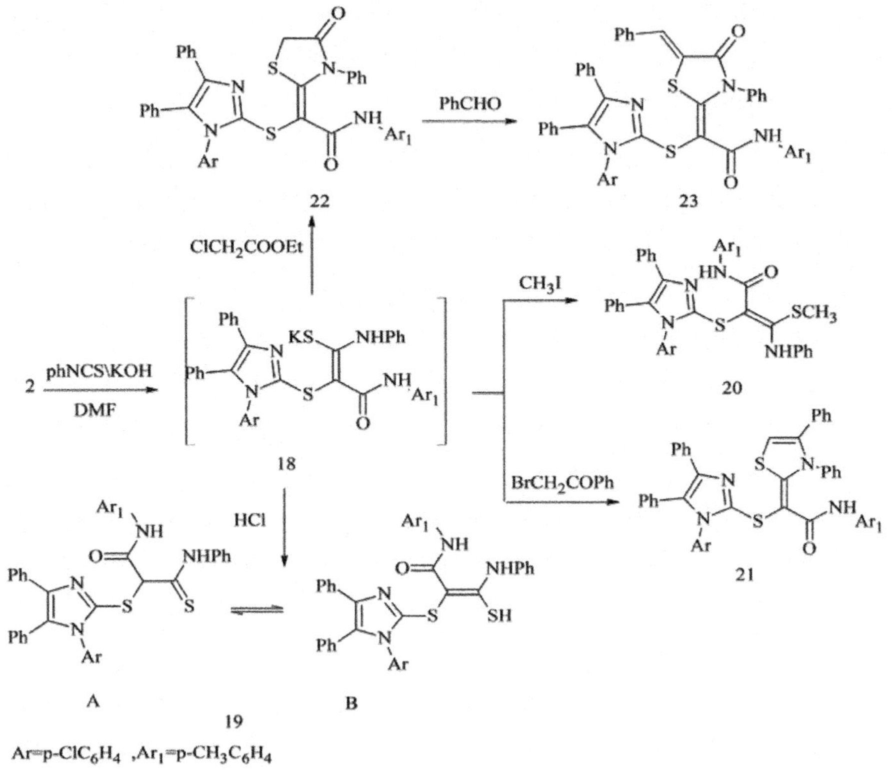

Ar=p-ClC$_6$H$_4$,Ar$_1$=p-CH$_3$C$_6$H$_4$

Scheme 4: Formation of compounds 18-22.

Diazotization of p-chloroaniline followed by coupling with active methylene group in compound 2 in pyridine yielded the hydrazone form 6 rather than the azo form 7 based on spectral data [10] .The 1H-NMR spectrum of compound 6 recorded in (DMSO-d6) revealed a signal at δ8.58 ppm, which could be attributed to hydrazone NH group(Scheme 1).

Reaction of compound 2 with carbon disulphide and one equivalent potassium hydroxide in dimethylformamide(DMF) gave non-isolated intermediate potassium salt 8.Treatment of the non-isolable potassium salts 8 with dilute hydrochloric acid [11] afforded the corresponding 2-(1H-imidazol-2-ylthio)-2-(p-tolylcarbamoyl)ethanedithioic acid 9.The assignment of structure 9 was based on both elemental analysis and spectral data.The IR spectrum displayed absorptions band at 3231cm^{-1}(NH) and 1271 cm^{-1} (C=S).The mass spectrum showed the molecular ion peak at m/z586 corresponding to the molecular formula C$_{31}$H$_{24}$ClN$_3$OS$_3$.On the other hand, treatment of intermediate salt 8 with phenacyl bromide [12] to give the corresponding 3-[benzoylmethyl-thio]-2-[1H-imidazol-2-ylthio]-N-p-tolyl-3-thioxo-propaneamide derivatives

10.The 1H-NMR spectra of 10 in (DMSO-d6)revealed a single at δ 4.39 ppm and δ 4.57ppm assigned to the CH_2 and CH protons respectively(Scheme 2).

While the reaction 2 with carbon disulphide in the presence of two equivalent potassium hydroxide in DMF to give non-isolated potassium salt 11, which was allowed to react with halogenated compounds namely ethylchloroacetate[13] and methyl iodide [14] afforded the corresponding 3,3-bis[(ethoxycarbonyl)methylthio]-N-p-toly- lacrylamide 12 and 3,3-bis-(methylthio)-N-p-tolylacrylamide 13 respectively.We suggest a mechanism forthe for- mation of 12 in which the intermediate I is obtained first, then elimination of potassium chloride (Figure 1).

The structure of synthesis compound 12 and 13 ware elucidated on the basis of the elemental analysis and spectral data. For example, 1H-NMR spectra in (DMSO-d6) of 12 displayed two multiple at 1.13-1.23 and 4.04-4.13 for ethoxy protons of two carboethoxy group and two single at 4.33 and 4.37 ppm for two methylene protons. On the other hand, 1H-NMR spectrum (DMSO-d6) of 13 showed single signal at δ 2.28 ppm for 6 protons of two similar methyl protons. The mass spectrum of compound 13 showed a molecular ion peak at m/z 614 corresponding to a molecular formula $C_{33}H_{28}ClN_3OS_3$(Scheme 2).

Moreover, condensation of 13 with hydrazine hydrate afforded the corresponding 3, 3-dihydrazino-N-p-toly- lacrylamide derivative 14.The structure of 14 was identified as the reaction product on the basis of its elemental analysis and spectral data.The 1H-NMR spectrum of 14 showed a multiple signals in the region at δ 7.09-7.56 ppm corresponding to the aromatic protons together with the NH proton, two single signals at δ 4.84 ppm and 4.89 corresponding to the two NH_2 protons, and another two single signals at δ 10.33 and δ 10.51 ppm assignable to two NH protons. Mass spectrum of 14 showed a molecular ion peak at m/z 582 corresponding to a molecular formula $C_{31}H_{28}ClN_7OS$. In addition, the condensation of 13 with suitable amine namely o-phenylenedi- amine, and o-aminophenol [15] in refluxing absolute ethanol to afford the corresponding 2-(1H-benzimidazol- 2-yl)- and 2-(1,3-benzoxazol-2-yl)-N-p-tolylacetamide15a, b respectively(Scheme 3).The structures of compounds 15a, b were established and confirmed by their elemental analysis and spectral data (see experimental section). The formation of 15 a, b were assumed to proceed through nucleophilic attack of the two -NH2 group in o-phenylenediamine,orNH,OH groups ino-aminophenol to the ethylenic double bond in the compound 13 followed by elimination of two moles of methyl mercaptan(Figure 2).

S,S-acetals13 was converted to corresponding S,N-acetals by reacting with appropriate primary.Thus, reaction of 13 with p-chloroaniline afforded ketene N,S-acetals16(Scheme 3).The assignment of the structure of 16 was basedon spectral data. The IR spectrum of 16 showed absorption bands at 3246, 3191 cm^{-1} for two NH. Its 1H-NMR spectrum (DMSO-d6) of 16 showed single signal at δ2.39 ppm for SCH$_3$ protons, δ 7.09-7.62 ppm corresponding to the aromatic protons together with the NH proton and single signals at δ 10.29 ppm assignable to NH. The mass spectrum of 16 showed a molecular ion peak at m/z 693 corresponding to the molecular formula C$_{38}$H$_{30}$Cl$_2$N$_4$OS$_2$.

Furthermore, the reaction of 13 with Glycine in ethanol containing triethylamine [16] afforded the corresponding 1H-pyrrole-2-carboxylic acidderivatives 17(Scheme 3).The assignment of the structure of 17was

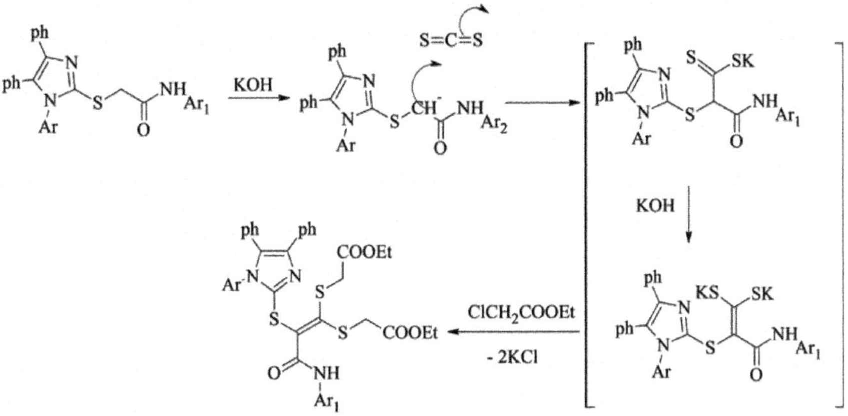

Figure 1: Proposed mechanism formation of 3,3-bis[(ethoxycarbonyl)methylthio]-N-p-tolylacrylamide derivatives 12.

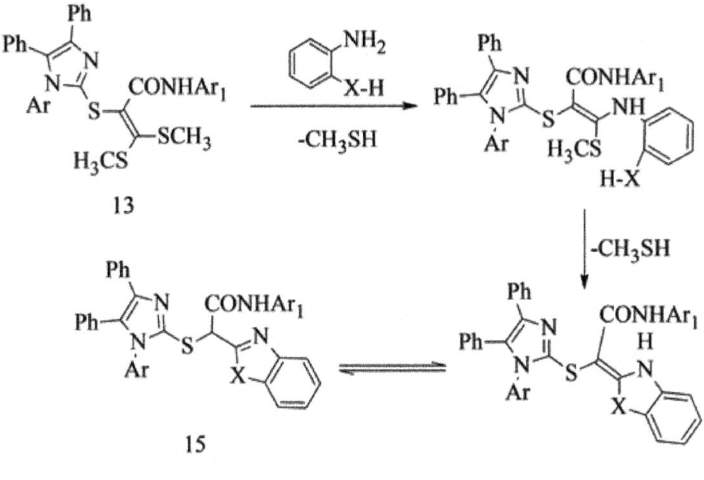

Figure 2: The proposed mechanism formation 15a, b.

Based on spectral data.The IR spectrum of 17 showed absorption bands at 3436 cm^{-1} (OH) and 3247, 3184 cm^{-1} (NH). Its 1H-NMRspectrum (DMSO-d6) single signal at δ2.29 ppm for SCH$_3$protons, δ7.09-7.47 ppm corresponding to the aromatic protons together with the 2NH protons.The mass spectrum of 17 showed a molecular ion peak at m/z623corresponding to the molecular formula C$_{34}$H$_{27}$Cl N$_4$O$_2$S$_2$.

2-(1H-imidazol-2-ylthio)-N-p-tolylacetamide2 was utilized as a key intermediate for the synthesis of thiocarbamoyl derivative 19 via its reaction with phenyl isothiocyanate.Thus, reaction 2 with phenyl isothiocyanate in DMF in the presence of an equimolar amount of potassium hydroxide yielded the non-isolable intermediate potassium sulphide salt 18. Acidification of the potassium salt 18 with dilute hydrochloric acid afforded the corresponding thiocarbamoyl derivative 19, which can exist in two tautomericthione-thiol forms A and B (Scheme4). Assignment of the product 19 was based on elemental analysis and spectral data.1H-NMR spectrumdisplayed multiple signals at δ 7.05 - 7.59 ppm for aromatic protons and two single signals at δ 10.40 and 10.20 ppm assignable to two NH protons. Mass spectrum showed a molecular ion peak at m/z 645 corresponding to a molecular formula C$_{37}$H$_{29}$ClN$_4$OS$_2$.

Treatment of the non-isolable potassium sulfide salt 18 with methyl iodide [17] afforded the ketene N, S- acetal20. The structure of 20 was established on the basis of its elemental analysis and spectral data. Its IR spectrum showed absorption bands at 3246, 3191cm^{-1} due to two NH groups.On addition 1H NMR spectrum (DMSO-d6) displayed single signal at δ2.54 ppm for SCH$_3$. The mass spectrum showed a molecular ion peak at m/z 659corresponding to a molecular formula C$_{38}$H$_{31}$ClN$_4$OS$_2$(Scheme 4).

On the other hand, reaction of 18 with phenacyl bromide and ethyl chloroacetate [18] afforded 2-(3, 4-di- phenyl-1,3-thiazol-2(3H)-ylidene)- and 2-(4-oxo-3-phenyl-1,3-thiazolidin-2-ylidene)-N-p-tolylacetamide 21 and 22 respectively.The structures of compounds 21 and 22 were established and confirmed by their elemental ana- lysis and spectral data.The 1H-NMR spectrum of 21 showed a multiple signals in the region at δ7.05-7.59 ppm corresponding to the aromatic protons together with the H-5 protons of the thiazole ring and a single signal at δ 10.40 ppm for NH proton. Mass spectrum of 21 revealed a molecular ion peak at m/z 745 corresponding to a molecular formula C$_{45}$H$_{33}$ClN$_4$OS$_2$. The IR spectrum of 22 showed absorption bands at 1726 cm^{-1} due to CO ofthiazolidinone ring. The 1H-NMR spectrum of 22 showed a single signal equivalent to two protons at δ 4.13ppm which represent the CH$_2$protons of the thiazolidinone ring.The Claisene Schmidt condensation of thiazolidin-5-one 22 with benzaldehyde [19] in DMF and in the presence of a catalytic amount of piperidine afforded arylidene derivatives 23(Scheme

4).The structures of latter products were confirmed based on elemental analysis and spectral data (see experimental section).

Antimicrobial Activity

Synthesized compounds 2, 3, 6, 13, 15a, 15b, 17, 20, 21, 22 and 23 were evaluated for antibacterial and antifungal activities.

Antibacterial Activity

Synthesized compounds 2, 3, 6, 13, 15a, 15b, 17, 20, 21, 22 and 23 were screened for their antibacterialactivities in vitro against Gram-positive namely Staphylococcus aureus(RCMB0100010) and Bacillus subtilis (RCMB 010067) andGram- negative Pseudomonas aeuroginosa (RCMB010043) and Escherichia coli (RCMB 010052).Ampicillin and gentamicin were used as references to evaluate the potency of the tested compounds.The inhibitory effects of the synthetic compounds against these organisms are given inTable 1,Figure 3.

In general, most of the tested compounds revealed better activity against the Gram-positive bacteria rather than the Gram-negative bacteria. Compounds 3, 13 and 15a exhibited excellent antibacterial activity against the tested organisms while compounds15b, 17, 20, 21 and 23 showed moderate antibacterial activity against the tested organisms and compound 2, 6 and 22 showed weak antibacterial activity against the tested organisms. In addition, all test compounds were found to be inactive against Pseudomonas aeuroginosa (RCMB010043).

Antifungal Activity

The newly synthesized compounds 2, 3, 6, 13, 15a, 15b, 17, 20, 21, 22 and 23 were screened for their antifungal activities in vitro against, Aspergill usfumigatus(RCMB02568),Syncephalastrumracemosum(RCMB   05922), Geotrichumcandidum(RCMB05097) and Candida albicans (RCMB05036). Amphotericin B was used as standards to evaluate the potency of the tested compounds.The inhibitory effects of the synthetic compounds against these organisms are given inTable 2,Figure 4.

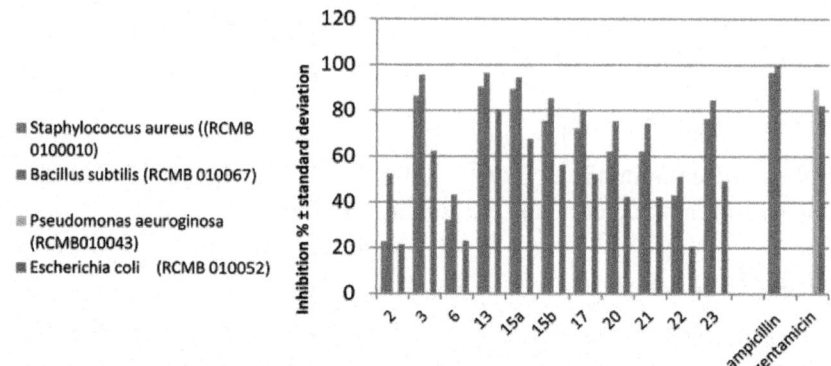

Figure 3: Graphical representation of the antibacterial activity of tested compounds compared to ampicillin and gentamicin.

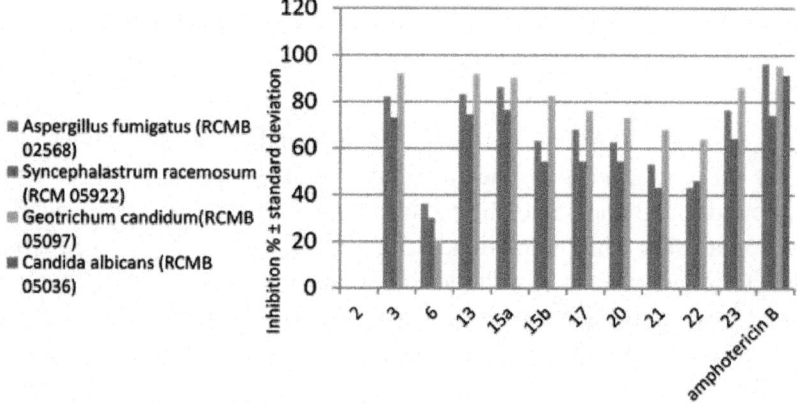

Figure 4. Graphical representation of the antifungal activity of tested compounds compared to amphotericin B.

Compounds 13, 15a exhibited excellent antifunger activity, which is better than the amphotericin B against Syncephalastrumracemosum (RCMB05922), while its strong antifunger activity againstAspergillusfumig atus(RCMB 02568) and Geotrichumcandidum(RCMB05097)is comparable to amphotericin B. The compounds 3, 15b, 17, 20, 21, 22 and 23 showed strong moderate activity againstAspergillusfumigatus(RCMB02568),Synceph alastrumracemosum(RCMB05922) and Geotrichumcandidum(RCMB05097) compared to amphotericin B against. While the compounds 6 weak antifungal activity against Aspergillusfumigatus (RCMB02568), Syncephalastrumracem osum(RCMB05922), Geotrichumcandidum(RCMB05097).On the other hand, compound 2 inactive againstall organism. Furthermore, all test compounds were found to be inactive against Candida albicans(RCMB 05036).

Table 1: Antibacterial evaluation of the some synthesized compounds.

| Comp. No. | Inhibition % ± standard deviation | | | |
| | Gram-positive bacteria | | Gram-negative bacteria | |
	Staphylococcus aureus ((RCMB 0100010)	Bacillus subtilis (RCMB 010067)	Pseudomonas aeuroginosa (RCMB 010043)	Escherichia coli (RCMB 010052)
2	22.63 ± 0.25	52.31 ± 0.37	NA	21.36 ± 0.44
3	86.32 ± 0.44	95.41 ± 0.25	NA	62.31 ± 0.25
6	32.12 ± 0.63	43.25 ± 0.42	NA	22.96 ± 0.16
13	90.31 ± 0.43	96.25 ± 0.53	NA	80.36 ± 0.33
15a	89.35 ± 0.15	94.32 ± 0.42	NA	67.58 ± 0.53
15b	75.44 ± 0.44	85.34 ± 0.58	NA	56.23 ± 0.19
17	72.13 ± 0.44	80.12 ± 0.63	NA	52.34 ± 0.25
20	62.14 ± 0.44	75.24 ± 0.63	NA	42.36 ± 0.25
21	62.14 ± 043	74.32 ± 0.53	NA	42.32 ± 0.25
22	43.12 ± 0.44	51.21 ± 0.25	NA	20.63 ± 0.33
23	76.52 ± 0.2	84.63 ± 0.3	NA	49.32 ± 0.3
Reference drugs				
Ampicillin	96.52 ± 0.2	99.65 ± 0.3		
Gentamicin			89.23 ± 0.1	82.14 ± 0.3

Table 2: Antifungal evaluation of the some synthesized compounds.

| Comp. No. | Inhibition % ± standard deviation | | | |
	Aspergillus fumigatus (RCMB 02568)	Syncephalastrum racemosum (RCMB 05922)	Geotrichum candidum (RCMB 05097)	Candida albicans (RCMB 05036)
2	NA	NA	NA	NA
3	82.21 ± 0.53	73.21 ± 0.44	92.14 ± 0.58	NA
6	36.21 ± 0.44	30.21 ± 0.37	20.36 ± 0.25	NA
13	83.25 ± 0.25	74.52 ± 0.25	92.13 ± 0.38	NA
15a	86.32 ± 0.33	76.52 ± 0.25	90.43 ± 0.34	NA
15b	63.22 ± 0.34	54.25 ± 0.25	82.68 ± 0.58	NA
17	68.23 ± 0.39	54.32 ± 0.58	76.24 ± 0.58	NA
20	62.52 ± 0.39	54.32 ± 0.16	73.25 ± 0.58	NA
21	53.21 ± 0.58	43.20 ± 25	68.30 ± 0.35	NA
22	43.25 ± 0.25	46.24 ± 0.58	64.25 ± 0.17	NA
23	76.52 ± 0.25	64.31 ± 0.44	86.23 ± 0.63	NA
Amphotericin B	96.25 ± 0.1	74.25 ± 0.2	95.36 ± 0.2	91.29 ± 0.1

CONCLUSION

In this paper, we report the synthesis of 2-(1H-imidazol-2-ylthio)-N-p-tolylacetamide2.The active methylene moiety of compound 2 was allowed to react with $CS_2$and/or phenyl isothiocyanate in dimethylformamide in the presence of potassium hydroxide and yielded the non-isolable intermediate potassium sulphide salt 8, 11and 18, which is used as intermediate to synthesis series of novel substituted imidazole derivatives in good yield. Synthesized compounds 2, 3, 6, 13, 15a, 15b, 17, 20, 21, 22 and 23 were evaluated for antibacterial and antifungal activities.Most of the tested compounds revealed better activity against the Gram-positive rather than the Gram-negative bacteria. Compound 13 exhibited excellent antibacterial activity against Staphylococcus aureus(RCMB0100010),Bacillus subtilis (RCMB 010067) and Escherichia coli (RCMB 010052).Compounds 13, 15a exhibited excellent antifunger activity, which is better than the amphotericin B against Syncephal astrumracemosum(RCMB05922).

REFERENCES

1.   Dieter, R.K. (1986) α-Oxo Ketene Dithioacetals and Related Compounds: Versatile Three-Carbon Synthons.Tetrahedron, 42, 3029-3096.http:// dx.doi.org/10.1016/S0040-4020(01)87376-2

2.   Junjappa, H.,Ila, H. and Asokan, C.V.(1990) α-Oxoketene-S,S-, N,S- and N,N-acetals: Versatile Intermediates in Organic Synthesis.Tetrahedron, 46, 5423-5506.http://dx.doi.org/10.1016/S0040-4020(01)87748-6

3.   Khali, M.A.,Sayed, S.M. and Raslan, M.A. (2012)Reactivity of 2-Cyano-N-(4-(1-methyl-1H-benzo[d]imidazol-2-yl)-3-(methylthio)-1-phenyl-1H-pyrazol-5-yl)acetamide:A Facile Synthesis of Pyrazole, Thiazole, 1,3,4-Thiadiazole and PolysubstitutedThiophene Derivatives. American Journal ofOrganic Chemistry, 2, 161-170.http://dx.doi.org/10.5923/j.ajoc.20120206.06

4.   Abdelhamid, A.O and Afifi, M.A.(2010) Synthesis of Some New Thiazoles and Pyrazolo[1,5-a]pyrimidines Containing an Antipyrine Moiety.Synthetic Communications, 40, 1539-1550.http://dx.doi.org/10.1080/00397910903100726

5.   Kaya, E.G.,Ozbilge, H.and Albayrak, S.(2009)Determination of Effect of Gentamicin against Staphylococcus aureus by Using Microbroth Kinetic System.AnkemDerg, 23, 110-114.

6.   Sofy, A.R.,Hmed, A.A.,Sharaf, A.M. and El-Dougdoug, K.A.(2014) Structural Changes of Pathogenic Multiple Drug Resistance Bacteria Treated with T. vulgarisAqueous Extract.Nature and Science,12, 83-88.

7.   Caldeira, E.M., Osório, A.,Oberosler, E.L.,Vaitsman, D.S.,Alviano, D.S. and Nojima, M.D.G.(2013) Antimicrobial and Fluoride Release Capacity of Orthodontic Bonding Materials.Journal of Applied Oral Science, 21, 327-334. http://dx.doi.org/10.1590/1678-775720130010

8.   Shabaan, M.,Taher, A. and Osman, E.O.(2011) Synthesis of Novel 3,4-Dihydroquinoxalin-2(1H)-one Derivatives.European Journal of Chemistry, 3, 365-371.

9.   Voskiene, A.,Mickevicius, V. and Mikulskiene, G.(2007) Synthesis and Structural Characerization of Products Condensation4-Carboxy-1-(4-styrylcarbonylphenyl)-2-pyrrolidinones   with   Hydrazines.

ARKIVOC:Archive for Organic Chemistry,2007,303-314.http://dx.doi.org/10.3998/ark.5550190.0008.f29

10. Mohare, R.M.,Fleit, D.H.and Sakka, O.K. (2011) Novel Synthesis of Hydrazide-Hydrazone Derivatives and Their Utilization in the Synthesis of Coumarin, Pyridine, Thiazole and Thiophene Derivatives with Antitumor Activity.Molecules,16, 16-27.http://dx.doi.org/10.3390/molecules16010016

11. El-Bayouki, K.A., Basyounia, W.M., Mohamed, Y.A., Aly, M.M. and Abbas, S.Y.(2011) Novel 4(3H)-Quinazolinones Containing Biologically ActiveThiazole, Pyridinone and Chromene of Expected Antitumor and Antifungal Activities.European Journal of Chemistry,2, 455-462. http://dx.doi.org/10.5155/eurjchem.2.4.455-462.171

12. Elgemeie, G.H., Elghandour, A.H.,Ali, H.A. and Hussein, A.M.(2002) Novel 2-Thioxohydantoin Ketene Dithioacetals:Versatile Intermediates for Synthesis of Methylsulfanylimidazo[4,5-c]pyrazole and Methylsulfanylpyrrolo[1, 2-c]imi- dazoles.Synthetic Communications, 32, 2245-2253.http://dx.doi.org/10.1081/SCC-120005435

13. Wang, Y., Dong, D., Yang, Y., Huang, J., Ouyang, Y.and Liu, Q. (2007) A Facile and Convenient One-Pot Synthesis of PolysubstitutedThiophenes from 1, 3-Dicarbonyl Compounds in Water.Tetrahedron,63, 2724-2728. http://dx.doi.org/10.1016/j.tet.2006.12.090

14. Khalil, A.M., Berghot, M.A. andGoudal, M.A.(2009) Synthesis and Antibacterial Activity of Some New Heterocycles Incorporating Phthalazine. European Journal of Medicinal Chemistry, 44, 4448-4454. http://dx.doi.org/10.1016/j.ejmech.2009.06.003

15. Elgemeie, G.H., Elghandour, A.H. and AbdElaziz, G.W.(2003)Novel Synthesis of Heterocyclic Ketene N,N-, N,O-, and N,S-Acetals Using CyanoketeneDithioacetals. Synthetic Communications, 33, 1659-1664. http://dx.doi.org/10.1081/SCC-120018927

16. Sommena, G.L.,Comelb, A.and Kirsch, G. (2005)An Easy Access to Variously Substituted Pyrroles Starting from Ketene Dithioacetals. Synthetic Communications, 35, 693-699.http://dx.doi.org/10.1081/SCC-200050365

17. Bondock, S., Rabie, R., Etman, H.A.and Fadda, A.(2008) Synthesis and Antimicrobial Activity of Some New Heterocycles Incorporating Antipyrine Moiety. European Journal of MedicinalChemistry, 43, 2122-2129. http://dx.doi.org/10.1016/j.ejmech.2007.12.009

18. Bondock, S., Fadaly, W.and Metwally, M.A.(2010) Synthesis and Antimicrobial Activity of Some New Thiazole, Thiophene and Pyrazole

Derivatives Containing Benzothiazole Moiety. European Journal of Medicinal Chemistry, 45, 3692-3701.http://dx.doi.org/10.1016/j.ejmech.2010.05.018

19. Rao, R.M., Reddy, G.N. and Sreeramulu, J.(2011) Synthesis of Some New Pyrazolo-Pyrazole Derivatives Containing Indoles with Antimicrobial Activity. Der Pharma Chemica, 3, 301-309.

Chapter 8

PREPARATION OF POLYFUNCTIONALLY SUBSTITUTED PYRIDINE-2(1H)-THIONE DERIVATIVES AS PRECURSORS TO BICYCLES AND POLYCYCLES

Fathi A. Abu-Shanab[1,2], Sayed A. S. Mousa[1], Sherif M. Sherif[3], Mohamed I. Hassan[1]

[1]Department of Chemistry, Faculty of Science, Al-Azhar University, Assiut, Egypt

[2]Department of Chemistry, Faculty of Science, Gazan University, Gazan, KSA

[3]Department of Chemistry, Faculty of Science, Cairo University, Giza, Egypt

ABSTRACT

Reaction of acetylacetone with 1 mole of dimethylformamide dimethyl acetal (DMFDMA) affords enamine 2a which reacts with cyanothioacetamide to give pyridinethione 3a. Pyridinethione 3a reacts with methyl iodide, halogenated compounds, aromatic aldehyde and malononitrile/ele- mental sulfur to yiled compounds 7-10 respectively. Reactions of thioether 7 in ethanolic K_2CO_3, 1 mole DMFDMA and 4-(dimethylamino)benzaldehyde give compounds 11, 13, 14 respectively. Enaminone 12 can be prepared by reaction of compound 11 with DMFDMA. We have demonstrated some reactions in order to show the potential usefulness of the prepared compounds for the preparation of new bipyridyl compounds 15, 16, 18, bicyclic compounds 17 and uncommon tricyclic compounds 20, 21, 22 and 23 respectively using DMFDMA.

INTRODUCTION

Formamide acetals are useful reagents in organic synthesis; [1] [2] their main application has been used for functional group transformations [3] , but they may also be regarded as one-carbon synthons in the construction of carbon skeletons. One type of reaction, which is potentially valuable for the future purpose, is the reaction of N,N'-dimethylformamide dimethyl acetal

(DMFDMA) with 1,3-dicarbonyl compounds 1 to give enamines 2 [2] [4] (Figure 1).

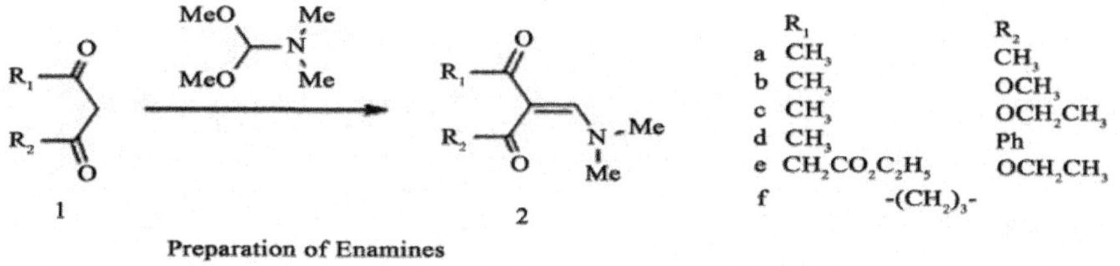

	R₁	R₂
a	CH₃	CH₃
b	CH₃	OCH₃
c	CH₃	OCH₂CH₃
d	CH₃	Ph
e	CH₂CO₂C₂H₅	OCH₂CH₃
f	-(CH₂)₃-	

Preparation of Enamines

Figure 1: Preparation of Enamines.

We have reported that enamines 2 were used as precursors in the synthesis of pentasubstituted pyridines 3-6 [5] -[8] (Figure 2).

The treatment of acetyacetone (1a) with dimethyl formamide dimethylacetal (DMFDMA) in dry DMF under nitrogen and stirring over night afforded the corresponding enamine 2a which on treatment with cyanothioacetamide and sodium hydride in dry DMF (in situ) afforded pyridine-2(1H)-thione (3a) [6] , when the emamine 2a was treated with cyanothioacetamide in ethanol and pepridine as a base afforded the pyridine-2(1H)-thione (5a) [7] [12] (Figure 3).

RESULTS AND DISCUSSION

In conjunction of this work, we report here the reaction of acetylacetone 1a with one mole of N,N'-dimetylfor- mamide dimethyl acetal (DMFDMA) in dry dioxane gave the corresponding enamine 2a. The treatment of this enamine (in situ) with cyanothioacetamide in ethanol in the presence of sodium ethoxide under reflux gave 5- acetyl-6-methyl-2-thioxo-1,2-dihydropyridine-3-carbonitrile 3a with a very good yied [7] , Scheme 1.

We have found that the prepared compound 3a included three functional groups which are thioamido group, cyano group and acetyl group. These functional groups can be used for the preparation of bicyclic or polycyclic compounds of biological interest. Thus, some illustrative reactions designed to demonstrate the potential usefulness of 5-acetyl-6-methyl-2-thioxo-1,2-dihydropyridine-3-carbonitrile 3a for further heterocyclic synthesis. Therefore, the reaction of 5-acetyl-6-methyl-2-thioxo-1,2-dihydropyridine-3-carbonitrile 3a with methyl iodide in alcoholic sodium hydroxide afforded the corresponding thioether derivative 7, which in turn is a good intermediate for the preparation of further heterocyclic compounds of biological interest. The structure of the isolated compound 7 is confirmed by spectral analysis. The IR spectrum shows the disappearance of (NH) group. Also, the ¹H NMR spectrum shows the disappearance of the thioamide proton and the appearance of a singlet signal corresponding to (SCH$_3$) at δ_H = 2.63 ppm. Also, the mass spectrum shows the molecular ion peak at m/e 206 which corresponding to the molecular formula (C$_{10}$H$_{10}$N$_2$OS). The reaction of 5-acetyl-6-methyl-2-thiox-o-1,2-dihydropyridine-3-carbonitrile 3a with ethyl chloroacetate or chloroacetamides in ethanolic sodium ethoxide afforded the corresponding 5-acetyl-3-amino-6-methylthieno[2,3-b]pyridine derivatives 8a-c in a good yield. The structure of the isolated compounds is confirmed by elemental and spectral analysis. The IR spectrum shows the disappearance of cyano group and appearance of amino group at υ_{max} = 3427, 3328 cm^{-1} in compound 8a as example beside the other functional groups. Also, the mass spectra show the molecular ion peaks fit to all compounds 8a-c. Also, the ^{1}H NMR spectra show signals fit to the structure of all compounds 8a-c. The presence of acetyl group in 5-acetyl-6-methyl-2-thioxo-1,2-dihydropyridine-3-carbonitrile 3a is useful for the preparation of fused heterocyclic compounds. So that the reaction of compound 3a with aldehydes like 4-(dimethylamino) benzaldehyde and 4-methylbenzaldehyde in ethanolic sodium hydroxide afforded the corresponding chalcones 9a,b. The structure of the isolated chalcones is confirmed by elemental analysis as well as spectral analysis. The mass spectra show the molecular ion peak fit to all compounds 9a,b. As an example compound 9a shows the molecular ion peak at m/e 323 which corresponding to the molecular formula (C$_{18}$H$_{17}$N$_3$OS). Also, the ^{1}H NMR spectra of these compounds 9a,b show the disappearance of the signal corresponding to the methyl of acetyl group and the appearance of two doublets signals corresponding to the two protons of double bond of chalcone. Finally, 5-acetyl-6-methyl-2-thioxo-1,2-dihydropyridine-3-carbonitrile 3a was treated with malononitrile and sulfur element (Gewald's reaction) in ethanol in the presence of triethylamine as a base to afford 5-(5-amino-4- cyanothiophen-3-yl)-6-methyl-2-thioxo-1,2-dihydropyridine-3-carbonitrile 10 in a good yield, Scheme 1. The IR spectrum of compound 10 shows the appearance of amino group at υ_{max} = 3435, 3350 cm^{-1} beside the other functional groups.

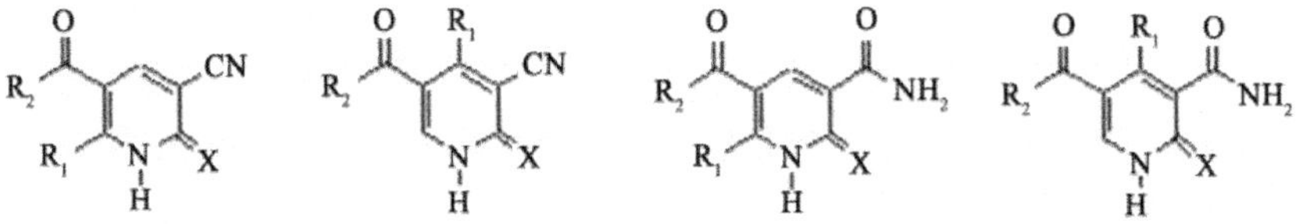

X = O,S,C(CN)$_2$

Defferent polysubstituted pyridines have been prepared

Figure 2: Different polysubstituted pridines have been prepared.

Figure 3: Tetrasubstituted pyridinethione have been prepared.

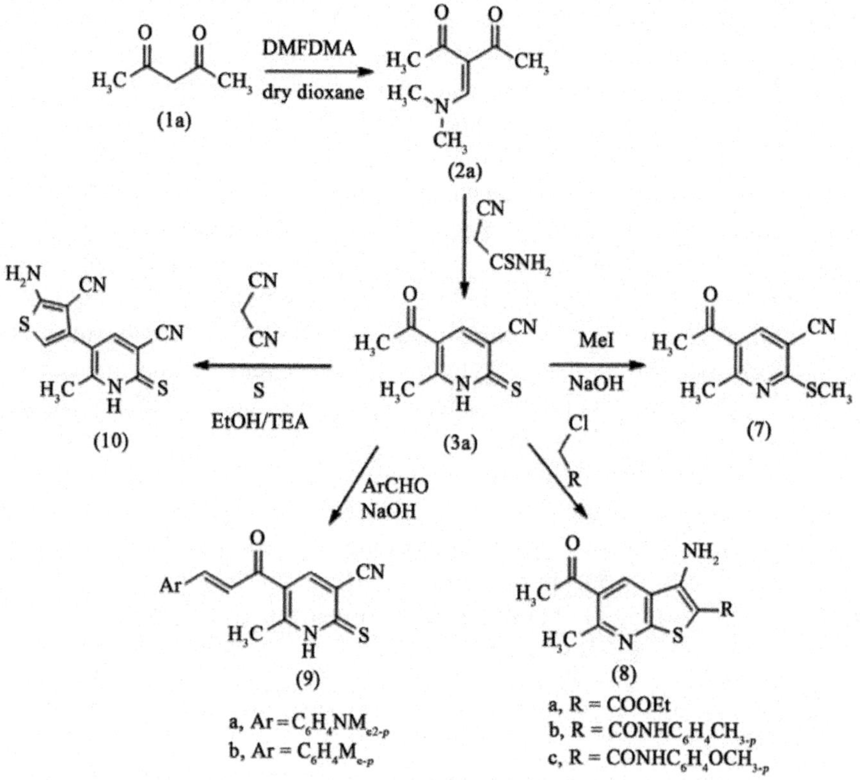

Scheme 1: Synthesis of pyridine (1H)-thione derivative 3a and its reactions with MeI, α- chloroketones, aldehydes and malononitrile.

Also, ^{1}H NMR spectrum of compound 10 shows singlet signal at $\delta_H =$ 2.45 ppm corresponding to methyl group and singlet signal at $\delta_H = 6.95$ ppm corresponding to amino group and singlet signal at $\delta_H = 7.07$ ppm corresponding to CH thiophene ring and singlet signal at $\delta_H = 7.2$ ppm corresponding to CH pyridine ring.

5-Acetyl-6-methyl-2-(methylthio)nicotinonitrile **7** can be used as intermediate for further preparation of heterocyclic compounds. So that compound **7** was treated with potassium carbonate in ethanol to afford 5-acetyl-2- ethoxy-6-methylnicotinonitrile **11**. This compound was formed by nucleophilic substitution of SMe by OEt group. The structure of the isolated compound is confirmed by elemental and spectral analyses. The mass spectrum shows the molecular ion peak at m/e 204 corresponding to the molecular formula ($C_{11}H_{12}N_2O_2$). Also, the ^1H NMR spectrum shows the disappearance of SMe signal and appearance of two signals; a triplet at $\delta_H = 1.43$ ppm and a quartet at $\delta_H = 4.54$ ppm corresponding to the OEt moiety, in addition to the rest of signals corresponding to the other protons in the molecule. Compound **11** was reacted with N,N›-dimetylformamide dimethyl acetal (DMFDMA) in dry xylene to give the corresponding enamine **12** in a good yield. The mass spectrum of compound **12** shows the molecular ion peak at m/e 259 which corresponding to the molecular formula ($C_{14}H_{17}N_3O_2$). Also, the ^1H NMR spectrum of compound **12** shows the disappearance of the singlet signal which is related to the methyl of acetyl group and the appearance of two singlet signals at $\delta_H = 2.68$ and 3.04 ppm corresponding to the two methyl groups of NMe$_2$ moiety. Consequently the ^1H NMR spectrum shows the appearance of two doublets at $\delta_H = 6.25$ ppm and 7.87 ppm corresponding to the two protons of the enamine double bond.

Enamine **13** can be prepared in a good yield by reaction of 5-acetyl-6-methyl-2-thioxo-1,2-dihydropyri-dine- 3-carbonitrile **3a** with two moles of N,N›-dimetylformamide dimethylacetal (DMFDMA) in dry xylene or by the reaction of 5-acetyl-6-methyl-2-(methylthio)nicotinonitrile **7** with one mole of N,N›-dimetylformamide dimethylacetal (DMFDMA) in dry xylene. The structure of the isolated compound is confirmed by elemental and spectral analysis. Whereas the mass spectrum shows the molecular ion peak at m/e 261 which corresponding to the molecular formula ($C_{13}H_{15}N_3OS$). Also, the ^1H NMR spectrum of it shows the disappearance of the singlet signal which is related to the methyl of acetyl group and appearance of two singlet signals at $\delta_H = 2.62$ and 2.64 ppm corresponding to the two methyl groups of NMe$_2$ moiety. Consequently, the ^1H NMR spectrum shows the appearance of two doublets at $\delta_H = 5.28$ ppm and 7.75 ppm corresponding to the two protons of double bond of enamine.

Chalcone **14** can either be prepared by the reaction of compound **7** with (4-(dimethylamino)benzaldehyde) in ethanolic sodium hydroxide or by treatment of compound **9a** with methyl iodide in ethanolic sodium hydroxide.

The mass spectrum of compound 14 shows the molecular ion peak at m/e 337 corresponding to the molecular formula ($C_{19}H_{19}N_3OS$). Also, the ^{1}H NMR spectrum of compound 14 shows singlet signal at $\delta_H = 2.62$ ppm corresponding to methyl group and singlet signal at $\delta_H = 2.66$ ppm corresponding to SCH$_3$ and two singlet signal at $\delta_H = 2.9, 3.04$ ppm corresponding to NMe$_2$ moiety and appearance of some signals of other protons in molecule.

For preparation of bipyridyl derivatives, we have carried out the reaction of chalcones 5-(3-(4-(dimethyla- mino)phenyl)acryloyl)-6-methyl-2-thioxo-1,2-dihydropyridine-3-carbonitrile 9a and 5-(3-(4-(dimethylamino) phenyl) acryloyl)-6-methyl-2-(methylthio)nicotinonitrile 14 with malononitrile dimmer [9] in acetic acid and ammonium acetate afforded the corresponding bipyridyl derivatives 6-(dicyanomethylene)-4-(4-(dimethylami- no)phenyl)-2›-methyl-6›-thioxo-1,1›,6,6›-tetrahydro-[2,3›-bipyridine]-5,5›-dicarbonitrile 15 and 6-(dicyanomethy- lene)-4-(4-(dimethylamino)phenyl)-2›-methyl-6›-(methylthio)-1,6-dihydro-[2,3›-bipyridine]-5,5›-dicarbonitrile

16 respectively. The reaction proceeds by Michael addition followed by cyclization through condensation as shown in Scheme 2.

The compound 16 can also be obtained by the reaction of 6-(dicyanomethylene)-4-(4-(dimethylami-no)   phenyl)-2›-methyl-6›-thioxo-1,1›,6,6›-tetrahydro-[2,3›-bipyridine]-5,5›-dicarbonitrile 15 with methyl iodide in alcoholic sodium hydroxide Scheme 2.

The structure of the isolated compounds 15 and 16 is established by elemental and spectral analysis. Whereas the mass spectra of these compounds show the molecular ion peaks at m/e 435 corresponding to the molecular formula ($C_{24}H_{17}N_7S$), and at m/e 449 corresponding to the molecular formula ($C_{25}H_{19}N_7S$) for 15 and 16 respectively. The IR spectra of both compounds 15 and 16 show the disappearance of the carbonyl group and the appearance of NH group. Also, the[1]H NMR spectra of these compounds show signals fit to structures 15 and 16.

For further preparation of heterocyclic compounds [10] , we carried out the following reactions. The reaction of enamine 13 with excess hydrazine hydrate in ethanol afforded 6-methyl-5-(1H-pyrazol-3-yl)-1H-pyrazolo[3, 4-b] pyridin-3-amine 17 in a good yield as shown in Scheme 3. The IR spectrum of compound 17 shows the disappearance of the cyano group and the appearance of NH$_2$ and NH groups at υ_{max} at 3405 cm^{-1}, 3329 cm^{-1} and 3136 cm^{-1} respectively. Also, the mass spectrum of compound 17 shows the molecular ion peak at m/e 214 corresponding to the molecular formula ($C_{10}H_{10}N_6$). Also, the ^{1}H NMR spectrum of compound 17 shows signals fit to the structure.

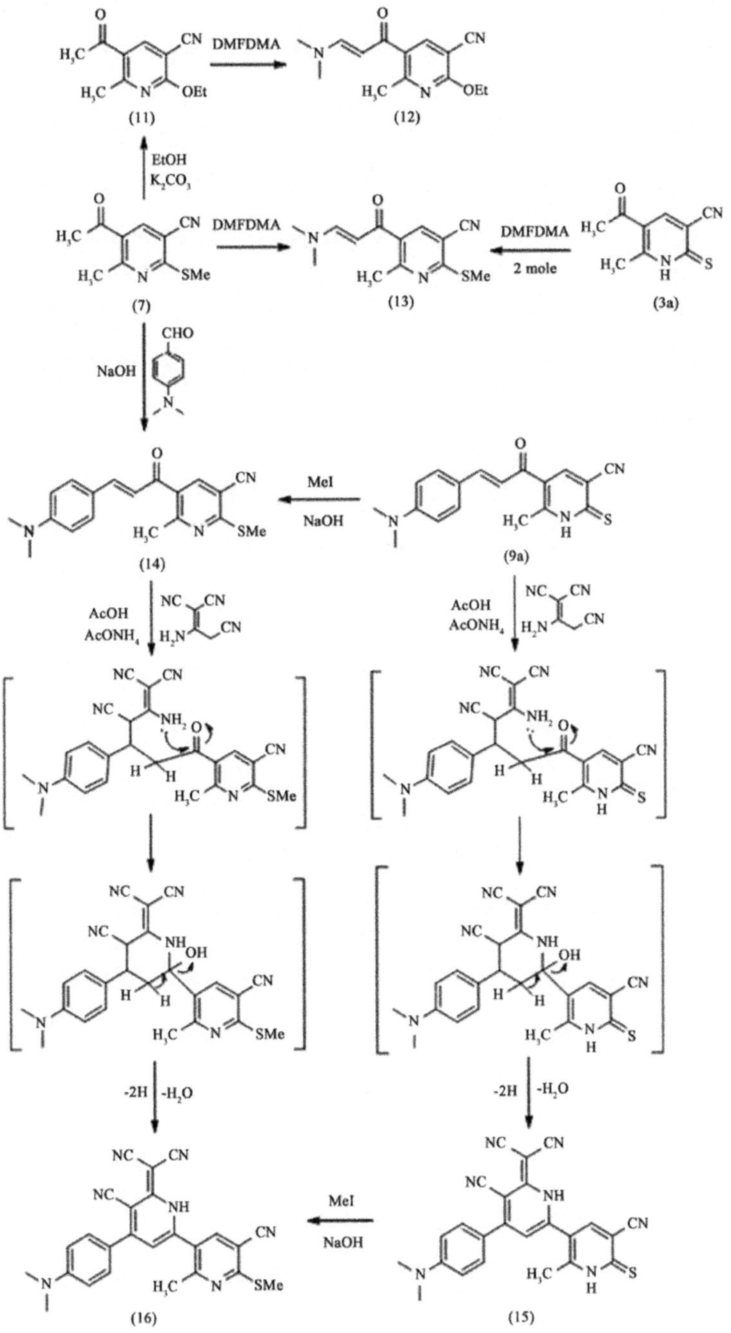

Scheme 2: Reactions of tetrasubstitutedpridine 7 with DMFDMA, alcholic K_2CO_3 and p-N, N- dimethylaminobezaldehyde.

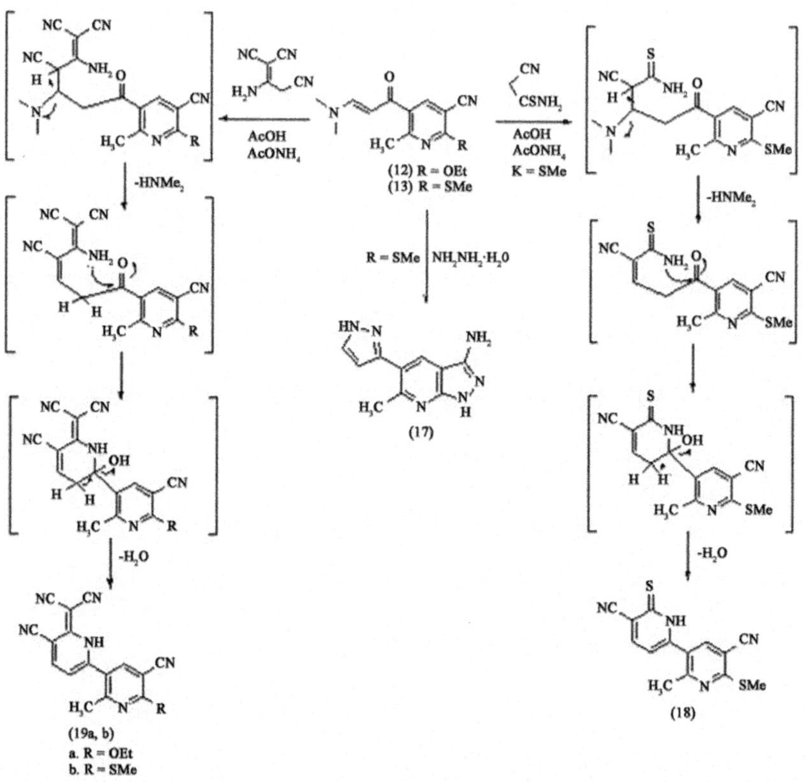

Scheme 3: Reactions of tetrasubstitutedpridine (12,13) with cyanothioacetamide, hydrazine hydrate and malononitrile dimer.

Also, the enamine 13 is treated with cyanothioacetamide in acetic acid and ammonium acetate afforded 2'- methyl-6'-(methylthio)-6-thioxo-1,6-dihydro-[2,3'-bipyridine]-5,5'-dicarbonitrile 18. The reaction is started by Micheal addition of cyanothioacatamide on the double bond followed by elimination of dimethylamine ($HNMe_2$) and cyclization with the carbonyl group. The structure of the isolated compound 18 is confirmed by elemental and spectral analysis. The IR spectrum of compound 18 shows the disappearance of carbonyl group and appearance of NH group at υ_{max} at 3428 cm^{-1}. The mass spectrum of compound 18 shows the molecular ion peak at m/e 298 corresponding to the molecular formula ($C_{14}H_{10}N_4S_2$). Also, the ^{1}H NMR spectrum of compound 18 shows the disappearance of protons of NMe_2 moiety and appearance of NH proton beside the other protons.

Another type of bipyridyl derivatives 19a,b can be prepared by the reaction of the enamines 12 and 13 with malononitrile dimmer in acetic acid and ammonium acetate. This reaction proceeds by Michael addition of malononitrile dimmer, followed by elimination of dimethylamine (HNMe$_2$) and cyclization through condensation of amino group with carbonyl group as shown in Scheme 3. The mass spectrum of compound 19a shows the molecular ion peak at m/e 328 corresponding to the molecular formula (C$_{18}$H$_{12}$N$_6$O), and compound 19b shows the molecular ion peak at m/e 330 corresponding to the molecular formula (C$_{17}$H$_{10}$N$_6$S). The IR spectra of the compounds 19a,b shows the disappearance of the carbonyl group and the appearance of NH group beside the other groups. Also, the ^{1}H NMR spectra of compounds 19a,b show the disappearance of protons of NMe$_2$moiety and appearance of NH proton beside the other protons.

The tricyclic heterocyclic compounds are biologically interest compounds. They are examples of uncommon ring system [11] [12] . Therefore we are interested for the preparation of this type of heterocyclic compound. Thus 5-acetyl-3-amino-6-methyl-N-(p-tolyl)benzo[b]thiophene-2-carboxamide 8b is reacted with N,N'-dime- tylformamide dimethyl acetal (DMFDMA) in dry dioxane afforded 8-acetyl-7-methyl-3-(p-tolyl)pyrido [3',2':4,5] thieno[3,2-d] pyrimidin-4(3H)-one 20. The IR spectrum of compound 20 shows the disappearance of (NH$_2$) and (NH) groups. The mass spectrum of compound 20 shows the molecular ion peak at m/e 349 which corresponding to the molecular formula (C$_{19}$H$_{15}$N$_3$O$_2$S). Also, the ^{1}H NMR spectrum of compound 20 shows the appearance of two singlet signals at δ_H = 8.43 ppm, and 8.52 ppm corresponding to two protons of pyrimidinone and pyridine rings respectively beside other signals for other protons.

For further reaction of 5-acetyl-3-amino-6-methyl-N-substituted[b] thiophene-2-carboxamide 8b,c it reacted with nitrous acid in acetic acid afforded the tricyclic compounds 21a,b in a good yield as shown in Scheme 4. The structures of the compounds 21a,b are established by elemental and spectral analysis. Whereas the IR spectra of both compounds 21a,b show the disappearance of the bands corresponding to (NH$_2$) and (NH) groups. The mass spectrum of the compound 21a as an example shows the molecular ion peak at m/e 350 corresponding to molecular formula (C$_{18}$H$_{14}$N$_4$O$_2$S).

Also, the ^{1}HNMR spectra of compounds 21a,b show the disappearance of the signals which corresponding to (NH$_2$) and (NH) groups beside the appearance the other signals for other groups.

We have found that the prepared tricyclic compounds 20 and 21a,b contain acetyl group which is very important for the preparation of new heterocyclic compounds. So that the reaction of 21a with malononitrile and sulphur element in ethanol and triethylamine (Geweld reaction) afforded 2-amino-4-(7-methyl-4-oxo-3-(p- tolyl)-3,4-dihydropyrido[3',2':4,5]thieno[3,2-d][1,2,3]triazin-8-yl)thiophene-3-carbonitrile 22. The IR spectrum of compound 22 shows the disappearance of the carbonyl group of acetyl moiety and the appearance of amino and cyano groups at υ_{max} at 3427 cm^{-1} and 2208 cm^{-1} respectively. Also, the mass spectrum of this compound 22 shows the molecular ion peak at m/e 430 which corresponding to the molecular formula $(C_{21}H_{14}N_6OS_2)$.

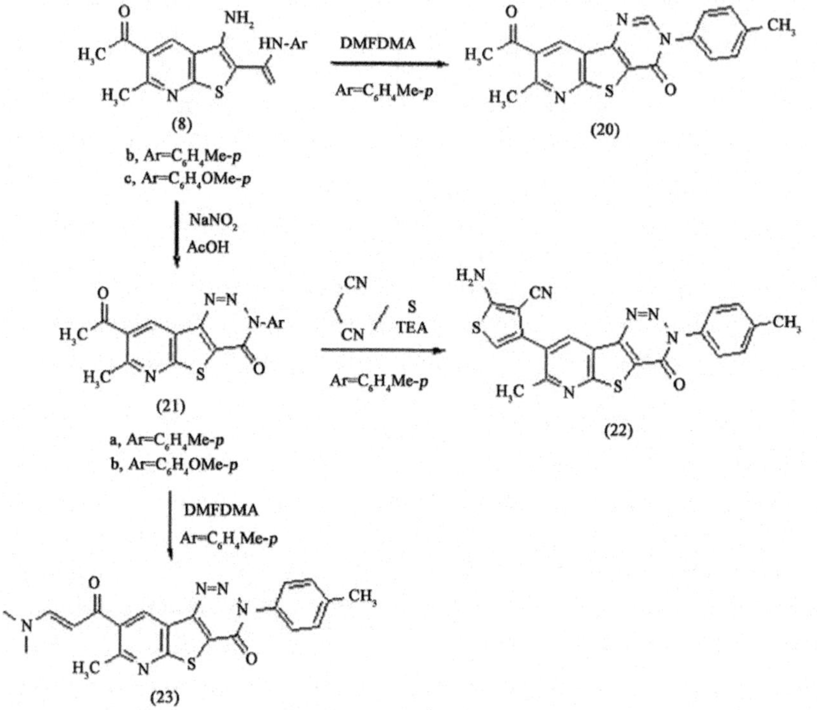

(8)

b, Ar=C$_6$H$_4$Me-p
c, Ar=C$_6$H$_4$OMe-p

NaNO$_2$
AcOH

(20)

(21)

a, Ar=C$_6$H$_4$Me-p
b, Ar=C$_6$H$_4$OMe-p

(22)

DMFDMA
Ar=C$_6$H$_4$Me-p

(23)

Scheme 4: Reactions of thinopyridine derivatives (8b,c) with DMFDMA and sodium nitrile in acetic.

Also, the compound 21a is treated with N,N'-dimetylformamide dimethyl acetal (DMFDMA) in dry xylene afforded the corresponding enamine 8-(3-(dimethylamino)acryloyl)-7-methyl-3-(p-tolyl)pyrido[3',2':4,5] thie- no[3,2-d][1,2,3]triazin-4(3H)-one 23 in a good yield, Scheme 4. The mass spectrum of compound 23 shows the molecular ion peak at m/e 405 corresponding to molecular formula $(C_{21}H_{19}N_5O_2S)$. Also, the ^{1}H NMR spectrum of compound 23 shows the disappearance of the methyl of acetyl moiety and appearance instead of it two singlet signals at $\delta_H = 3.63$ ppm and 3.67 ppm corresponding to (NMe$_2$) moiety. Also, it shows the appearance of two doublet signals at $\delta_H = 5.42$ ppm and 7.82 ppm respectively corresponding to the double bond protons of enaminone moiety beside signals for other protons.

EXPERIMENTAL

All melting points are uncorrected. IR spectra were recorded on a Perkin-Elmer 17,100 FTIR spectrometer as KBr disks. NMR spectra were recorded on a Varian Gemini (400 MHz) spectrometer for solutions in $CDCl_3$ or DMSO-d_6 with tetramethylsilane (TMS) as an internal standard unless otherwise. Mass spectra were obtained on Finnigan 4500 (low resolution) spectrometers using electron impact (EI) at Micro-analytical Center Cairo University Giza Egypt.

Preparation of 5-acetyl-3-cyano-6-methylpyridine-2(1H)-thione (3a):

A mixture of acetylacetone (1a) (1.0 g, 10 mmol), dry dioxane (10 mL) and N,N-dimethylformamide dimethyl acetal (1.19 g, 10 mmol) was stirred under dry condition at room for 24 h. In a second flask, a mixture of dry ethanol (15 mL), and sodium metal (0.46 g, 20 mmol) was left under stirring for 10 min. Then cyanothioacetamide (1.0 g, 10 mmol) was added to the mixture. The mixture was left for further 10 min. The contents of the second flask were transferred into the first flask, and the resulting mixture was stirred for further 1 h, followed by converting stirring into reflux for 4 h. After cooling, the mixture was poured onto acidified ice/cold water. The product was recovered by filtration and recrystallised from ethanol as orange crystals (76%), Mp. 232° - 233°, similar to be published before [7] and mixed Mp.

Preparation of 5-acetyl-6-methyl-2-(methylthio)nicotinonitrile (7)

Mixture of 5-acetyl-6-methyl-2-thioxo-1,2-dihydropyridine-3-carbonitrile 3a (1.92 g, 10 mmol) in ethanol as solvent and sodium hydroxide (0.4 g, 10 mmol) with stirring for 1 hr., and add methyl iodide (0.62 ml, 10 mmol) with stirring until precipitate formed. The product was recovered by filtration and recrystallised from ethanol as white crystals (1.52 g, 74%), Mp. 140°C - 142°C; ^{1}H-NMR ($CDCl_3$): δ = 2.54 (3H, s, CH_3 py.), 2.63 (3H, s, SCH_3), 2.77 (3H, s, CH_3CO), 8.07 (1H, s, CH py.); IR (KBr) υ 2227 (CN), 1685 cm^{-1} (C=O); MS (EI)$^+$: m/z 206 M$^+$; Anal. Calcd for $C_{10}H_{10}N_2OS$ (206.27): C, 58.23; H, 4.89; N, 13.58. Found: C, 58.03; H, 4.73; N, 13.41.

General procedure for the preparation of compounds 8a-c

In dry flask, a mixture 5-acetyl-6-methyl-2-thioxo-1,2-dihydropyridine-3-carbonitrile 3a (1.92 g, 10 mmol) and a-chloro compounds (10 mmol) in ethanol and sodium ethoxide (20 mmol) was left under reflux for two hours. The mixture was left for cooling and poured onto ice cold water. The solid product was recovered by filtration and recrystallised from the proper solvent.

Ethyl 5-acetyl-3-amino-6-methylthieno[2,3-b]pyridine-2-carboxylate (8a): Obtained using ethyl 2-chloroa- cetate (1.06ml, 10 mmol). The product was recrystallised from acetic acid as yellow crystals (2.16 g, 77.7%), Mp.

220°C - 222°C; ^{1}H-NMR (DMSO): δ = 1.25 (3H, t, CH_3 ethyl), 2.6 (3H, s, CH_3 py.), 2.66 (3H, s, CH_3CO), 4.25 (2H, q, CH_2 ethyl), 7.29 (2H, s, NH_2), 8.95 (1H, s, CH py.); IR (KBr) υ 3427, 3328 (NH_2), 1679 cm^{-1} (C=O); MS (EI)$^+$: m/z 278 M$^+$; Anal. Calcd for $C_{13}H_{14}N_2O_3S$ (278.33): C, 56.10; H, 5.07; N, 10.06, Found: C, 55.96; H, 4.94; N, 9.97.

5-Acetyl-3-amino-6-methyl-N-(p-tolyl)thieno[2,3-b]pyridine-2-carboxamide (8b): Obtained using 2-chloro- N-(p-tolyl)acetamide (1.83 g, 10 mmol). The product was recrystallised from ethanol as yellow crystals (2.7 g, 79%), Mp. 218°C - 220°C; ^{1}H-NMR (DMSO) δ = 2.26 (3H, s, CH_3 Ar), 2.64 (3H, s, CH_3 py.), 2.73 (3H, s, CH_3CO), 7.12 (2H, d, Ar), 7.47 (2H, s, NH_2), 7.55 (2H, d, Ar), 9.04 (1H, s, CH py.), 9.4 (1H, s, NH); IR (KBr) υ 3428 , 3312 (NH_2, NH), 1685 cm^{-1} (C=O); MS (EI)$^+$: m/z 339 M$^+$; Anal. Calcd for $C_{18}H_{17}N_3O_2S$ (339.42): C, 63.70; H, 5.05; N, 12.38, Found: C, 63.56; H, 4.93; N, 12.15.

5-Acetyl-3-amino-N-(4-methoxyphenyl)-6-methylthieno[2,3-b]pyridine-2-carboxamide (8c):

Obtained using 2-chloro-N-(4-methoxyphenyl)acetamide (1.99 g, 10 mmol). The product was recrystallised from ethanol as yellow crystals (2.8 g, 79%), Mp. 240°C - 242°C; ^{1}H-NMR (DMSO) δ = 2.65 (3H, s, CH_3 py.), 2.73 (3H, s, CH_3CO), 3.76 (3H, s, CH_3O), 6.9 (2H, d, Ar), 7.45 (2H, s, NH_2), 7.56 (2H, d, Ar), 9.04 (1H, s, CH py.), 9.4 (1H, s, NH); IR (KBr) υ 3428, 3310, 3251 (NH_2, NH), 1680 cm^{-1} (C=O); MS (EI)$^+$: m/z 355 M$^+$; Anal. Calcd for $C_{18}H_{17}N_3O_3S$ (355.42): C, 60.83; H, 4.82; N, 11.82, Found: C, 60.76; H, 4.73; N, 11.69.

General procedure for the preparation of compounds 9a,b

A mixture of 5-acetyl-6-methyl-2-thioxo-1,2-dihydropyridine-3-carbonitrile 3a (1.92 g, 10 mmol) in ethanol as solvent in presence of sodium hydroxide (0.4 g, 10 mmol) with aromatic aldehydes (10 mmol) with stirring for 2hr. then poured onto ice, cold water and acidified with conc. hydrochloric acid until the precipitate was formed. The solid product was recovered by filtration and recrystallised from ethanol.

5-(3-(4-(Dimethylamino)phenyl)acryloyl)-6-methyl-2-thioxo-1,2-dihydropyridine-3-carbonitrile (9a):

Obtained using 4-(dimethylamino)benzaldehyde (1.49g, 10 mmol). Mp. 140°C - 142°C as yellow crystals (2.45 g, 76%); ^{1}H-NMR (CDCl$_3$) δ = 2.65 (3H, s, CH_3), 2.79 (6H, s, NMe$_2$), 6.70 (2H, d, Ar), 7.06 (1H, d, CH chalcone), 7.74 (2H, d, Ar), 7.85 (1H, d, CH chalcone), 8.07 (1H, s, CH py.), 13.2 (1H, br., NH); IR (KBr) υ 3437 (NH), 2225 (CN), 1685 cm^{-1} (C=O); MS (EI)$^+$: m/z 323

M$^+$; Anal. Calcd for $C_{18}H_{17}N_3OS$ (323.42): C, 66.85; H, 5.30; N, 12.99, Found: C, 66.69; H, 5.17; N, 12.86.

6-Methyl-2-thioxo-5-(3-(p-tolyl)acryloyl)-1,2-dihydropyridine-3-carbonitrile (9b): Obtained using 4-methyl- benzaldehyde (1.2 g, 10 mmol). Mp. = 240°C - 242°C as yellow crystals (2.2 g, 74.8%); ^{1}H-NMR (DMSO) δ = 2.31 (3H, s, CH$_3$ Ar), 2.58 (3H, s, CH$_3$ py.), 7.24 (2H, d, Ar), 7.49 (1H, d, CH chalcone), 7.6 (1H, d, CH chalcone), 7.69 (2H, d, Ar), 8.58 (1H, s, CH py.), 13 (1H, br., NH); IR (KBr) υ 3434 (NH), 2231 (CN), 1659 cm^{-1} (C=O); MS (EI)$^+$: m/z 294 M$^+$; Anal. Calcd for $C_{17}H_{14}N_2OS$ (294.38): C, 69.36; H, 4.79; N, 9.52, Found: C, 69.19; H, 4.68; N, 9.45.

5-(5-Amino-4-cyanothiophen-3-yl)-6-methyl-2-thioxo-1,2-dihydropyridine-3-carbonitrile (10)

In dry flask, a mixture 5-acetyl-6-methyl-2-thioxo-1,2-dihydropyridine-3-carbonitrile 3a (1.92 g, 10 mmol), malononitrile (0.66 g, 10 mmol) and sulfur (0.32 g, 10 mmol) in ethanol and few drops of triethylamine as base was left under reflux for three hours. The mixture was left for cooling then poured onto ice cold water. The product obtained was recrystallised from a mixture of ethanol/DMF (3:1) as brown crystals (1.9 g, 69.8%), Mp. > 300°C; ^{1}H-NMR (DMSO) δ = 2.45 (3H, s, CH$_3$), 6.95 (2H, s, NH$_2$), 7.07 (1H, s, CH thiophene), 7.2 (1H, s, CH py.); IR (KBr) υ 3435, 3350 (NH$_2$), 3250 (NH), 2210 cm^{-1} (CN); Anal. Calcd for $C_{12}H_8N_4S_2$ (272.35): C, 52.92; H, 2.96; N, 20.57, Found: C, 52.85; H, 2.92; N, 20.40.

5-Acetyl-2-ethoxy-6-methylnicotinonitrile (11)

In dry flask, a mixture of 5-acetyl-6-methyl-2-(methylthio)nicotinonitrile 7 (2.06 g, 10 mmol) in ethanol and potassium carbonate was left under reflux for 3hr. after cooling the mixture was poured onto ice cold water. The product was recovered and recrystallised from EtOH/H$_2$O (1:1) as yellowish crystals (1.6 g, 78%), Mp. 78°C - 80°C; ^{1}H-NMR (CDCl$_3$) δ = 1.43 (3H, t, CH$_3$ ethyl), 2.66 (3H, s, CH$_3$), 3.04 (3H, s, CH$_3$CO), 4.54 (2H, q, CH$_2$ ethyl), 7.85 (1H, s, CH py.); IR (KBr) υ 2228 (CN), 1688 cm^{-1} (C=O); MS (EI)$^+$: m/z 204 M$^+$; Anal. Calcd for $C_{11}H_{12}N_2O_2$ (204.23): C, 64.69; H, 5.92; N, 13.72, Found: C, 64.51; H, 5.83; N, 13.54.

5-(3-(Dimethylamino)acryloyl)-2-ethoxy-6-methylnicotinonitrile (12)

In dry flask a mixture of 5-acetyl-2-ethoxy-6-methylnicotinonitrile 11 (2.04 g, 10 mmol) in dry xylene as solvent and N,N›-dimethylformamide dimethyl acetal (DMFDMA) (1.32 ml, 10 mmol) was left under reflux for 2 hr., cool and the solvent was evaporated. The product was recovered and recrystallised from EtOH/H$_2$O (1:1) as yellow crystals (1.9 g, 73.3%), Mp. 68°C - 70°C; ^{1}H-NMR (CDCl$_3$) δ = 1.3 (3H, t, CH$_3$ ethyl), 2.62 (3H, s, CH$_3$),

2.68, 3.04 (6H, 2s, NMe$_2$), 4.58 (2H, q, CH$_2$ethyl), 6.25 (1H, d, CH), 7.87 (1H, d, CH), 8.2 (1H, s, CH py.); IR (KBr) υ 2230 (CN), 1684 cm^{-1}(C=O); MS (EI)$^+$: m/z 259 M$^+$; Anal. Calcd for C$_{14}$H$_{17}$N$_3$O$_2$ (259.31): C, 64.85; H, 6.61; N, 16.20, Found: C, 64.56; H, 6.47; N, 16.11.

5-(3-(Dimethylamino)acryloyl)-6-methyl-2-(methylthio)nicotinonitrile (13)

(A) In dry flask, a mixture of 5-acetyl-6-methyl-2-(methylthio) nicotinonitrile 7 (2.06 g, 10 mmol) in dry xylene as solvent and N,N›-dimethylformamide dimethyl acetal (DMFDMA) (1.32 ml, 10 mmol) was left under reflux for 2 hr., cool and poured in dry backer and the solvent was evaporated. The product was recovered and recrystallised from EtOH/H$_2$O (1:1) as yellow crystals (2 g, 76.6%), Mp. 100°C - 102°C; (B) In dry flask a mixture of 5-acetyl-6-methyl-2-thioxo-1,2-dihydropyridine-3-carbonitrile 3a (1.92 g, 10 mmol) in dry xylene as solvent and N,N'-dimethylformamide dimethyl acetal (DMFDMA) (2.64 ml, 20 mmol) was left under reflux for 2 hr., cool and poured in dry backer and the solvent was evaporated. The product was recovered and recrystallised from EtOH/H$_2$O (1:1) as yellow crystals (2.1 g, 80.4%), Mp. and mixed Mp. 100°C - 102°C; ^{1}H-NMR (CDCl$_3$) δ = 2.62, 2.64 (6H, 2s, NMe$_2$), 2.9 (3H, s, CH$_3$ py.), 3.15 (3H, s, SCH$_3$), 5.28 (1H, d, trans CH), 6.28 (1H, d, cis CH), 7.75 (1H, d, trans CH), 8.07 (1H, s, CH py.), 10.15 (1H, d, cis CH); IR (KBr) υ 2227 (CN), 1685 cm^{-1} (C=O); MS (EI)$^+$: m/z 261 M$^+$; Anal. Calcd for C$_{13}$H$_{15}$N$_3$OS (261.35): C, 59.74; H, 5.79; N, 16.08, Found: C, 59.63; H, 5.45; N, 15.8.

5-(3-(4-(Dimethylamino)phenyl)acryloyl)-6-methyl-2-(methylthio) nicotinonitrile (14)

(A) mixture of 5-(3-(4-(dimethylamino)phenyl)acryloyl)-6-methyl-2-thioxo-1,2-dihydropyridine-3-carboni- trile 9a (3.23 g, 10 mmol) in ethanol as solvent and sodium hydroxide (0.4g, 10mmol) with stirring for 1hr., and add methyl iodide (0.62 ml, 10 mmol) with stirring until precipitate was formed. The product was recovered by filtration and was purified by recrystallised from ethanol as yellow crystals (2.5 g, 74%), Mp. 160°C - 162°C; (B) mixture of 5-acetyl-6-methyl-2-(methylthio)nicotinonitrile 7 (2.06 g, 10 mmol) in ethanol as solvent in presence of sodium hydroxide (0.4 g, 10 mmol) with 4-(dimethylamino)benzaldehyde (1.49 g, 10 mmol) with stirring for 2 hr., until precipitate formed and dilute with water. The product was recovered by filtration and purified by recrystallised from ethanol as yellow crystals (2.4 g, 71%), Mp. and mixed Mp. 160°C - 162°C; ^{1}H-NMR (CDCl$_3$) δ = 2.62 (3H, s, CH$_3$), 2.66 (3H, s, SCH$_3$), 2.9, 3.04 (6H, 2s, NMe$_2$), 6.83 (2H, d, Ar), 7.46 (2H, d, Ar), 6.67 (1H, d, CH), 7.38 (1H, d, CH), 7.85 (1H, s, CH py.); IR (KBr) υ 2217 (CN), 1648 cm^{-1} (C=O); MS (EI)$^+$: m/z 337 M$^+$; Anal. Calcd for

$C_{19}H_{19}N_3OS$ (337.45): C, 67.63; H, 5.68; N, 12.45, Found: C, 67.49; H, 5.62; N, 12.48.

6-(Dicyanomethylene)-4-(4-(dimethylamino)phenyl)-2'-methyl-6'-thioxo-1,1',6,6'-tetrahydro-[2,3'-bipyridine]-5,5'-dicarbonitrile (15)

In dry flask, a mixture 5-(3-(4-(dimethylamino)phenyl)acryloyl)-6-methyl-2-thioxo-1,2-dihydropyridine-3- carbonitrile 9a (3.23 g, 10 mmol) and malononitrile dimmer (1.32 g, 10 mmol) in acetic acid and presence of ammonium acetate was left under reflux for three hours. The mixture was left for cooling and poured onto ice, cold water. The product was recovered by filtration and recrystallisation from ethanol as brown crystals (3.25 g, 74.7%), Mp. 260°C - 262°C; ^{1}H-NMR (DMSO) δ = 2.38 (3H, s, CH_3), 3.06 (6H, s, NMe_2), 6.83 (2H, d, Ar), 7.5 (1H, s, CH py.), 7.93 (2H, d, Ar), 8.21 (1H, s, CH py.), 11.93 (1H, br, NH), 12.4 (1H, br, NH); IR (KBr) υ 3334, 3207 (2NH), 2206 cm^{-1} (CN); MS (EI)$^+$: m/z 435 M$^+$; Anal. Calcd for $C_{24}H_{17}N_7S$ (435.51): C, 66.19; H, 3.93; N, 22.51, Found: C, 66.06; H, 3.78; N, 22.35.

6-(Dicyanomethylene)-4-(4-(dimethylamino)phenyl)-2'-methyl-6'-(methylthio)-1,6-dihydro-[2,3'-bipyridine]-5,5'-dicarbonitrile (16)

(A) In dry flask a mixture of 5-(3-(4-(dimethylamino)phenyl)acryloyl)-6-methyl-2-(methylthio)nicotinonitrile 14 (3.37 g, 10 mmol) and malononitrile dimmer (1.32 g, 10 mmol) in acetic acid acid and ammonium acetate was left under reflux for four hours, cool. The solid product was recovered by filtration and recrystallised from acetic acid as brown crystals (3.4 g, 76%), Mp. 220°C - 222°C; (B) mixture of 6-(dicyanomethylene)-4-(4-(di- methylamino)phenyl)-2'-methyl-6'-thioxo-1,1',6,6'-tetrahydro-[2,3'-bipyridine]-5,5'-dicarbonitrile (15) (4.35 g, 10 mmol) in ethanol as solvent in presence of sodium hydroxide (0.4 g, 10mmol) and methyl iodide (0.62 ml, 10 mmol) with stirring until precipitate formed. The product was recovered by filtration and recrystallised from acetic acid as brown crystals (3.2 g, 71.5%), Mp. and mixed Mp. 220°C - 222°C; ^{1}H-NMR (DMSO) δ = 2.61 (3H, s, CH_3), 2.65 (3H, s, SCH_3), 2.99, 3.01 (6H, 2s, NMe_2), 6.82 (2H, d, Ar), 7.09 (1H, s, CH py.), 7.73 (2H, d, Ar), 8.66 (1H, s, CH py.), 10.3 (1H, br, NH); IR (KBr) υ 3345 (NH), 2213 cm^{-1} (CN); MS (EI)$^+$: m/z 449 M$^+$; Anal. Calcd for $C_{25}H_{19}N_7S$ (449.54): C, 66.80; H, 4.26; N, 21.81, Found: C, 66.69; H, 4.18; N, 21.66.

6-Methyl-5-(1H-pyrazol-3-yl)-1H-pyrazolo[3,4-b]pyridin-3-amine (17)

In flask a mixture of 5-(3-(dimethylamino)acryloyl)-6-methyl-2-(methylthio)nicotinonitrile 13 (2.61 g, 10 mmol) and excess of hydrazine hydrate was left under reflux for four hours, cool. The solid product was recovered by filtration and recrystallised from ethanol as yellowish crystals (1.6 g, 75%), Mp. 260°C - 262°C; ^{1}H- NMR (DMSO) δ = 2.49 (3H, s, CH_3), 5.54

(2H, s, NH$_2$), 6.47 (1H, d, CH pyrazole), 7.8 (1H, s, CH py.), 8.3 (1H, d, CH pyrazole), 11.75 (1H, s, NH), 12.91 (1H, s, NH); IR (KBr) υ at 3405, 3329, 3136 cm^{-1} (NH$_2$, NH); MS (EI)$^+$: m/z 214 M$^+$; Anal. Calcd for C$_{10}$H$_{10}$N$_6$ (214.23): C, 56.07; H, 4.71; N, 39.23, Found: C, 55.85; H, 4.56; N, 39.16.

2›-Methyl-6›-(methylthio)-6-thioxo-1,6-dihydro-[2,3›-bipyridine]-5,5›-dicarbonitrile (18)

In dry flask a mixture of 5-(3-(dimethylamino)acryloyl)-6-methyl-2-(methylthio)nicotinonitrile 13 (2.61 g, 10 mmol) and cyanothioacetamide (1 g, 10 mmol) in acetic acid and ammonium acetate was left under reflux for four hours. Cool and poured the mixture onto ice cold water. The product was recovered by filtration and recrystallised from ethanol as brown crystals (2.3 g, 77.1%), Mp. 170°C - 172°C; ^{1}H-NMR (DMSO) δ = 2.63 (3H, s, CH$_3$), 2.65 (3H, s, SCH$_3$), 7.7 (1H, d, CH py.), 8 (1H, d, CH py.), 8.14 (1H, s, CH py.), 12.25 (1H, br, NH); IR (KBr) υ = 3428 (NH), 2221 cm^{-1}(CN); MS (EI)$^+$: m/z 298 M$^+$; Anal. Calcd for C$_{14}$H$_{10}$N$_4$S$_2$ (298.39): C, 56.35; H, 3.38; N, 18.78, Found: C, 56.12; H, 3.24; N, 18.58.

General procedure for the preparation of compounds 19a,b

In dry flask, a mixture of 5-(3-(dimethylamino)acryloyl)-2-ethoxy-6-methylnicotinonitrile 12 (2.59 g, 10 mmol) or 5-(3-(dimethylamino)acryloyl)-6-methyl-2-(methylthio)nicotinonitrile 13 (2.61 g, 10 mmol) and malononitrile dimmer (1.32 g, 10 mmol) in acetic acid and ammonium acetate was heated under reflux for four hours, cool. The solid product was recovered by filtration and recrystallised from ethanol

6-(Dicyanomethylene)-6›-ethoxy-2›-methyl-1,6-dihydro-[2,3›-bipyridine]-5,5›-dicarbonitrile (19a):

Obtained   using   5-(3-(dimethylamino)acryloyl)-2-ethoxy-6-methylnicotinonitrile 12. Mp. 200°C - 202°C as brown crystals (2.4 g, 73.1%); ^{1}H-NMR (DMSO) δ = 1.39 (3H, t, CH$_3$), 2.62 (3H, s, CH$_3$), 4.50 (2H, q, CH$_2$), 7.58 (1H, d, CH py.), 8.48 (1H, d, CH py.), 8.7 (1H, s, CH py.), 11.3 (1H, br, NH); IR (KBr) υ 3330 (NH), 2218 cm^{-1} (CN); MS (EI)$^+$: m/z 328 M$^+$; Anal. Calcd for C$_{18}$H$_{12}$N$_6$O (328.34): C, 65.85; H, 3.68; N, 25.60, Found: C, 65.71; H, 3.52; N, 25.43.

6-(Dicyanomethylene)-2'-methyl-6'-(methylthio)-1,6-dihydro-[2,3'-bipyridine]-5,5'-dicarbonitrile

(19b): Obtained using 5-(3-(dimethylamino)acryloyl)-6-methyl-2-(methylthio)nicotinonitrile 13. Mp. 190°C - 192°C as brown crystals (2.3 g, 69.7%); ^{1}H-NMR (DMSO) δ = 2.58 (3H, s, CH$_3$), 2.64 (3H, s, SCH$_3$), 6.5 (1H, d, CH py.), 8.2 (1H, d, CH py.), 8.69 (1H, s, CH py.), 11.31 (1H, br, NH); IR (KBr) υ 3340 (NH), 2212 cm^{-1} (CN); MS (EI)$^+$: m/z 330 M$^+$; Anal. Calcd for

$C_{17}H_{10}N_6S$ (330.37): C, 61.80; H, 3.05; N, 25.44, Found: C, 61.63; H, 2.89; N, 25.27.

8-Acetyl-7-methyl-3-(p-tolyl)pyrido[3',2':4,5]thieno[3,2-d]pyrimidin-4(3H)-one (20)

A mixture of 5-acetyl-3-amino-6-methyl-N-(p-tolyl)thieno[2,3-b] pyridine-2-carboxamide 8b (3.39 g, 10 mmol) in dry dioxane and DMFDMA (1.32 ml, 10 mmol) with stirring for 12 hrs. The product was recovered by filtration and recrystallised from acetic acid as gray crystals (2.6 g, 74.5%), Mp. 200°C - 202°C; ¹H-NMR (DMSO) δ 2.26 (3H, s, CH_3 Ar), 2.68 (3H, s, CH_3 py.), 2.69 (3H, s, CH_3CO), 7.16 (2H, d, Ar), 7.52 (2H, d, Ar), 8.43 (1H, s, CH pyrimidinone), 8.52 (1H, s, CH py.); IR (KBr) υ at 1691, 1649 cm⁻¹ (2C=O); MS (EI)⁺: m/z 349 M⁺; Anal. Calcd for $C_{19}H_{15}N_3O_2S$ (349.41): C, 65.31; H, 4.33; N, 12.03, Found: C, 65.19; H, 4.26; N, 11.95.

General procedure for the preparation of compounds 21a,b

A mixture of N-substituted-5-acetyl-3-amino-6-methylthieno[2,3-b] pyridine-2-carboxamide 8b,c (10 mmol) in acetic acid and sodium nitrite (1.38 g, 20mmol) with stirring for 1 hr. the precipitate was formed and dilute with water. The product was recovered by filtration and recrystallised from ethanol.

8-Acetyl-7-methyl-3-(p-tolyl)pyrido[3',2':4,5]thieno[3,2-d][1,2,3]triazin-4(3H)-one (21a): Obtained using 5- acetyl-3-amino-6-methyl-N-(p-tolyl) thieno[2,3-b]pyridine-2-carboxamide 8b (3.39g, 10 mmol). Mp. 170°C - 172°C as gray crystals (3 g, 85.7%); ¹H-NMR (DMSO) δ = 2.4 (3H, s, CH_3 Ar), 2.74 (3H, s, CH_3 py.), 2.77 (3H, s, CH_3CO), 7.4 (2H, d, Ar), 7.54 (2H, d, Ar), 9.17 (1H, s, CH py.); IR (KBr) υ 1700, 1687 cm⁻¹ (2C=O); MS (EI)⁺: m/z 350 M⁺; Anal. Calcd for $C_{18}H_{14}N_4O_2S$ (350.40): C, 61.70; H, 4.03; N, 15.99, Found: C, 61.56; H, 3.94; N, 15.78.

8-Acetyl-3-(4-methoxyphenyl)-7-methylpyrido[3',2':4,5]thieno[3,2-d] [1,2,3]triazin-4(3H)-one (21b): Obtained using 5-acetyl-3-amino-N-(4-methoxyphenyl)-6-methylthieno[2,3-b]pyridine-2-carboxamide 8c (3.55 g, 10 mmol). Mp. = 220°C - 222°C as gray crystals (2.9 g, 79.4%); ¹H-NMR (DMSO) δ = 2.74 (3H, s, CH_3 py.), 2.81 (3H, s, CH_3CO), 3.85 (3H, s, CH_3O), 7.15 (2H, d, Ar), 7.61 (2H, d, Ar), 9.26 (1H, s, CH py.); IR (KBr) υ 1687 cm⁻¹ (2C=O); Anal. Calcd for $C_{18}H_{14}N_4O_3S$ (366.40): C, 59.01; H, 3.85; N, 15.29, Found: C, 58.90; H, 3.76; N, 15.17.

2-Amino-4-(7-methyl-4-oxo-3-(p-tolyl)-3,4-dihydropyrido[3',2':4,5] thieno[3,2-d][1,2,3]triazin-8-yl)thiophene-3-carbonitrile (22):

In dry flask a mixture 8-acetyl-7-methyl-3-(p-tolyl)pyrido[3',2':4,5] thieno[3,2-d][1,2,3]triazin-4(3H)-one 21a (3.5g, 10 mmol), malononitrile

(0.66 g, 10 mmol) and elemental sulfer (0.32 g, 10mmol) in ethanol and few drops of triethylamine as base was heated under reflux for three hours. The mixture was left for cooling and poured onto ice cold water. The product was recovered by filtration and recrystallised from a mixture of ethanol/DMF (3:1) as brown crystals (3 g, 69.7%), M.p 260°C - 262°C; IR (KBr) υ 3427 (NH$_2$), 2208 (CN), 1683 cm^{-1} (C=O); MS (EI)$^+$: m/z 430 M$^+$; Anal. Calcd for C$_{21}$H$_{14}$N$_6$OS$_2$ (430.51): C, 58.59; H, 3.28; N, 19.52, Found: C, 58.43; H, 3.14; N, 19.36.

8-(3-(Dimethylamino)acryloyl)-7-methyl-3-(p-tolyl)pyrido[3›,2›:4,5] thieno[3,2-d][1,2,3]triazin-4(3H)-one (23):

In dry flask a mixture 8-acetyl-7-methyl-3-(p-tolyl)pyrido[3›,2›:4,5] thieno[3,2-d][1,2,3]triazin-4(3H)-one 21a (3.5 g, 10 mmol) and DMFDMA (1.32 ml, 10 mmol) in dry dioxane was left under reflux for two hours. The mixture was left for cooling and evaporates the solvent. The product was recovered by filtration and recrystallised from ethanol as brown crystals (2.9 g, 71.6%), Mp. 210°C - 212°C; ^{1}H-NMR (DMSO) δ = 2.39 (3H, s, CH$_3$ Ar), 2.66 (3H, s, CH$_3$ py.), 3.63, 3.67 (6H, 2s, NMe$_2$), 5.42 (1H, d, CH), 7.41 (2H, d, Ar), 7.54 (2H, d, Ar), 7.82 (1H, d, CH), 9.12 (1H, s, CH py.); IR (KBr) υ 1693, 16.44 cm^{-1} (2C=O); MS (EI)$^+$: m/z 405 M$^+$; Anal. Calcd for C$_{21}$H$_{19}$N$_5$O$_2$S (405.48): C, 62.21; H, 4.72; N, 17.27, Found: C, 62.08; H, 4.59; N, 17.11.

CONCLUSION

From the biological importance of pyridine-2(1H)-thione derivatives, we have used it in order for the preparation of biologically important bipyridyles, bi- and uncommon tricyclic compounds.

REFERENCES

1. Granik, V.G., Zhidkova, A.M. and Glushkov, R.G. (1977) Advances in the Chemistry of the Acetals of Acid Amides and Lactams. Russian Chemical Reviews, 46, 361.http://dx.doi.org/10.1070/RC1977v046n04ABEH002137

2. Abdulla, R.F. and Brinkmeyer, R.S. (1979) The Chemistry of Formamide Acetals. Tetrahedron, 35, 1675-1735.

3. Anelli, P.L., Brocchetta, M., Palano, D. and Visigalli, M. (1997) Mild Conversion of Primary Carboxamides into Carboxylic Esters. Tetrahedron Letters, 38, 2367-2368.http://dx.doi.org/10.1016/S0040-4039(97)00350-X

4. Malesic, M., Krbavcic, A., Golobic, A., Golic, L. and Stanovenik, B. (1997) The Synthesis and Transformation of Ethyl 2-(2-acetyl-2-

benzoyl-1-ethenyl)amino-3-dimethylaminopropenoate. A New Synthesis of 2,3,4-Trisubstituted Pyrroles. Journal of Heterocyclic Chemistry, 34, 1757-1762.

5.   Abu-Shanab, F.A., Elnagdi, M.H., Aly, F.M. and Wakefield, B.J. (1994) α,α-Dioxoketene Dithioacetals as Starting Materials for the Synthesis of Polysubstituted Pyridines. Journal of the Chemical Society, Perkin, 1, 1449-1452. http://dx.doi.org/10.1039/p19940001449

6.   Abu-Shanab, F.A., Redhouse, A.D., Thompson, J.R. and Wakefield, B.J. (1995) Synthesis of 2,3,5,6-Tetrasubstituted Pyridines from Enamines Derived from N,N-Dimethylformamide Dimethyl Acetal. Synthesis, 5, 557-560. http://dx.doi.org/10.1055/s-1995-3954

7.   Abu-Shanab, F.A., Aly, F.M. and Wakefield, B.J. (1995) Synthesis of Substituted Nicotinamides from Enamines De- rived from N,N-Dimethylformamide Dimethyl Acetal. Synthesis, 8, 923-925.

8.   Abu-Shanab, F.A., Hessen, A.M. and Mousa, S.A.S. (2007) Dimethylformamide Dimethyl Acetal in Heterocyclic Syn- thesis: Synthesis of Polyfunctionally Substituted Pyridine Derivatives as Precursors to Bicycles and Polycycles. Journal of Heterocyclic Chemistry, 44, 787-791. http://dx.doi.org/10.1002/jhet.5570440406

9.   Carboni, R.A., Conffman, D.D. and Howard, E.G. (1958) Cyanocarbon Chemistry. XI.[1]Malononitrile Dimer. Journal of the American Chemical Society, 80, 2838-2840.http://dx.doi.org/10.1021/ja01544a061

10. Abu-Shanab, F.A., Sherif, S.M. and Mousa, S.A.S. (2009) Dimethylformamide Dimethyl Acetal as a Building Block in Heterocyclic Synthesis. Journal of Heterocyclic Chemistry, 46, 801-827. http://dx.doi.org/10.1002/jhet.69

11. Melani, F., Cecchi, L., Colotta, V., Filacchini, G., Martini, C., Giannicini, G. and Lucacchini, A. (1989) Dipyrazolo[5,4- b:3',4'-d]pyridines. Synthesis, Inhibition of Benzodiazepine Receptor Binding and Structure-Activity Relationships. Farmaco, 44, 585-594.

12. Abu-Shanab, F.A.M., Mousa, S.A.S., Eshak, E.A., Sayed, A.Z. and Al-Harrasi, A. (2011) Dimethylformamide Dime- thyl Acetal (DMFDMA) in Heterocyclic Synthesis: Synthesis of Polysubstituted Pyridines, Pyrimidines, Pyridazine and Their Fused Derivatives. International Journal of Chemistry, 1, 207-214.

Chapter 9

OPTIMIZATION OF EXTRACTION CONDITIONS OF SOME PHENOLIC COMPOUNDS FROM WHITE HOREHOUND (MARRUBIUM VULGARE L.) LEAVES

Karim Bouterfas[1], Zoheir Mehdadi[1], Djamel Benmansour[2], Meghit Boumedien Khaled[3], Mohamed Bouterfas[4], Ali Latreche[1]

[1]Laboratory of Vegetal Biodiversity: Conservation and Valorization, Faculty of Life and Natural Sciences, Djillali Liabes University, Sidi Bel-Abbes, Algeria

[2]Laboratory of Statistics and Random Model, Faculty of Natural and Universe Sciences, Abou-Bekr Belkaid University, Tlemcen, Algeria

[3]Department of Biology, Faculty of Natural and Life Sciences, Université Djillali Liabes, Sidi Bel Abbes, Algeria

[4]Laboratory of Microscopy, Microanalysis of the Matter and Molecular Spectroscopy, Faculty of Exact Sciences, Djillali Liabes University, Sidi Bel-Abbes, Algeria

ABSTRACT

This research was aimed to optimize the extraction conditions of three phenolic compounds: total phenolics, flavonoids and condensed tannins, from White Horehound's leaves (Marrubium vulgare L.). Distilled water and different organic solvents such as: methanol, ethanol and acetone, were used, with various concentrations (20% - 80%, v/v), temperatures (20°C - 60°C) and extraction times (30 - 450 min). Results showed that the maximum total phenolics amounts (293.34 ± 14.60 mg gallic acid equivalent/g dry weigh), were obtained with 60% aqueous methanol at 25°C for 180 min; total flavonoids (79.52 ± 0.55 mg catechin equivalent/g dry weigh) with 80% aqueous methanol at 20°C for 450 min, and condensed tannins (28.15 ± 0.80 mg catechin equivalent/g dry weigh) with 60% aqueous acetone at 50°C and for 180 min. ANOVA test showed the significant effect (***P < 0.001) of the extraction conditions tested on phenolic compounds. The Principal Component

Analysis (PCA) exhibited the positive effect of low temperatures on total phenolics and flavonoids extraction, and the effect of high temperatures on the condensed tannins extraction. The Response Surface Methodology (RSM) provided predicted values of extraction conditions and maximum polyphenols amounts similar to those obtained experimentally.

INTRODUCTION

Polyphenols constitute one of the most common and widespread groups of substances in flowering plants, occurring in all vegetative organs, as well as in flowers and fruits. These molecules are involved in many physiological processes such as cell growth, root formation, seed germination and fruit ripening [1]. Moreover, these compounds are considered secondary metabolites involved in the chemical defense of plants against predators, pathogens, environmental stresses and in plant-plant interferences [2].

Nowadays, phenolic compounds represent a unique and a functional place, composed of bioactive products, present in plant-derived foods and beverages and included in the formulations of well-marketed cosmetic and parapharmaceutical products [3]. Furthermore, polyphenols exhibit various biological activities such as anti- cancer [4], antioxidant [5], antimicrobial [6] and anti-inflammatory activities [7]. Therefore, in recent years, the determination of phenolic compounds concentrations in fruits [8] [9], vegetables [10] and some aromatic and medicinal plants [11] [12] has been of increasing interest in the scientific community as well as among health professionals and business partners.

It is well known that the content of phenolic compounds could be influenced by environmental conditions, such as season [13] [14], sampling period and geographic origin [15], precipitations and temperatures [16], and soil type [17]. Additionally, there are several experimental factors that can influence the rate of extraction and the quality of extracted bioactive phenolic compounds. These factors include extraction method, solvent type used and concentration [18], particle size of medicinal plants, temperature and pH of extraction [19], extraction time [20], number of extractions repetition [1] and solvent-to-sample ratio [21].

In order to recover bioactive compounds from plant raw materials, extraction is widely used and it constitutes the first important step [22]. Different solvents and techniques are used for the extraction of polyphenols from plants [23] [24]. However, there is no one standard extraction method used to extract phenolic compounds from plant materials because of their complexity and their interaction with other bioactive compounds [25]. Furthermore, each plant material has its unique properties, in term of phenolic extraction; different plants may require different extraction conditions to achieve the optimum recovery of phenolic compounds [26] .

The family Lamiaceae includes aromatic and medicinal plants, which are used in traditional medicine, although Lamiaceae species are well known for their volatile oil content, their therapeutic activities and other properties. This reflects the existence of other chemical components, such as the polyphenols. Marrubium vulgare L. commonly known in Europe as "White horehound", and in the Mediterranean region as "Marute" or as "Merriouet" in Algeria, is a perennial herb of Lamiaceae family, naturalized in North and South America, and Western Asia [27] . In Algeria, M. vulgare is used in folk medicine to treat several digestive diseases, diarrhea, as well as diabetes, rheumatism, acute or chronic bronchitis, cough, asthma and other respiratory infections [28] . Earlier, phytochemical investigation of M. vulgare have led to the characterization of a very complex metabolic pattern, containing, among other secondary metabolites, diterpenes [29] , phenyl propanoids esters [30] , tannins [31] and flavonoids [32] . Several activities, traditionally attributed to M. vulgare, were approved by intensive modern research and clinical trials, such as hypoglycemic [33] , vasorelaxant and antihypertensive [29] , analgesic [34] , antidiabetic [35] , anti-inflammatory [36] , and antioxidant properties [37] .

However, to the best of our knowledge, optimizing the extraction of phenolic compounds from M. vulgare leaves, using different extraction conditions and response surface methodology (RSM), has not been reported yet. Hence, the purpose of the current study was to investigate the effects of different extracting conditions (organic solvent type, concentration of organic solvent, temperature and time) on the extraction of phenolic compounds (total phenolic content, TPC; total flavonoid content, TFC; and condensed tannins content, CTC) from M. vulgare leaves.

MATERIAL AND METHODS

Plant Material

Leaves of M. vulgare were collected in April 2012, from Tessala Mountain (north-western Algeria, semi-arid climate) at the level of a station which latitudinal coordinates 35°16'33''N, and longitudinal 0°46'27''W, altitude 596 m. The identification of plant specimen was done by Professor Z. Mehdadi and put at the Laboratory of Vegetal Biodiversity: Conservation and Valorization (Faculty of Natural and Life Sciences, Djillali Liabes University of Sidi Bel-Abbes, Algeria).

Upon arrival at the laboratory, samples were thoroughly rinsed with distilled water, dried in the dark for three weeks at room temperature and crushed with the cutting mill. The powdered samples were packaged into a linear-low-density polyethylene (LLDPE) film and stored in dark at room temperature for further experiments.

Extract Preparation

Two grams of leaf powder M. vulgare were extracted using 20 ml of extraction solvent with different concentrations, introduced in conical flask of capacity 100 ml sealed with parafilm and wrapped with aluminum foil to prevent solvent loss and exposure to light. Therefore, the mixture was stirred at 150 rpm in bath water with controlling temperature at a constant speed (level 8) for a particular duration. After completing the extraction process, the M. vulgare extract was filtered using Whatman No. 1 filter paper into amber bottle for analysis without storage overnight, in order to obtain a clear crude extract solution.

To determine the optimal conditions for phenolic extraction from M. vulgare leaves, the extraction conditions were set according to the experimental design described below:

- Organic solvents type: three kinds of organic solvents (methanol, ethanol and acetone) were selected. Distilled water was tested as control. The selection of the three extraction solvents is made firstly after several (about one hundred) referenced protocols in terms of quantification of phenolic compounds and secondly from the viewpoint of the availability of the chemicals used;
- Solvent concentrations: four concentrations (20%, 40%, 60% and 80%; v/v) were prepared in distilled water;
- Extraction temperatures: six temperatures (20°C, 25°C, 30°C, 40°C, 50°C and 60°C) were used;
- Extraction times: six different times (30, 90, 180, 270, 360 and 450 min) were chosen.

Total Phenolic Contents (TPC)

TPC of M. vulgare extracts was determined using Folin-Ciocalteu reagent, according to method suggested by Li et al. [38] and slightly modified by Chew et al. [39] . Crude extracts were diluted 50 times with deionized water prior to analysis. 1 ml of diluted extract was mixed with 1 ml of diluted Folin-Ciocalteu reagent (10 times diluted with deionized water). After incubating the mixture at room temperature for 4 min, 0.8 ml of 7.5% (w/v), sodium carbonate anhydrous solution was added. The mixture was then vortexed for 10 s and incubated in the dark at room temperature during 2 h. The absorbance of the mixture was measured against blank at 765 nm using UVi light spectrophotometer. Gallic acid with different concentrations (0-100-200-300-400-500 mg/l) was used to calibrate the standard curve. The calibration equation for gallic acid was $Y = 0.0042X - 0.0178$ ($R^2 = 0.9992$). Each crude extract was analyzed in triplicate and the results were expressed in milligrams of gallic acid equivalents per gram of dry weight (mg GAE/g DW). Data were expressed as mean ± standard deviation.

Total Flavonoids Contents (TFC)

TFC was determined using procedures described by Tan et al. [21]. The crude extract was diluted 10 times. An amount of 1.25 ml deionised water followed by 75 µl of 5% sodium nitrite (NaNO2), was added to 0.25 ml of diluted crude extract in an aluminium foil-wrapped 15 ml test tube. The mixture was left standing for 6 min before adding 150 µl of 10% (w/v) aluminium chloride (AlCl3). The mixture was left standing for 5 min before adding 0.5 ml of 1 M sodium hydroxide (NaOH) and 275 µl of deionised water. The tip of the test tube was covered with parafilm and then mixed using vortex mixer for approximately 10 s. The absorbance of the mixture was determined at 510 nm versus the prepared blank using Uvi light spectrophotometer. Catechin with different concentration (0 - 100 - 200 - 300 - 400 - 500 mg/l) was used for calibration. The calibration equation for catechin was calculated as follow: $Y = 0.0035X - 0.0062$ ($R^2 = 0.9995$). Each crude extract was analyzed in triplicate and the results were expressed in milligrams of catechin equivalents per gram of dry weight (mg CE/g DW). Data were expressed as mean ± standard deviation.

Total Flavonoid Contents (TFC)

TFC was determined using procedures described by Tan et al. [21]. The crude extract was diluted 10 times. An amount of 1.25 ml deionised water followed by 75 µl of 5% sodium nitrite (NaNO$_2$), was added to 0.25 ml of diluted crude extract in an aluminium foil-wrapped 15 ml test tube. The mixture was left standing for 6 min before adding 150 µl of 10% (w/v) aluminium chloride (AlCl3). The mixture was left standing for 5 min before adding 0.5 ml of 1 M sodium hydroxide (NaOH) and 275 µl of deionised water. The tip of the test tube was covered with parafilm and then mixed using vortex mixer for approximately 10 s. The absorbance of the mixture was determined at 510 nm versus the prepared blank using Uvi light spectrophotometer. Catechin with different concentration (0 - 100 - 200 - 300 - 400 - 500 mg/l) was used for calibration. The calibration equation for catechin was calculated as follow: $Y = 0.0035X - 0.0062$ ($R^2 = 0.9995$). Each crude extract was analyzed in triplicate and the results were expressed in milligrams of catechin equivalents per gram of dry weight (mg CE/g DW). Data were expressed as mean ± standard deviation.

Condensed Tannins Contents (TFC)

CTC assay was performed according to the method described by Chew et al. [39] . 0.5 ml undiluted crude extract was firstly mixed with 3 ml of vanillin reagent (4%, w/v, in absolute methanol), followed by addition of 1.5 ml of concentrated HCl (37%). After that, the mixture was stored in the dark at room temperature for 15 min. The absorbance of mixture was measured at 500 nm against blank using Uv light spectrometer. Each undiluted crude extract was measured in triplicate. Catechin with different concentrations (0 - 100 - 200 - 300 - 400 - 500 mg/l) was used for calibration of standard curve. The calibration equation for catechin was calculated using the formula $Y = 0.0021X - 0.0143$ ($R^2 = 0.997$). Each crude extract was analyzed in triplicate and the results were expressed in milligrams of catechin equivalents per gram of dry weight (mg CE/g DW). Data were expressed as mean ± standard deviation.

Statistical Analysis

In order to control the influence of extraction conditions (solvent type, solvent concentration, extraction time, and extraction temperature) on the mean concentrations of each phenolic compound, ANOVA, with more classification criteria, using Fisher's least significant difference test and the significant differences at the 5% level, were calculated. The difference was considered as not significant when $P > 0.05$, significant when $^*P \leq 0.05$, and highly significant for $^{**}P \leq 0.01$ and extremely significant for $^{***}P \leq 0.001$. These analyzes were performed using Minitab 16. Tukey's test was also performed for pair-wise comparisons at the 5% level.

To determine possible correlation among polyphenols concentrations and extraction conditions, PCA was used. This statistical tool is dedicated for data exploration, which allows the reduction of the number of quantitative variables to a small number of components. Moreover, using PCA, interrelationships between different variables could be seen, and detected and sample patterns, groupings, similarities or differences could be interpreted [40] . A matrix is prepared using XLSTAT 2012 software, by taking polyphenols concentrations as observations and extraction conditions as variables.

Optimal conditions for the extraction of phenolic compounds from M. vulgare leaves were obtained using response surface methodology (RSM) [41] [42] . This method is adopted for each type of phenolic compound with the solvent giving the highest experimental concentration. The independent variables studies were solvent concentration (X_1), extraction temperature (X_2) and extraction time (X_3); while the dependent variable (Y: response variable) measured was the contents of phenolic compound. To obtain the best combination of independent variables giving a maximum percentage of adjusted R^2, a best subsets regression is performed using Minitab 16 software. Thus, the regression equation of the model is obtained by an analysis of general regression. Finally, the optimum polyphenols contents and extraction conditions are obtained by introducing the regression equation in Maple 6 software. The experimental and predicted values of polyphenols contents and extraction conditions were compared, in order to determine the validity of the model.

RESULTS AND DISCUSSION

Results of total phenolic, total flavonoids, and condensed tannins amounts are shown in Figure 1and Figure 2. The influence of extraction's conditions used on phenolic compounds amounts obtained using ANOVA, with more classification criteria, is presented in Table 1.

Table 1: Analysis of variance (ANOVA) with more classification criteria of quantified phenolic compounds

Source	Degree of freedom	Sum of Squares	Mean squares	F-value	P-value
{1}Solvent type	2	40326.1^{TPC}	20163.1^{TPC}	262.15^{TPC}	$<0.0001^{***}$
		2326.12^{TFC}	1163.06^{TFC}	274.61^{TFC}	$<0.0001^{***}$
		43.14^{CTC}	21.57^{CTC}	UD^{CTC}	$<0.0001^{***}$
{2}Solvent concentration	4	107832.8^{TPC}	26958.2^{TPC}	350.49^{TPC}	$<0.0001^{***}$
		5132.83^{TFC}	1283.21^{TFC}	302.98^{TFC}	$<0.0001^{***}$
		257.24^{CTC}	64.31^{CTC}	UD^{CTC}	$<0.0001^{***}$
{3}Extraction time	5	45983.2^{TPC}	9196.6^{TPC}	119.57^{TPC}	$<0.0001^{***}$
		111.09^{TFC}	22.22^{TFC}	5.25^{TFC}	$<0.0001^{***}$
		592.16^{CTC}	118.43^{CTC}	UD^{CTC}	$<0.0001^{***}$
{4}Extraction temperature	5	553938.7^{TPC}	110787.7^{TPC}	1440.39^{TPC}	$<0.0001^{***}$
		39821.11^{TFC}	7964.22^{TFC}	1880.44^{TFC}	$<0.0001^{***}$
		10589.43^{CTC}	2117.89^{CTC}	UD^{CTC}	$<0.0001^{***}$
{1}*{2}	8	31097.1^{TPC}	3887.1^{TPC}	50.54^{TPC}	$<0.0001^{***}$
		1654.30^{TFC}	206.79^{TFC}	48.82^{TFC}	$<0.0001^{***}$
		10.91^{CTC}	1.36^{CTC}	UD^{CTC}	$<0.0001^{***}$
{1}*{3}	10	7922.1^{TPC}	792.2^{TPC}	10.30^{TPC}	$<0.0001^{***}$
		54.86^{TFC}	5.49^{TFC}	1.30^{TFC}	$=0.235^{NS}$
		2.59^{CTC}	0.26^{CTC}	UD^{CTC}	$<0.0001^{***}$
{1}*{4}	10	14801.5^{TPC}	1480.1^{TPC}	19.24^{TPC}	$<0.0001^{***}$
		2422.48^{TFC}	242.25^{TFC}	57.20^{TPC}	$<0.0001^{***}$
		0.00^{CTC}	0.00^{CTC}	UD^{CTC}	$<0.0001^{***}$

{2}*{3}	20	35664.5^{TPC}	1783.2^{TPC}	23.18^{TPC}	<0.0001***
		599.22^{TFC}	29.96^{TFC}	7.07TFC	<0.0001***
		100.97^{CTC}	5.05^{CTC}	UD^{CTC}	<0.0001***
{2}*{4}	20	48446.9^{TPC}	2422.3^{TPC}	31.49^{TPC}	<0.0001***
		7603.54^{TFC}	380.18^{TFC}	89.76^{TFC}	<0.0001***
		233.26^{CTC}	11.66^{CTC}	UD^{CTC}	<0.0001***
{3}*{4}	25	18241.9^{TPC}	729.7^{TPC}	9.49^{TPC}	<0.0001***
		767.42^{TFC}	30.70^{TFC}	7.25^{TFC}	<0.0001***
		92.06^{CTC}	3.68^{CTC}	UD^{CTC}	<0.0001***
{1}*{2}*{3}	40	8775.9^{TPC}	219.4^{TPC}	2.58^{TPC}	<0.0001***
		314.37^{TFC}	7.86^{TFC}	1.86^{TFC}	=0.003**
		1.26^{CTC}	0.03^{CTC}	UD^{CTC}	<0.0001***
{1}*{2}*{4}	40	18002.2^{TPC}	450.1^{TPC}	5.85^{TPC}	<0.0001***
		3061.91^{TFC}	76.55^{TFC}	18.07^{TFC}	<0.0001***
		0.00^{CTC}	0.00^{CTC}	0.00^{CTC}	<0.0001***
{1}*{3}*{4}	50	5440.4^{TPC}	108.8^{TPC}	1.41^{TPC}	=0.050*
		272.27^{TFC}	5.45^{TFC}	1.29^{TFC}	=0.116^{NS}
		0.00^{CTC}	0.00^{CTC}	UD^{CTC}	<0.0001***
{2}*{3}*{4}	100	46098.9^{TPC}	461.0^{TPC}	5.99^{TPC}	<0.0001***
		1176.72^{TFC}	11.77^{TFC}	2.78^{TFC}	<0.0001***
		176.67^{CTC}	1.77^{CTC}	UD^{CTC}	<0.0001***
{1}*{2}*{3}*{4}	200	15383.0	76.09	UD^{TPC}	<0.0001***
		847.06	4.24	UD^{TFC}	<0.0001***
		0.00	0.00	UD^{CTC}	<0.0001***

UD: undefined (the denominator of Fisher's test is null or undefined).

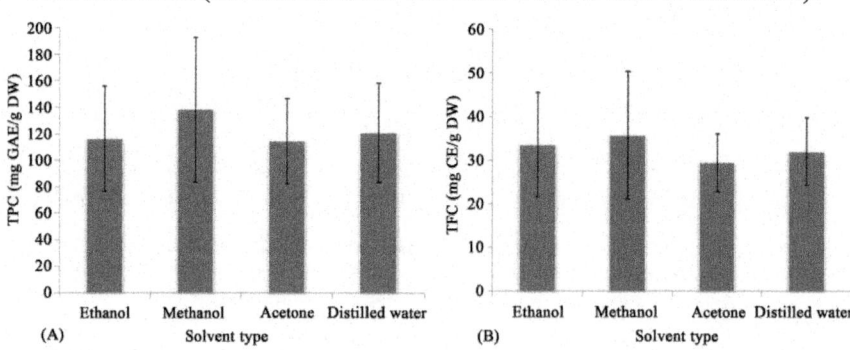

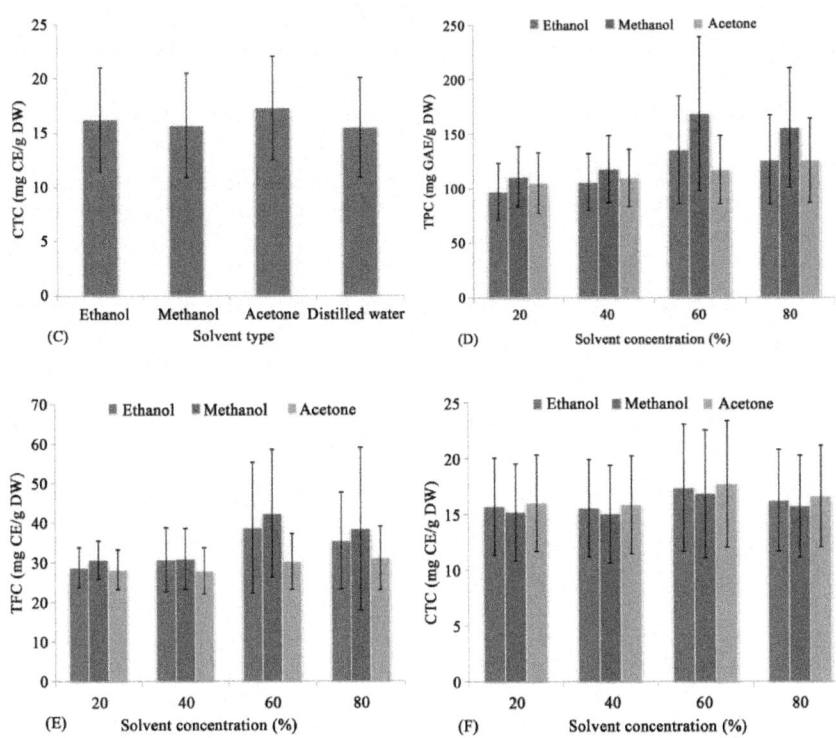

Figure 1: Effects of solvent type ((A), (B), (C)) and solvent concentration ((D), (E), (F)) on TPC, TFC and CTC extracted from M. vulgare leaves.

Effect of Solvent Type on Extraction of Phenolic Compounds

The choice of extraction solvents is important for complex food samples. They allow determining the amount and the type of phenolic compounds to extract. Organic solvents, particularly acetone, ethanol and methanol are the most commonly used in polyphenols extraction from botanical materials [43] .

Our results showed that methanol (137.79 ± 54.62 mg GAE/g) was significantly high effective (***P < 0.001) when compared with all other solvent systems used in extracting TPC from M.

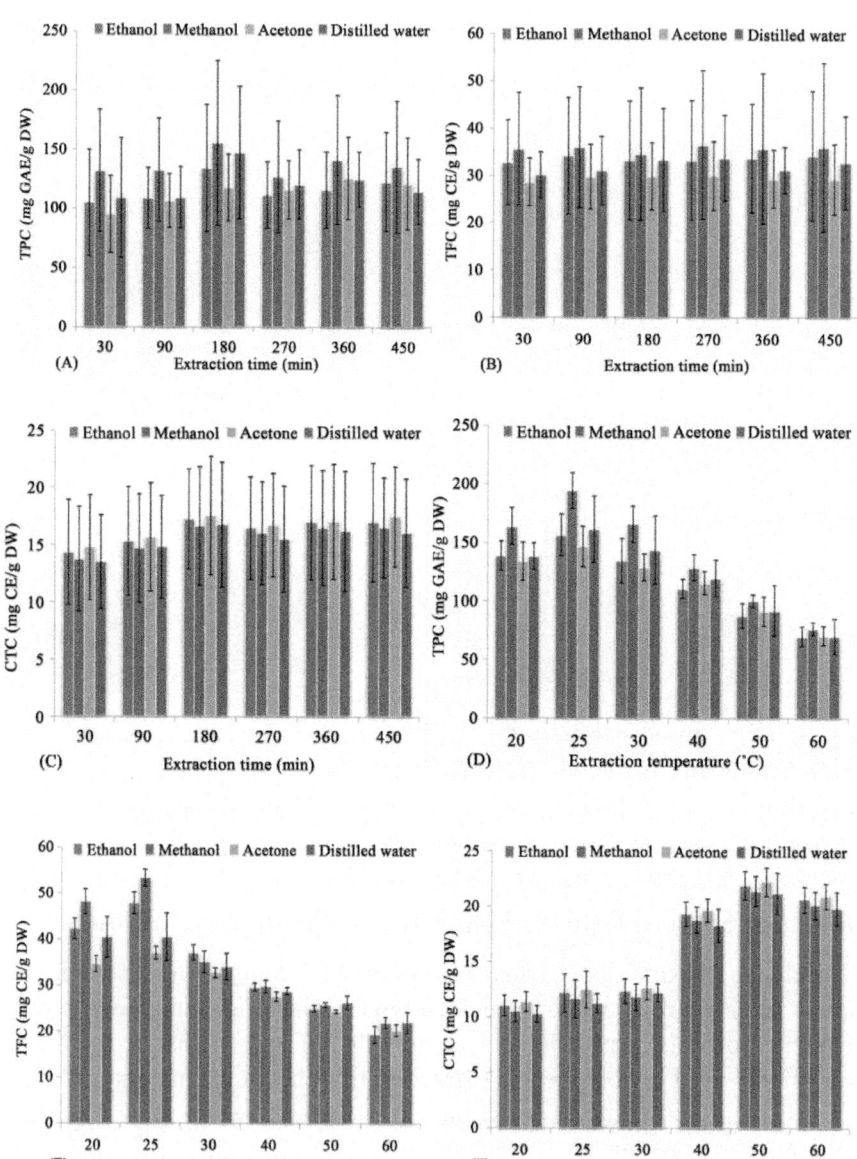

Figure 2: Effects of extraction time ((A), (B), (C)) and extraction temperature ((D), (E), (F)) on TPC, TFC and CTC extracted from M. vulgare leaves.

vulgare leaves, as shown in Figure 1(A). Methanol has been considered as an ideal solvent for the TPC extraction from green walnut fruits [44], Zingiber officinalis leaves [45], Tamarix aphylia leaves [46] and Artemisia annua leaves [47]. Several studies confirmed that TPC depends on organic solvent polarity and methanol is one of the most suitable solvents for the extraction of TPC from plants [48] [49] . Mohammedi and Atik [46] reported that the high polarity of methanol, estimated at 6.6, compared to other organic solvents, plays a key role in increasing phenolic solubility. However, Kong et al. [50] found that TPC in methanolic extracts, obtained from Pouteria campechiana fruit parts, was about 23 % to 45 % lower than the ethanolic extracts.

Methanol permitted to obtain the highest extraction levels of TFC (35.64 ± 14.61 mg CE/g) from M. vulgare leaves (Figure 1(B)). Furthermore, a significant difference (***P < 0.001) was found between the type of organic solvents used. Similar results were observed for Moringa oleifera and Aloe barbadensis leaves [51] , Euphorbia helioscopia leaves, steams and flowers [52] . Our results agree with previous studies, which found that ethanol was less effective than methanol for extracting TFC [53] [54] . In contrast, Settharaksa et al. [55] have found that using water to extract solvent from some Thai medicinal plants, was better than using ethanolic and methanolic solvents to extract TFC. Ghasemzadeh et al. [45] reported that extraction of TFC from Zingiber officinale leaves, using methanol, was about 3 times higher when using acetone and 4 times higher when using hexane. Spigno et al. [56] suggested that high level of TPC obtained using methanol, could be explained by the fact that this solvent allows a good solubility of flavonoids' hydroxyl groups.

A maximum level of CTC (17.24 ± 4.77 mg CE/g) was reached using acetone, as shown in Figure 1(C). There was a high significant difference between the four tested solvents (***P < 0.001). Our results agree with those of Trabelsi et al. [18] who found that maximum CTC level, from Limoniastrum monopetalum leaves, was obtained when acetone solvent was used. Similarly, in other studies performed by Wina et al. [57] and Mazandarani et al. [58] , acetone was found to be the most effective solvent extracting CTC from Acacia mangium barks and Onosma dichroanthum roots, respectively. According to Antwi-Boasiako and Animapauh [59] , the four solvents tested in the current study, especially acetone and methanol were the best extraction solvents of CTC from the barks of three tropical hardwoods. Additionally, Uma et al. [60] showed that acetone was the most effective solvent for extraction of condensed tannins as tannins have a relatively high molecular weight.

Effect of Solvent Concentration on Extraction of Phenolic Compounds

Mixtures of alcohols with different proportions of water have shown to be more effective in extracting phenolic compounds compared to mono-component solvent system [25] . Addition of small quantity of water to organic solvent usually leads to a more polar medium, which facilitates the polyphenols extraction [56] .

We observed that maximum rate of TPC (168.41 ± 70.60 mg GAE/g), extracted from M. vulgare leaves, was obtained when aqueous methanol was used at 60% (Figure 1(D)). However, there were high significant differences (***P < 0.001) in TPC among the various concentrations. Our results are in line with those found by Chan et al. [61] and Chew et al. [39] who found that 60% aqueous methanol gave the best effectiveness extracting TPC from Citrus hystrix peels and Centella asiatica leaves, respectively. Yilmaz and Toledo [62] reported that aqueous mixtures of methanol, ethanol or acetone were better than a mono-component solvent for the extraction of TPC from Muscadine seeds.

The highest value of TFC (42.40 ± 16.22 mg CE/g), as shown in Figure 1(E), was obtained, when we used 60% of aqueous methanol and was significantly different (***P < 0.001) compared to the other studied concentrations. In contrast, aqueous methanol at 80% was the best concentration for Calendula officinalis flowers [63] and aqueous ethanol at 80% from the leaves of Limoniastrum monopetalum [18] , Bauhinia monandra [64] and Callicarpa nudiflora [41] . Musa et al. [65] confirmed that the mixture of an aqueous solvent (distilled water) with an organic solvent (methanol and ethanol) improves the flavonoids yield comparing to water or organic solvent used separately.

Sixty percent (60%) acetone in water gave the highest value of CTC, which was 17.68 ± 5.69 mg CE/g (Figure 1(F)). Moreover, there were high significant (***P < 0.001) differences in CTC values among the different concentrations used. Our data are in agreement with those reported by Downey and Hanlin [66] on grape skin. In other studies, it has been noticed that 80% aqueous acetone was the best concentration for CTC extraction from Limonium densiflorum shoots [67] , 60% aqueous methanol from Cichorium intybus L. roots, leaves, stems and seeds [68] , and 80% aqueous ethanol from the Chinese chestnut [69] .

Effect of Extraction Time on Extraction of Phenolic Compounds

Extraction time represents another key parameter in optimizing the phenolic compounds extraction. In the literature, this parameter might be as short as few minutes or long up to 24 hours, depending on the phenolic compounds present in samples [70] .

In this study, extraction time showed a significant effect (***P < 0.001) on the extraction of TPC from M. vulgare leaves. As shown in Figure 2(A), the highest value of TPC (155.16 ± 69.83 GAE/g) was recorded using methanol for 180 min. Chan et al. [61] demonstrated that 180 min represented the best extraction time for TPC from the peels of Citrus hystrix. In other studies, the best extraction time of TPC was estimated to 18 hours for black tea, Camellia sinensis [49] , 90 min from grape pomace extracts [71] , 45 min from Areca catechu seeds [72] and Azadirachta indica leaves [70] . Increased time of extraction beyond 180 min (270, 360 and 450 min) induced a loss in TPC. Dent et al. [73] highly recommended that the extraction time, should not exceed 3 h for the extraction of TPC from Salvia officinalis leaves. Several authors stated that more the extraction time is long, less the content of polyphenols is obtained. This could be the result of loss of phenolic compounds, via oxidation, which might polymerize into insoluble compounds [26] [74] . Therefore, extraction time of TPC depends, not only, on maceration or agitation times, but also on several factors such as filtration time or the time spent during the evaporation of solvents [75] .

The maximum of TFC concentration (39.95 ± 17.90 mg CE/g) was obtained using methanol for 450 min as illustrated in Figure 2(B). Extraction time showed significant effect ($^{***}P < 0.001$) in TFC. However, 3 hours was considered to be the best extraction time for TFC from Callicarpa nudiflora leaves [46] and from some thyme varieties [76] . Additionally, 30 min has been shown to be the most favorable extraction time for TFC from Gynura medica leaves [77] .

180 min was the longest extraction time, using acetone, for CTC (Figure 2(C)) with a maximum rate of 17.51 ± 5.18 CE/g. A high significant difference ($^{***}P < 0.001$) was obtained in CTC among different extraction times. Our results agree with those obtained by Zhekova and Pavlov [76] on some thyme varieties. In other works, the best extraction time was of 120 min for mangosteen fruits [78] , 150 min for Parkia clappertoniana husks [79] , 80 min for Cichorium intybus different organs [67] , and 20 min for Punica granatum peels [80] .

Effect of Extraction Temperature on Extraction of Phenolic Compounds

The effectiveness of extraction process of phenolic compounds is largely regulated by different experimental parameters particularly by the extraction temperature [81] . An increase of temperature is mainly due to an increase of the diffusion rate and solubility of the extracted substances. On the other hand, it should be taken into account, that some important biological active substances, such as TPC are damaged at high temperatures [82] . The extraction of TPC, as shown in Figure 2(D), was optimal when methanol was used at 25°C. In this case, 194.16 ± 15.45 mg GAE/g DW were obtained with high significant difference (***P < 0.001) and between different temperatures. Similar to our finding, 25°C was the most optimal extraction temperature for TPC from Lawsonia inermis leaves [60] and Azadirachta indica leaves [70] . Extension of extraction temperature beyond 25°C led to an important decrease in TPC. However, it should be noticed that increasing the extraction temperature, beyond certain values, might promote possible concurrent decomposition of phenolic compounds, which were already mobilized at lower temperature. Furthermore, this elevation of temperature can even lead to the breakdown of phenolic compounds that remained in the plant matrix [83] [84] . Hence, heating may affect the polyphenolic composition in many cases; therefore, high-temperature drying should be avoided as much as possible [2] . Usually, TPC extraction is used at room temperature (»25°C) to avoid the degradation of phenolic compounds [25] . However, in other experiments, the optimal extraction temperature for TPC was found more elevated and estimated to 40°C for Citrus hystrix peels [61] , 100°C for Areca catechu seeds [72] , 65°C for Centella asiatica leaves [39] and Orthosiphon stamineus stems and leaves [85] , and 90°C for Moringa oleifera leaves [86] .

A maximum of TFC rate (53.31 ± 1.83 CE/g) was obtained at 25°C using methanol as solvent (Figure 2(E)). In the current investigation we found that extraction temperature significantly (***P < 0.001) affected the TFC extraction. However, in other researches, the highest extraction temperature was fixed at 90 °C for Callicarpa nudiflora leaves [41] and 60°C from some thyme varieties [76] . Like TPC, the extension of extraction temperature higher than 25°C led to an important decrease in TFC. However, the temperature conditions during the extraction procedures of flavonoids have to be carefully adjusted because of the possibility of thermal degradation of flavonoid derivatives, especially hydroxyl groups [87] [88] . In addition, mild heating was also found to have the ability to soften the plant tissues, to weaken the cell wall integrity, and thus to favor the release of bound phenolic compounds [56] [89] .

As shown in Figure 2(F), the maximum of CTC rate (22.18 ± 1.27 CE/g) was reached when acetone was used as a solvent and when the extraction temperature was about 50°C. We observed a significant difference (***P < 0.001) in CTC between the extraction temperatures. Our results agree with those of Zam et al. [80] on Punica granatum peels. In other studies, the best extraction temperature was slightly elevated compared to those we obtained: 80°C for mangosteen fruits [78] , 70°C Parkia clappertoniana husks [79] and 60°C for some thyme varieties [73] . Unlike the obtained results for TPC and TFC, the extension of extraction temperature beyond 25°C led to a significant increase in CTC. Al-Farsi and Lee [90] reported that elevated temperature could stimulate the CTC extraction by increasing both diffusion coefficient and solubility of condensed tannins in extraction solvent. Besides that, intense heat from solvent allowed the release of cell wall phenolics and bounded phenolics by breaking down cellular constituents [91] and consequently, increasing the phenolic yield in extract. Moreover, Juntachote et al. [89] reported that elevated extraction temperature would increase, on one hand, the mass transfer of condensed tannins, and on the other hand, it would reduce the solvent viscosity and surface tension and, promotes the extraction of phenolic compounds.

Principal Component Analysis (PCA)

PCA aimed to diminish the size of data collected into a reduced number of components to examine the possible grouping of phenolic compounds according to the different extraction conditions. The first factor, PC1 presented 85.55% of variance accounted for, whereas, the second one, PC2 presented 10.86%. With the two first PCs, the explained variance accumulated was of 96.41%. This great value means that nearly all the variance contained in the original data was explained by just using the first new coordinates. By layering variables projection circle on the observations scatter plot, two groups were obtained as shown in the PCA plot (Figure 3):

- Group 1 (Gr1) on the positive side of PC1, formed by TPC and TFC with 0.932 and 0.958 as contributions, respectively. These phenolic compounds are related with ethanolic, methanolic and aqueous extracts at low temperatures 20°C and 25°C; ET1, ET2, MT1, MT2, DWT1, DWT2 with respective contributions of 2.285; 3.040; 3.578; 4.596; 2.253 and 2.618;

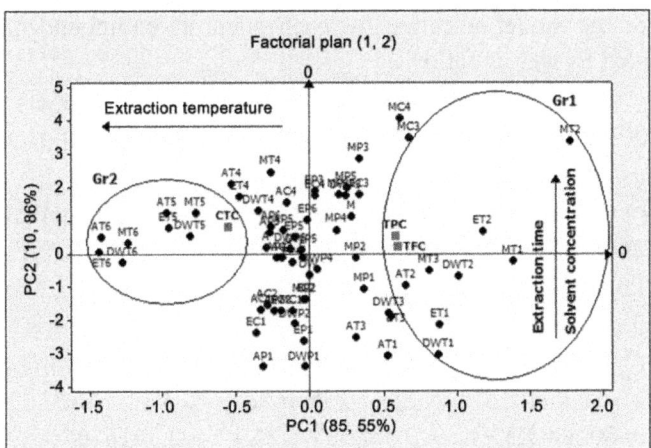

Figure 3. Principal component plot of PC1 and PC2. Gr: group; PC1: first principal component; PC2: second principal component; M: methanol; E: ethanol; DW: distilled

water; A: acetone, C: concentration in % (1: 20; 2: 40; 3: 60; 4: 80); T: temperature in °C (1: 20; 2: 25; 3: 30; 4: 40; 5: 50; 6: 60); P: time in min (1: 30; 2: 90; 3: 180; 4: 270; 5: 360; 6: 450).

- Group 2 (Gr2) on the negative side of PC1, constituted by CTC with a contribution of −0.883. This phenolic compound was associated to ethanolic, acetonic and aqueous extracts at high temperatures 50°C and 60°C; ET5, ET6, AT5, AT6, DWT5, and DWT6 with contributions of 2.473; 3.705; −2.506; −3.661; −2.089 and −3.290, respectively.

- Moreover, considering the contributions of observations and variables on PC1, we defined two gradients:

- A horizontal gradient, moving from right to left of PC1, formed by the extraction temperature used, which explains the position of TPC and TFC right of PC1, correlated with low temperatures and the location of CTC to the left of PC1 in conjunction with high temperatures;

- A vertical gradient from the bottom to the top of PC1, constituted by the organic solvent concentration and the extraction time. This gradient explains the position of the three phenolic compounds measured at the top of PC1, correlated with high solvent concentrations (60% for TPC and CTC, 80% for TFC) and high times (180 min for TPC and CTC, 270 min for TFC).

Response Surface Methodology (RSM)

The Response Surface Methodology (RSM) was plotted to study the effects of extraction conditions on phenolic compounds extraction from M. vulgare leaves. 3 D response surface plots are illustrated in Figure 4. The regression equation of the model selected, for each phenolic compound measured and their adjusted R^2 was as follow:

- TPC = 323.205 − 827.094 X1 − 4.2516 X2 + 0.015029 X3 + 843.115 X12 + 17.1341 X1X2 − 13.5293 X2X12 − 0.0615286 X1X22, with adjusted R2 = 82.03%,

- TFC = 32.4239 + 31.9683 X1 + 0.42576 X2 + 0.0308808 X3 + 98.4907 X12 − 0.0108914 X22 − 2.96617 X1X2 − 0.00163763 X3X2 − 0.00122396 X1X3X2 + 0.0702683 X3X12 − 2.22016 X2X12 + 2.10557e-5 X3X22 + 0.0492972 X1X32, with adjusted R2 = 77.61%,

- CTC = −2.19999 + 7.81089 X1 + 0.603279 X2+ 0.01403 X3 − 12.395 X12 − 0.00405521 X22 − 2.96468e − 5 X32 − 0.220734 X1X2 + 0.00680008 X3X12 + 0.388604 X2X12 + 2.36161e − 7 X2X32, with adjusted R2 = 90.62%.

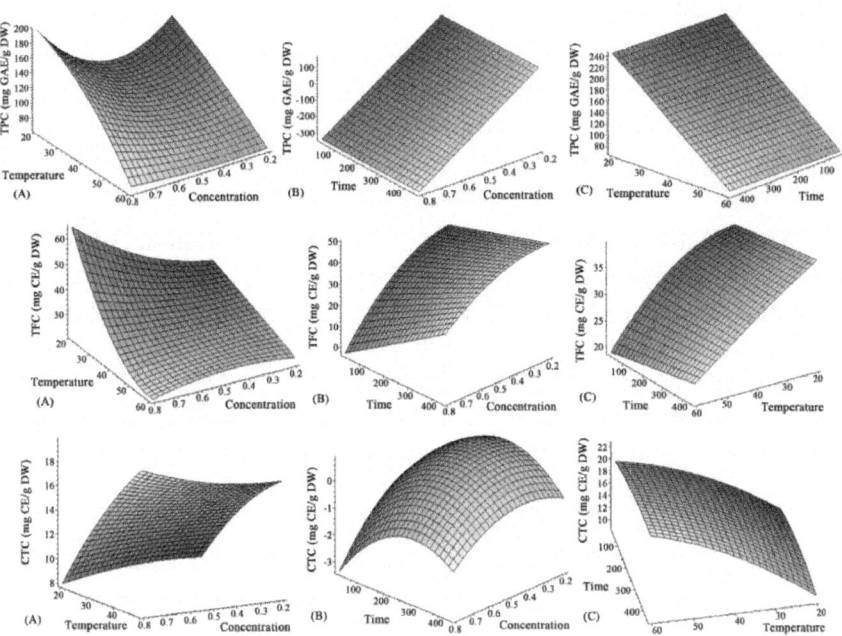

Figure 4: The 3D response surface plots representing the effects of (A) solvent concentration and extraction temperature; (B) solvent concentration and time extraction and (C) extraction temperature and extraction time on the extraction of phenolic compounds from M. vulgare leaves.

The comparison between the maximum experimental values of phenolic compounds and extraction conditions, and the predicted ones (Table 2), showed that the two sets of values were close. Indeed, the maximum value of TFC (experimental: 79.52 mg CE/g; predicted: 79.46 mg CE/g) and CTC (experimental: 28.15 mg CE/g, predicted: 20.81 mg CE/g) were obtained by the same experimental and predicted extraction conditions (TFC: aqueous methanol 80% at 20°C and for 450 min; CTC: 60% at 50°C and for 180 min). Concerning TPC, the maximum values of experimental and predicted contents and extraction conditions were different. The experimental values were of 293.34 mg GAE/g, obtained with aqueous methanol 60% at 25°C and for 180 min. However, the predicted ones were of 204.75 mg GAE/g, obtained with aqueous methanol 80% at 20°C during 167 min. Both, accounting experimental and predicted results indicated that the experimental model was valid. This implied that there was a high fit degree between the values observed in experiment and those predicted from the regression model. Hence, the response surface modeling could be applied effectively to predict extraction of phenolic compounds from M. vulgare leaves.

CONCLUSIONS

The results obtained in our study indicate that the experimental conditions tested (organic solvent type, solvent concentration, extraction time and extraction temperature) influence notably the values of TPC, TFC and CTC extracted from M. vulgare leaves, with extremely significant differences ([***]P < 0.001).

We were able to define the optimum extraction conditions too, for obtaining higher values of some phenolic compounds from M. vulgare leaves. Indeed, the optimum level of TPC (293 ± 14.60 mg GAE/g) was obtained using 60% aqueous methanol at 25°C for 180 min, TFC (79.52 ± 0.55 mg CE/g) using 80% aqueous methanol at 20°C for 450 min, and CTC (28.15 ± 0.80 mg CE/g) with 80% aqueous acetone at 50°C for 450 min. These levels highlight the richness of M. vulgare in these secondary metabolites. We conclude that this species remains poorly studied compared to other species belonging to the same taxonomic family (Lamiaceae) as thyme and sage.

Our results provide some confirmations on the effect of the variability of extraction procedures and assay on the amount of phenolic compounds recorded and reinforce previous works in this context.

This work is far from exhaustive as other experimental conditions can be tested, as the solid-to-solvent ratio, the number of extractions and the nature of the extraction method. Also, other extraction solvents (ethylene glycol, acetic acid and ethyl acetate), fractional extraction for many times or multistage extraction can be taken into consideration in our future works.

The impact of environmental conditions as the geographical region, altitude, exposure and the harvest season on the levels of polyphenols in M. vulgare is a promising perspective to this study and is still under investigation in our laboratory.

ACKNOWLEDGEMENTS

This research is a part of the Algerian National Project Research (entitled "Flora of Tessala Mountains: inventory, valorization and conservation", code 1/u22/397), financially supported by the Algerian Ministry of High

Table 2: Optimum values of experimental and predicted polyphenols contents and extraction conditions

	Polyphenols concentrations		Solvent type		Solvent concentration (X_1)		Extraction temperature (X_2)		Extraction time (X_3)	
	EV	PV	EV	PV	EV	PV	EV	PV	EV	PV
TPC	293.34	204.75	Methanol		60	80	25	20	180	167
TFC	79.52	79.46	Methanol		80	80	20	20	450	450
CTC	28.15	20.81	Acetone		60	60	50	50	180	180

EV: experimental values; PV: predicted values.

Education and Scientific Research. Also, the authors are grateful to Dr. Amel Latifi, Bacterial Chemistry Laboratory (Marseille, France), for valuable editorial assistance during the drafting of this manuscript.

REFERENCES

1. Khoddami, A., Wilkes, M.A. and Roberts, T.H. (2013) Techniques for Analysis of Plant Phenolic Compounds. Molecules, 18, 2328-2375. http://dx.doi.org/10.3390/molecules18022328

2. Tsao, R. (2010) Chemistry and Biochemistry of Dietary Polyphenols. Nutrients, 2, 1231-1246. http://dx.doi.org/10.3390/nu2121231

3. Ferrazzano, G.F., Amato, I., Ingenito, A., Zarrelli, A., Pinto, G. and Pollio, A. (2011) Plant Polyphenols and Their Anti-Cariogenic Properties: A Review. Molecules, 16, 1486-1507.http://dx.doi.org/10.3390/molecules16021486

4. Berghe, W.V. (2012) Epigenetic Impact of Dietary Polyphenols in Cancer Chemoprevention: Lifelong Remodeling of Our Epigenomes. Pharmacological Research, 65, 565-576. http://dx.doi.org/10.1016/j.phrs.2012.03.007

5. Kazeem, M.I., Akanji, M.A., Hafizur, R.M. and Choudhary, M.I. (2012) Antiglycation, Antioxidant and Toxicological Potential of Polyphenol Extracts of Alligator Pepper, Ginger and Nutmeg from Nigeria. Asian Pacific Journal of Tropi- cal Biomedicine, 2, 727-732.http://dx.doi.org/10.1016/S2221-1691(12)60218-4

6. Konaté, K., Hilou, A., Mavoungou, J.F., Lepengué, A.N., Souza, A., Barro, N., Datté, J.Y., M'Batchi, B. and Nacoul- ma, O.G. (2012) Antimicrobial Activity of Polyphenol-Rich Fractions from Sida alba L. (Malvaceae) against Cotrimoxazol-Resistant Bacteria Strains. Annals of Clinical Microbiology and Antimicrobials, 11, 1-6.http://dx.doi.

org/10.1186/1476-0711-11-5

7.   Lolayekar, N. and Shanbhag, C. (2012) Polyphenols and Oral Health. RSBO, 9, 74-84.

8.   Dragovic-Uzelac, V., Levaj, B., Mrkic, V., Bursac, D. and Boras, M. (2007) The Content of Polyphenols and Carote- noids in Three Apricot Cultivars Depending on Stage of Maturity and Geographical Region. Food Chemistry, 102, 966-975.http://dx.doi.org/10.1016/j. foodchem.2006.04.001

9.   Ignat, I., Volf, I. and Popa, V.I. (2011) A Critical Review of Methods for Characterisation of Polyphenolic Compounds in Fruits and Vegetables. Food Chemistry, 126, 1821-1835.http://dx.doi.org/10.1016/j. foodchem.2010.12.026

10. D'Archivio, M., Filesi, C., Di Benedetto, R., Gargiulo, R., Giovannini, C. and Masella, R. (2007) Polyphenols, Dietary Sources and Bioavailability. Annali dell›Istituto Superiore di Sanità, 43, 348-361.

11. N'Guessan, A.H.O., Déliko, C.E.D., Mamyrbékova-Békro, J.A. and Békro, Y.A. (2011) Teneurs en composés phéno- liques de 10 plantes médicinales employées dans la tradithérapie de l'hypertension artérielle, une pathologie émergente en Côte d'Ivoire. Revue de Génie Industriel, 6, 55-61.

12. Okuda, T. and Ito, H. (2011) Tannins of Constant Structure in Medicinal and Food Plants—Hydrolyzable Tannins and Polyphenols Related to Tannins. Molecules, 16, 2191-2217. http://dx.doi.org/10.3390/ molecules16032191

13. Cosmulescu, S. and Trandafir, I. (2011) Seasonal Variation of Total Phenols in Leaves of Walnut (Juglans regia L.). Journal of Medicinal Plants Research, 5, 4938-4942.

14. Generalic, I., Skroza, D., Surjaka, J., Mozinab, S.S. , Ljubenkovc, I., Katalinic, A., Simate, V. and Katalinic, V. (2012) Seasonal Variations of Phenolic Compounds and Biological Properties in Sage (Salvia officinalis L.). Chemistry & Biodiversity, 9, 441-456.http://dx.doi.org/10.1002/ cbdv.201100219

15. Raal, A., Orav, A., Pussa, T., Valner, C., Malmiste, B. and Arak, E. (2012) Content of Essential Oil, Terpenoids and Polyphenols in Commercial Chamomile (Chamomilla recutita L. Rauschert) Teas from Different Countries. Food Chemistry, 131, 632-638.http://dx.doi.org/10.1016/j. foodchem.2011.09.042

16. Ghasemi, K., Ghasemi, Y., Ehteshamnia, A., Nabavi, S.M. , Nabavi, S.F. , Ebrahimzadeh, M.A. and Pourmorad, F. (2011) Influence of

Environmental Factors on Antioxidant Activity, Phenol and Flavonoids Contents of Walnut (Juglans regia L.) Green husks. Journal of Medicinal Plants Research, 5, 1128-1133.

17. Bruni, R. and Sacchetti, G. (2009) Factors Affecting Polyphenol Biosynthesis in Wild and Field Grown St. John's Wort (Hypericum perforatum L. Hypericaceae/Guttiferae). Molecules, 14, 682-725. http://dx.doi.org/10.3390/molecules14020682

18. Trabelsi, N., Megdiche, W., Ksouri, R., Falleh, H., Oueslati, S., Soumaya, B., Hajlaoui, H. and Abdelly, C. (2010) Solvent Effects on Phenolic Contents and Biological Activities of the Halophyte Limoniastrum monopetalum Leaves. LWT-Food Science and Technology, 43, 632-639.

19. Gironi, F. and Piemonte, V. (2011) Temperature and Solvent Effects on Polyphenol Extraction Process from Chestnut Tree Wood. Chemical Engineering Research and Design, 89, 857-862. http://dx.doi.org/10.1016/j.cherd.2010.11.003

20. Falleh, H., Ksouri, R., Lucchessi, M.E., Abdelly, C. and Magné, C. (2012) Ultrasound-Assisted Extraction: Effect of Extraction Time and Solvent Power on the Levels of Polyphenols and Antioxidant Activity of Mesembryanthemum edule L. Aizoaceae Shoots. Tropical Journal of Pharmaceutical Research, 11, 243-249.http://dx.doi.org/10.4314/tjpr.v11i2.10

21. Tan, P.W., Tan, C.P. and Ho, C.W. (2011) Antioxidant Properties: Effects of Solid-to-Solvent Ratio on Antioxidant Compounds and Capacities of Pegaga (Centella asiatica). International Food Research Journal, 18, 557-562.

22. Santana, C.M. , Ferrera, Z.S. , Padrón, M.E.T. and Rodríguez, J.J.S. (2009) Methodologies for the Extraction of Phenolic Compounds from Environmental Samples: New Approaches. Molecules, 14, 298-320. http://dx.doi.org/10.3390/molecules14010298

23. Stalikas, C.D. (2010) Phenolic Acids and Flavonoids: Occurrence and Analytical Methods. Methods in Molecular Biology, 610, 65-90. http://dx.doi.org/10.1007/978-1-60327-029-8_5

24. Jace, D.E. and Shahidul, I. (2012) Effects of Extraction Procedures, Genotypes and Screening Methods to Measure the Antioxidant Potential and Phenolic Content of Orange-Fleshed Sweetpotatoes (Ipomoea batatas L.). American Journal of Food Technology, 7, 50-61. http://dx.doi.org/10.3923/ajft.2012.50.61

25. Garcia-Salas, P., Morales-Soto, A., Segura-Carretero, A. and Fernández-Gutiérrez, A. (2010) Phenolic-Compound-Ex- traction Systems for

Fruit and Vegetable Samples. Molecules, 15, 8813-8826. http://dx.doi. org/10.3390/molecules15128813

26. Chirinos, R., Rogez, H., Campos, D., Pedreschi, R. and Larondelle, Y. (2007) Optimisation of Extraction Conditions of Antioxidant Phenolic Compounds from Mashua (Tropaeolum tuberosum Ruíz & Pavón) Tubers. Journal of Separation and Purification Technology, 55, 217-225. http://dx.doi.org/10.1016/j.seppur.2006.12.005

27. Kanyonga, P.M., Faouzi, M.A., Meddah, B., Mpona, M., Essassi, E.M. and Cherrah, Y. (2011) Assessment of Methanolic Extract of Marrubium vulgare for Antiinflammatory, Analgesic and Anti-Microbiologic Activities. Journal of Chemical and Pharmaceutical Research, 3, 199-204.

28. Belhattab, R. and Larous, L. (2006) Essential oil Composition and Glandular Trichomes of Marrubium vulgare L. Growing Wild in Algeria. Journal of Essential Oil Research, 18, 369-373. http://dx.doi.org/10.108 0/10412905.2006.9699116

29. El Bardai, S., Morel, N., Wibo, M., Fabre, N., Liabres, G., Lyoussi, B. and Quetin-Leclercq, J. (2003) The Vasorelaxant Activity of Marrubenol and Marrubiin from Marrubium vulgare.Planta Medica , 69, 75-77. http:// dx.doi.org/10.1055/s-2003-37042

30. Sahpaz, S., Garbacki, N., Tits, M. and Bailleul, F. (2012) Isolation and Pharmacological Activity of Phenylpropanoid Esters from Marrubium vulgare. Journal of Ethnopharmacology, 79, 389-392. http://dx.doi. org/10.1016/S0378-8741(01)00415-9

31. Kurbatova, N.V., Muzychkina R.A., Mukhitdinov, N.M. and Parshina, G.N. (2013) Comparative Phytochemical Investigation of the Composition and Content of Biologically Active Substances in Marrubium vulgare and Marrubium alternidens. Chemistry of Natural Compounds, 39, 501-502. http://dx.doi.org/10.1023/B:CONC.0000011128.64886.f4

32. Nawwar Mahmoud , A.M., El-Mousallamy, A.M.D., Barakat, H.H. , Buddrus, J. and Linscheid, M. (1989) Flavonoid Lactates from Leaves of Marrubium vulgare. Phytochemistry, 28, 3201-3206. http://dx.doi. org/10.1016/0031-9422(89)80307-3

33. Vergara-Galicia, J., Aguirre-Crespo, F., Tun-Suarez, A., Crespo, A.A., Estrada-Carrillo, M., Jaimes-Huerta, I., Flores- Flores, A., Estrada-Soto, S. and Ortiz-Andrade, R. (2012) Acute Hypoglycemic Effect of Ethanolic Extracts from Marrubium vulgare. Phytopharmacology, 3, 54-60.

34. Meyre-Silva, C., Yunes, R.A. , Schlemper, V., Campos-Buzzi, F. and Cechinel-Filho, V. (2005) Analgesic Potential of Marrubiin Derivatives,

a Bioactive Diterpene Present in Marrubium vulgare L. (Lamiaceae). Il Farmaco, 60, 312-326.http://dx.doi.org/10.1016/j.farmac.2005.01.003

35. Boudjelal, A., Henchiri, C., Siracusa, L., Sari, M. and Ruberto, G. (2012) Compositional Analysis and in Vivo Anti- Diabetic Activity of Wild Algerian Marrubium vulgare L. Infusion. Fitoteparia, 2, 286-292.

36. De Jesus, R.A. and Cechinel Filho, V. (2000) Analysis of the Antinociceptive Properties of Marrubiin Isolated from Marrubium vulgare L. Phytomedicine, 7, 111-115.http://dx.doi.org/10.1016/S0944-7113(00)80082-3

37. Pukalskas, A., Rimantas Venskutonis, P., Salido, S., De Waard, P. and Van Beek., T.A. (2012) Isolation, Identification and Activity of Natural Antioxidants from Horehound (Marrubium vulgare L.) Cultivated in Lithuania. Food Chemistry, 130, 695-701.http://dx.doi.org/10.1016/j. foodchem.2011.07.112

38. Li, H., Wong , C., Cheng, K. and Chen , F. (2008) Antioxidant Properties in Vitro and Total Phenolic Contents in Methanol Extracts from Medicinal Plants. Lebensmittel-Wissenschaft und -Technologie, 41, 385-390.

39. Chew, K.K. , Ng, S.Y. , Thoo, Y.Y. , Khoo, M.Z. , Wan Aida, W.M. and Ho, C.W. (2011) Effect of Ethanol Concentration, Extraction Time and Extraction Temperature on the Recovery of Phenolic Compounds and Antioxidant Capacity of Centella asiatica Extracts. International Food Research Journal, 18, 571-578.

40. Ceto, X., Gutiérrez, J.M., Gutiérrez, M., Céspedes, F., Capdevila, J., Manguez, S., Jiménez-Jorquera, C. and Del Valle, M. (2012) Determination of Total Polyphenol Index in Wines Employing a Voltammetric Electronic Tongue. Analytica Chimica Acta, 732, 172-179. http://dx.doi.org/10.1016/j.aca.2012.02.026

41. Liao , L., Yin, X. and Wang, Z. (2012) Optimization of Total Flavonoid Extraction in Callicarpa nudiflora Hook. et Arn. Using Response Surface Methodology. Journal of Medicinal Plants Research, 6, 5038-5047.

42. Zheng, N., Wang, Z., Shi, Y. and Lin , J. (2012) Evaluation of the Antifungal Activity of Total Flavonoids Extract from Patrinia Villosa Juss and Optimization by Response Surface Methodology. African Journal of Microbiology Research, 6, 586-593.http://dx.doi.org/10.5897/ AJMR11.1393

43. Naczk, M. and Shahidi, F. (2004) Extraction and Analysis of Phenolics in Food. Journal of Chromatography, 1054, 95-111. http://dx.doi. org/10.1016/j.chroma.2004.08.059

44. Jakopic, J., Veberic, R. and Stampar, F. (2009) Extraction of Phenolic

Compounds from Green Walnut Fruits in Different Solvents. Acta Agriculturae Slovenica , 93, 11-15.http://dx.doi.org/10.2478/v10014-009-0002-4

45. Ghasemzadeh, A., Jaafar, H.Z.E. and Rahmat, A. (2011) Effects of Solvent Type on Phenolics and Flavonoids Content and Antioxidant Activities in Two Varieties of Young Ginger (Zingiber officinale Roscoe) Extracts. Journal of Medicinal Plants Research, 5, 1147-1154.

46. Mohammedi, Z. and Atik, F. (2011) Impact of Solvent Extraction Type on Total Polyphenols Content and Biological Activity from Tamarix aphylla (L.) Karst. International Journal of Pharma and Bio Sciences, 2, 609-615.

47. Iqbal, S., Younas, U., Chan, K.W. , Zia-Ul-Haq, M. and Ismail, M. (2012) Chemical Composition of Artemisia annua L. Leaves and Antioxidant Potential of Extracts as a Function of Extraction Solvents. Molecules, 17, 6020-6032.http://dx.doi.org/10.3390/molecules17056020

48. Lapornik, B., Prosek, M. and Wondra, A.G. (2005) Comparison of Extracts Prepared from Plant By-Products Using Different Solvents and Extraction Time. Journal of Food Engineering, 71, 214-222. http://dx.doi.org/10.1016/j.jfoodeng.2004.10.036

49. Turkmen, N., Sari, F. and Velioglu, S. (2006) Effect of Extraction Solvents on Concentration and Antioxidant Activity of Black and Black Mate Polyphenols Determined by Ferrous Tartrate and Folin-Ciocalteu Methods. Food Chemistry, 99, 838-841.http://dx.doi.org/10.1016/j.foodchem.2005.08.034

50. Kong , K.W. , Khoo, H.E . , Prasad, N.K. , Chew, L.Y. and Amin, I. (2013) Total Phenolics and Antioxidant Activities of Pouteria campechiana Fruit Parts. Sains Malaysiana , 42, 123-127.

51. Sultana, B., Anwar, F. and Ashraf, M. (2009) Effect of Extraction Solvent/Technique on the Antioxidant Activity of Selected Medicinal Plant Extracts. Molecules, 14, 2167-2180.http://dx.doi.org/10.3390/molecules14062167

52. Ben Mohamed Maoulainine, L., Jelassi, A., Hassen, I. and Ould Mohamed Salem Ould Boukhari, A. (2012) Antioxidant Proprieties of Methanolic and Ethanolic Extracts of Euphorbia helioscopia (L.) Aerial Parts. International Food Research Journal, 19, 1125-1130.

53. Pérez, M.B., Calderón, N.L. and Croci, C.A. (2007) Radiation-Induced Enhancement of Antioxidant Activity in Extracts of Rosemary (Rosmarinus officinalis L.). Food Chemistry, 104, 585-592. http://dx.doi.org/10.1016/j.foodchem.2006.12.009

54. Karimi, E., Oskoueian, E., Hendra, R. and Jaafar, H.Z.E. (2010) Evaluation

of Crocus sativus L. Stigma Phenolic and Flavonoid Compounds and Its Antioxidant Activity. Molecules, 15, 6244-6256. http://dx.doi.org/10.3390/molecules15096244

55. Settharaksa, S., Jongjareonrak, A., Hmadhlu, P., Chansuwan, W. and Siripongvutikorn, S. (2012) Flavonoid, Phenolic Contents and Antioxidant Properties of Thai Hot Curry Paste Extract and Its Ingredients as Affected of pH, Solvent Types and High Temperature. International Food Research Journal, 19, 1581-1587

56. Spigno, G., Tramelli, L. and De Faveri, D.M. (2007) Effects of Extraction Time, Temperature and Solvent on Concentration and Antioxidant Activity of Grape MarcPhenolics. Journal of Food Engineering, 81, 200-208.

57. Wina, E., Susana, I.W.R. and Tangendjaja, B. (2010) Biological Activity of Tannins fromAcacia mangium Bark Extracted by Different Solvents. Media Peternakan, 33, 103-107.http://dx.doi.org/10.5398/medpet.2010.33.2.103

58. Mazandarani, M., Zarghami Moghaddam, P., Zolfaghari, M.R., Ghaemi, E.A. and Bayat, H. (2012) Effects of Solvent Type on Phenolics and Flavonoids Content and Antioxidant Activities in Onosma dichroanthum Boiss. Journal of Medicinal Plants Research, 6, 4481-4488. http://dx.doi.org/10.5897/JMPR11.1460

59. Antwi-Boasiako, C. and Animapauh, S.O. (2012) Tannin Extraction from the Barks of Three Tropical Hardwoods for the Production of Adhesives. Journal of Applied Sciences Research, 8(6), 2959-2965.

60. Uma, D.B., Ho, C.W. and Aida, W.M.W. (2010) Optimization of Extraction Parameters of Total Phenolic Compounds from Henna (Lawsonia inermis) Leaves. Sains Malaysiana, 39, 119-128.

61. Chan, S.W., Lee, C.Y., Yap, C.F., Wan Aida, W.M. and Ho, C.W. (2009) Optimisation of Extraction Conditions for Phenolic Compounds from Limau Purut (Citrus hystrix) Peels. International Food Research Journal, 16, 203-213.

62. Yilmaz, Y. and Toledo, R.T. (2005) Oxygen Radical Absorbance Capacities of Grape/Wine Industry Byproducts and Effect of Solvent Type on Extraction of Grape Seed Polyphenols. Journal of Food Composition and Analysis, 19, 41- 48.http://dx.doi.org/10.1016/j.jfca.2004.10.009

63. Butnariu, M. and Coradini, C.Z. (2012) Evaluation of Biologically Active Compounds from Calendula officinalis Flowers Using Spectrophotometry. Chemistry Central Journal, 6, 35-42. http://dx.doi.org/10.1186/1752-153X-6-35

64. Fernandes, A.J.D., Ferreira, M.R.A., Randau, K.P., De Souza, T.P. and Soares, L.A.L. (2012) Total Flavonoids Content in the Raw Material and Aqueous Extractives from Bauhinia monandra Kurz (Caesalpiniaceae). The Scientific World Journal, 2012, Article ID: 923462.

65. Musa, K.H., Abdullah, A., Jusoh, K. and Subramaniam, V. (2011) Antioxidant Activity of Pink-Flesh Guava (Psidium guajava L.): Effect of Extraction Techniques and Solvents. Food Analytical Methods, 4, 100-107. http://dx.doi.org/10.1007/s12161-010-9139-3

66. Downey, M.O. and Hanlin, R.L. (2010) Comparison of Ethanol and Acetone Mixtures for Extraction of Condensed Tannin from Grape Skin. South African Journal of Enology and Viticulture, 31, 154-159.

67. Medini, F., Ksouri, R., Falleh, H., Megdiche, W., Trabelsi, N. and Abdelly, C. (2011) Effects of Physiological Stage and Solvent on Polyphenol Composition, Antioxidant and Antimicrobial Activities of Limonium densiflorum. Journal of Medicinal Plants Research, 5, 6719-6730.

68. Shad, M.A., Nawaz, H., Rehman, T., Ahmad , H.B. and Hussain, M. (2012) Optimization of Extraction Efficiency of Tannins from Cichorium intybus L.: Application of Response Surface Methodology. Journal of Medicinal Plants Research, 6, 4467-4474.

69. Zhao, S., Liu, J.Y. , Chen , S.Y. , Shi, L.L. , Liu, Y.J. and Ma, C. (2011) Antioxidant Potential of Polyphenols and Tannins from Burs of Castanea mollissima Blume. Molecules, 16, 8590-8600. http://dx.doi.org/10.3390/molecules16108590

70. Hismath, I., Wan Aida, W.M. and Ho, C.W. (2011) Optimization of Extraction Conditions for Phenolic Compounds from Neem (Azadirachta indica) Leaves. International Food Research Journal, 18, 931-939.

71. Franco , D., Sineiro, J., Rubilar, M., Sanchez , M., Jerez, M., Pinelo, M., Costoya, N. andJosé Nunez , M. (2008) Poly- phenols from Plant Materials: Extraction and Antioxidant Power. Electronic Journal of Environmental, Agricultural and Food Chemistry, 7, 3210-3216.

72. Sardsaengjun, C. and Jutiviboonsuk, A. (2010) Effect of Temperature and Duration Time on Polyphenols Extract of Areca catechu Linn. Seeds. Thai Pharmaceutical and Health Science Journal, 5, 14-17.

73. Dent, M., Dragovi-Uzelac, V., Peni, M., Brncic, M., Bosiljkov, T. and Levaj, B. (2013) The Effect of Extraction Solvents, Temperature and Time on the Composition and Mass Fraction of Polyphenols in Dalmatian Wild Sage (Salvia officinalis L.) Extracts. Food Technology and Biotechnology, 51, 84-91.

74. Naczk, M. and Shahidi, F. (2006) Phenolics in Cereals, Fruits

and Vegetables: Occurrence, Extraction and Analysis. Journal of Pharmaceutical and Biomedical Analysis, 41, 1523-1542. http://dx.doi.org/10.1016/j.jpba.2006.04.002

75. Garcia-Marquez, E., Roman-Guerrero, A., Perez-Alonso, C. and Cruz-Sosa, F. (2012) Effect of Solvent-Temperature Extraction Conditions on the Initial Antioxidant Activity and Total Phenolic Content of Muitle Extracts and Their Decay upon Storage at Different pH. Revista Mexicana de Ingeniería Química, 11, 1-10.

76. Zhekova, G. and Pavlov , D. (2012) Influence of Different Factors on Tannins and Flavonoids Extraction of Some Thyme Varieties Representatives of Thymol, Geraniol and Citral Chemotype. Agricultural Science and Technology, 4, 148-153.

77. Liu, W., Yu, Y., Yang, R., Wan, C., Xu, B. and Cao, S. (2010) Optimization of Total Flavonoid Compound Extraction from Gynura medica Leaf Using Response Surface Methodology and Chemical Composition Analysis. International Journal of Molecular Sciences, 11, 4750-4763. http://dx.doi.org/10.3390/ijms11114750

78. Moosophin, K., Wetthaisong, T., Seeratchakot, L. and Kokluecha, W. (2010) Tannin Extraction from Mangosteen Peel for Protein Precipitation in Wine. KKU Research Journal, 15, 377-385.

79. Mustapha, M.B. , Adefisan, H.A. and Olawale, A.S. (2012) Studies of Tannin Extract Yield from Parkia clappertoniana's Husk. Journal of Applied Sciences Research, 8, 65-68.

80. Zam, W., Bashour, G., Abdelwahed, W. and Khayata, W. (2012) Effective Extraction of Polyphenols and Proanthocyanidins from Pomegranate's Peel. International Journal of Pharmacy and Pharmaceutical Sciences, 4, 675-682.

81. Druzynska, B., Stepniewska, A. and Wolosiak, R. (2007) The Influence of Time and Type of Solvent on Efficiency of the Extraction of Polyphenols from Green Tea and Antioxidant Properties Obtained Extracts. Acta Scientiarum Polo- norum. Technologia Alimentaria, 6, 27-36.

82. Jokic, S., Velic, D., Bilic, M., Bucic-Kojic, A., Planinic, M. and Tomas, S. (2010) Modelling of the Process of Solid- Liquid Extraction of Total Polyphenols from Soybeans. Czech Journal of Food Sciences, 28, 206-212.

83. Perva-Uzunalic, A., Skerget, M., Knez, Z., Weinreich, B., Otto, F. and Grucher, S. (2006) Extraction of Active Ingredients from Green Tea (Camellia sinensis), Extraction Efficiency of Major Catechins and Caffeine. Food Chemistry, 96, 597-605.http://dx.doi.org/10.1016/j.

foodchem.2005.03.015

84. Akowuah, G.A. , Mariam A. and Chin, J. H. (2009) The Effect of Extraction Temperature on Total Phenols and Antioxidant Activity of Gynura procumbens Leaf. Pharmacognosy Magazine, 5, 81-85.

85. Chew, K.K. , Thoo, Y.Y. , Khoo, M.Z. , Wan Aida, W.M. and Ho, C.W. (2011) Effect of Ethanol Concentration, Extraction Time and Extraction Temperature on the Recovery of Phenolic Compounds and Antioxidant Capacity of Orthosiphon stamineus Extracts. International Food Research Journal, 18, 427-1435.

86. Naeem, S., Ali , M. and Mahmood, A. (2012) Optimization of Extraction Conditions for the Extraction of Phenolic Compounds from Moringa oleifera Leaves. Pakistan Journal of Pharmaceutical Sciences, 25, 535-541.

87. Davidov-Pardo, G., Arozarena, M.R.I. and Marin-Arroyo, M.R. (2011) Stability of Polyphenolic Extracts from Grape Seeds after Thermal Treatments. European Food Research and Technology, 232, 211-220. http://dx.doi.org/10.1007/s00217-010-1377-5

88. Biesaga, M. and Pyrzynska, K. (2013) Stability of Bioactive Polyphenols from Honey during Different Extraction Methods. Food Chemistry, 136, 46-54.http://dx.doi.org/10.1016/j.foodchem.2012.07.095

89. Juntachote, T., Berghofer, E., Bauer, F. and Siebenhandl, S. (2006) The Application of Response Surface Methodology to the Production of Phenolic Extracts of Lemon Grass, Galangal, Holy Basil and Rosemary . International Journal of Food Science and Technology, 41, 121-133. http://dx.doi.org/10.1111/j.1365-2621.2005.00987.x

90. Al-Farsi, M.A. and Lee , C.Y. (2007) Optimization of Phenolics and Dietary Fibre Extraction from Date Seeds. Food Chemistry, 108, 977-985.http://dx.doi.org/10.1016/j.foodchem.2007.12.009

91. Wang, J., Sun, B., Cao, Y., Tian, Y. and Li, X. (2008) Optimisation of Ultrasound-Assisted Extraction of Phenolic Compounds from Wheat Bran. Food Chemistry, 106, 804-810.http://dx.doi.org/10.1016/j.foodchem.2007.06.062

Chapter 10

NOVEL SYNTHESIS APPROACH AND ANTIPLATELET ACTIVITY EVALUATION OF 6-ARYLMETHYLENEAMINO-2-ALKYLSULFONYLPYRIMIDIN-4(3H)-ONE DERIVATIVES AND ITS NUCLEOSIDES

Abdulrahman G. Alshammari[1], Abdel-Rhman B. A. El-Gazzar[1,2]

[1]Department of Chemistry, Faculty of Science, Al-Imam Mohammad Ibn Saud Islamic University (IMSIU), Riyadh, KSA

[2]Photochemistry Department, Heterocyclic & Nucleosides Unit, National Research Centre, Cairo, Egypt

ABSTRACT

A new and efficient procedure has been designed for the preparation of 6-arylmethylene-amino-2-alkyl sulfonylpyrimidine. The first alkylthio group was introduced into the pyrimidine ring by S-alkylation. The introduction of the second one was successfully achieved using the phosphorous oxychloride method to afford 4-chloro-2-alkylthiopyrimidines. Subsequent nucleophilic displacement by the corresponding alkylamines followed by glycoside bromide addition conveniently gave a series of the target compounds. Thus, the two same or different alkylamino groups were easily introduced into the pyrimidine ring through the two different approaches. The human anti-platelet aggregation activity of the newly synthesized compounds was also described.

INTRODUCTION

Pyrimidines play an essential role in several biological processes and have considerable chemical and pharmacological importance in terms that the pyrimidine ring can be found in nucleoside antibiotics, anti-bacterial and cardiovascular [1-9]. Pyrimidine derivatives, especially alkylthio-substituted pyrimidines [10], and thieno pyrimidines [11], have attracted much attention because of their quite high anti-platelet aggregation activity as inhibitors for P2-receptor family. As related works, there have been active attempts to develop the antagonist of P2Y receptors (which mediate platelet aggregation induced by adenosine diphosphate, ADP) by employing adenine nucleotide derivatives containing two phosphate groups, i.e., adenosine-3',5'-bis-phosphate analogues, as P2Y1 receptor antagonists [12,13] and 4-alkoxyl-2-alkylthio-6-aminopyrimidine derivatives as P2Y12 receptor antagonists [14]. The evaluated anti-platelet aggregation ability of a series of the synthesized pyrimidine derivatives proves their potential as lead compounds to develop a new series of P2Y12 antagonists [14]. Furthermore, the results appear to suggest the importance of the chemical structure of alkyl-thio-substituents and of the presence of a free amino group for the activity. As for adenine nucleotide analogues against P2T receptor, there's effective enhancement of the activity by N-mono-alkylation at the 6-position of the adenine moiety [15]. Considering such findings on the structure of the antagonist candidates, we achieved one hypothesis that N-monoalkylation of pyrimidine compounds might also increase anti-platelet aggregation activity. Furthermore, to date, there are few reports on the synthesis and evaluation of dialkyl(aryl) thio-substituted pyrimidines as platelet aggregation inhibitors. Hence, we designed to introduce another thio-nucloside group into the pyrimidine ring. As it is well known, the introduction of alkyl/arylthio groups into the pyrimidine ring is commonly achieved through either the alkylation of thiol groups or the nucleophilic substitution of halogens by alkylmer-captides. However, these processes must be conducted under harsh conditions, especially in the absence of activating substituents. In the present study, we successfully introduced the two same or different alkyl amino groups into the pyrimidine ring, giving the intermediate 2- and 4-alkylamino pyrimidines. Thus far, there has been no report on the application of the diazotizatione-alkylthionation reaction to aminopyrimidine derivatives. The subsequent nucleophilic displacement of the chloro group in C-4-position by the corresponding amines affords a series of 4-alkylamino- 2-thioxopyrimidines. Herein, we describe the details of the convenient synthesis and the evaluation results of all the synthesized compounds as human platelet aggregation inhibitors.

RESULTS AND DISCUSSION

Chemistry

In a previous work we synthesized a series of substituted pyrimidines and substituted thieno[2,3-d] pyrimidine compounds [16,17] which presented a noticeable platelet antiaggregating power [18,19]. The most potent activity was exhibited by the thieno pyrimidinone derivatives. These thienopyrimidine compounds present a similar scaffold to the agents cited above. Several studies have shown that the presence of a quinazoline skeleton substituted in position 4 by vaious substituted anilino groups potentially increased the EGFR inhibitory effect [20,21]. According to the results observed with the former pyrimidine derivatives as platelet antiaggregating agents, the substitution of these compounds at the 4 position could lead to new PDGFR or EGFR pathway inhibitors. Our synthetic strategy to the final 2- and 4-alkylamino pyrimidine nucleosides was first to synthesize the glycosyl donor and then to condense with pyrimidine derivative bases. The pyrimidine bases were synthesized via the key intermediate 2 and 3 from 6-amino thiouracil and aromatic aldehydes followed by alkylation (Scheme 1).

The reactions of the starting 6-[(4-chlorophenyl)-meethylene]amino-2-methylthio-pyrimidin-4(1H)-one (3) with β-D-glycofuranosyl bromide were carried out in acetone in the presence of potassium carbonate under stirring at room temperature until the precipitation of sodium bromide had ceased.

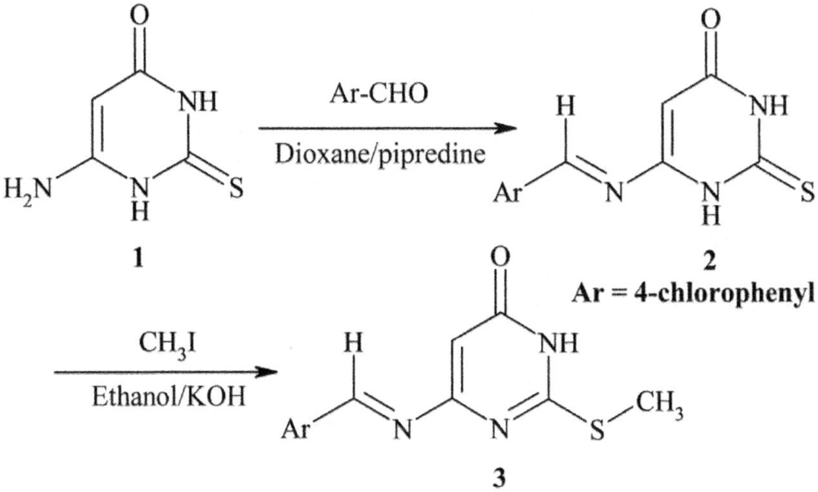

Scheme 1: Synthesis the starting materials.

The reaction mixture was filtered off; the solvent was removed under reduced pressure, and the solid formed was purified by using flash chromatography afforded only one product in each case (as evidenced by TLC). The isolated products were identified as N-(2',3',5'-tri-O-acetyl/benzoyl-β-D-glycofuranosyl)-6-[4-(aryl)-methylene]amino-2-(methylthio) pyrimidin-4(1H)-one (4a,b). Deacetylation of the later nucleosides 4a,b using saturated ammonia solution in methanol at room temperature afforded the corresponding deacetylated N-(β-D-glyco furanosyl)-6-{[4-(chlorophenyl)-methylene]amino}-2-(methylthio)-pyrimidin-4(1H)-one (5a,b) respectively.

In the other way, the latter 6-[4-(chlorophenyl)-methylene]amino-2-(methylthio) pyrimidine series were oxidized in acetic acid solution containing hydrogen per-oxide at room temperature, affording 6-[4-(chloro phenyl)- methylene]amino-2-(methylsulfonyl)pyrimidine (6). Under the same conditions [16] the conden-sation of 2-(methylsulfonyl)-pyrimidine (6) with β-D-glyco furanosyl bromide afforded   N-(2',3',5'-tri-O-acetyl-β-D-glycofu-ranosyl)-6-[4-(chlorophenyl) methylene]amino-2-(me-thylsulfonyl)-pyrimidin-4(1H)-one (7a,b) which give respectively, the deacetylated N-(β-D-glycofurano syl)-6-[4- (chlorophenyl)-methylene]amino-2-(methylthio)- pyrimidin-4(1H)-one (8a,b), (Scheme 2). Some newer 6-[4-(aryl)-methylene]amino-2-alkyl-amino-pyrimidin- 4(1H)-one, have been synthesized for the further diversification in our pyrimidine nucleosides core motif to generate alkylamino group at C-2 position via a two step protocol. The Fusion of 2-methythiopyrimidine 3 with secondary alkyl amines namely (morpholine, piprazine and N-methyl piprazine) at 180°C in sand bath without using any solvent, affording the 6-[4-(aryl)-methylene] ami-no-2-alkylamino-pyrimidin-4(1H)-one (9a-c), (Scheme 3).

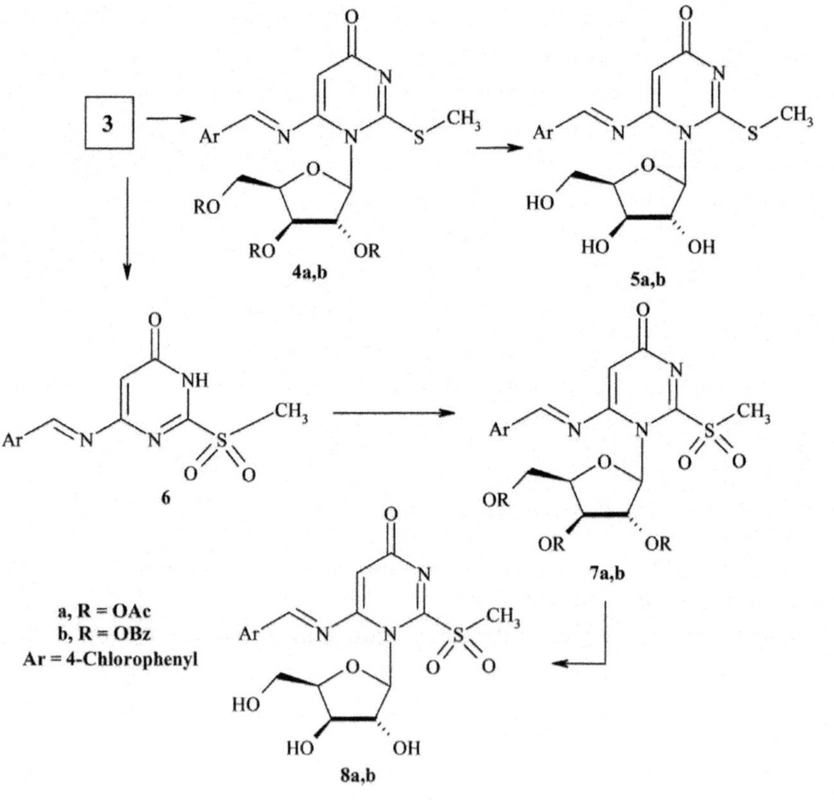

Scheme 2: Synthesis of Sulfone N-Nucleosides.

These lactol derivatives were transformed in to N-nucleosides derivatives (11a-f) by the reaction 2,3,5-tri-O-acetyl-β-D-glyco furanosyl bromide in quantitative yield.

Our recent interest was centered on the 4 position of pyrimidine nucleobases as potential anti-platelet agents. We modified the above mentioned uracils by introducing a alkylamino moiety and reported the synthesis of 4-(alka mino)-N-glycofuranosylpyrimidines 16a-f, 18a-f (Scheme 4) and studied there anti-platelet activity against. Thus, the phosphorus oxychloride in dry condition modified the pyrimidin-4-one (2) to 4-chloro pyrimidine (13), the latter is a key to introduced the alkyamino goups namely morpholine, piprazine, and N-methyl piprazine at 4-position in pyrimidines (14a-c), The synthesis was initiated by formation of potassium salts of the starting compounds 14a-c. Potassium salt uracils reacted with protected sugar in the presence of acetone at room temperature to afford 4-alkyamino-pyrimidine acetylated S-nucleosides (15a-f) that were separated by flash colum chromate graphy in good yields.

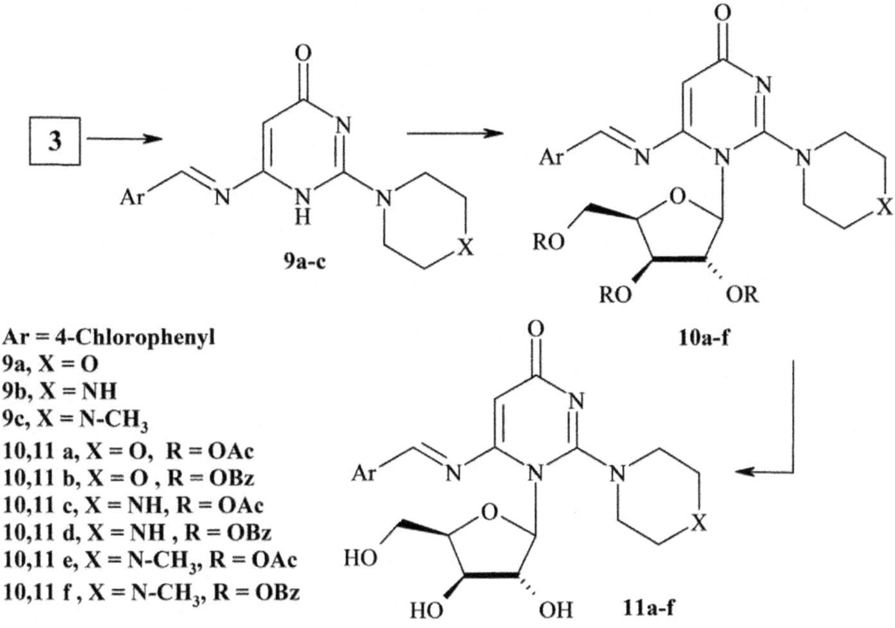

Ar = 4-Chlorophenyl
9a, X = O
9b, X = NH
9c, X = N-CH₃

10,11 a, X = O, R = OAc
10,11 b, X = O , R = OBz
10,11 c, X = NH, R = OAc
10,11 d, X = NH , R = OBz
10,11 e, X = N-CH₃, R = OAc
10,11 f , X = N-CH₃, R = OBz

Scheme 3: Synthesis of Nucleosides.

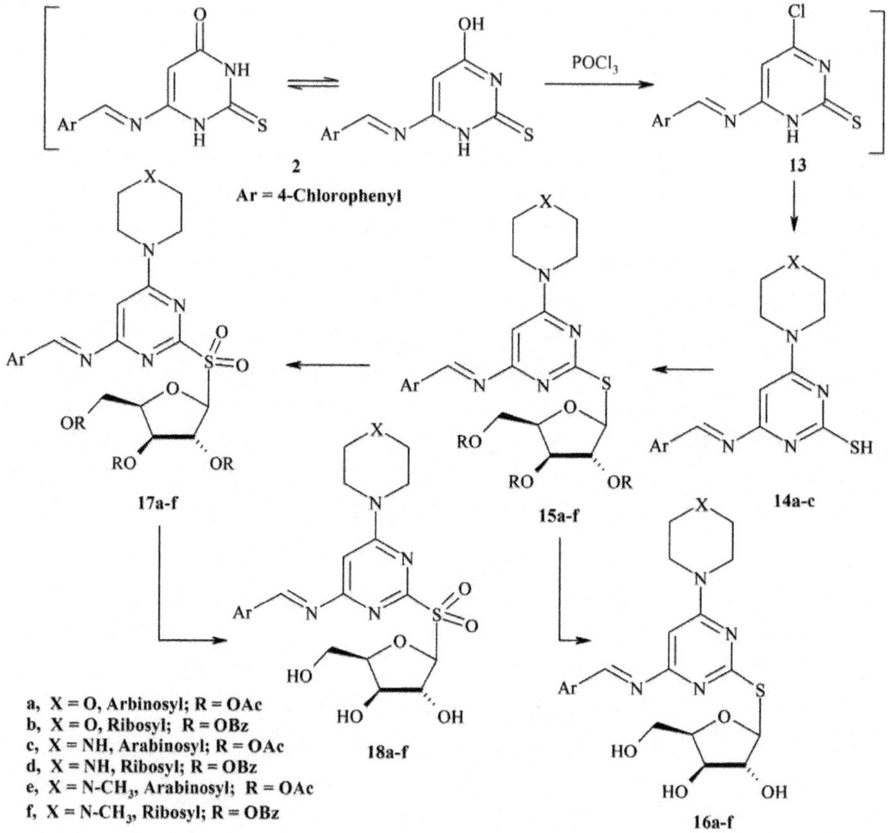

Scheme 4: Synthesis of sulfone S-Nucleosides.

Treatment of glycosides 15a-f with methanolic ammonia at room temperature afforded the nucleosides 16a-f. Moreover, the acetylated S-nucleosides (15a-f) was oxidized in a mixture of acetic acid and hydrogen peroxide at room temperature affording the acetylated 6-[4-(aryl)-methylene] amino-2-(Nglycofuranosyl sulf one)-4-alkylamino-pyrimidine (17af), respectively in good yield (55% - 64%). Thin layer chromatography (chloro form: methanol, 8:2) indicated the formation of the pure compounds. The structures of 17a-f were confirmed by elemental analysis and spectral data (IR, ^{1}H NMR, ^{13}C.NMR) (cf. Exp.). The ^{1}H NMR spectrum of compound 17a as an example, showed the anomeric proton of the glycoside moiety as a doublet at d 6.81 ppm with a coupling constant J $_{1'-2'}$ = 3.69 Hz indicating β-configuration of the anomeric center. The other protons of the glycoside ring resonated at d 4.00 - 5.40 ppm, while the three acetoxy groups appeared as three singlets at d 1.94 - 2.15 ppm.

Deacetylation of acetylated nucleosides 17a-f using saturated solution of ammonia in methanol at room temperature afforded the corresponding deacetylated nucleosides 18a-f respectively. The structures of free nucleosides 18 have been established on the basis of their spectral data and elemental analyses. Thus, the ^{1}H NMR spectrum of 18a showed the anomeric proton as a doublet at d 6.84 ppm. The signals of the other five glycoside protons appeared at d 3.94 - 5.73 ppm, while the signals that disappear on rapid exchange with D_2O are observed at d 5.13. - 5.47 ppm, were assigned as the three hydroxyl groups.

Antiplatelet Aggregation Activity

The antiplatelet aggregation activity of the derivatives is listed in **Table 1**. All the tested derivatives effectively inhibited platelet aggregation higher than 50% at 100 μm concentration. The majority of the compounds showed lower IC50 values than that of aspirin and among them, compounds 18b, 18d and 18f exhibited comparable IC_{50} values to that of indomethacin, a potent inhibitor of the enzyme COX. The results show that the tested compounds have a similar activity profile to those of previously studied pyrimidine derivatives, that is they inhibited AA-induced platelet aggregation more effectively than once ADP was used as inducer. However, compound 16f showed satisfying activity with IC_{50} value of 18.1 μm against ADP-induced platelet aggregation; notably that this derivatives proved to be effective to inhibit platelet aggregation induced by both ADP and AA. Of the most potent derivatives, compounds 18f and 18b with IC50 values of 2.2 μm and 3.8 μm both contain five membered electron-rich heterocyclic rings and this would suggest that the existence of these ring systems can be an important factor affecting the antiplatelet activity.

This has been confirmed the previous studies [22], as in their study the most potent derivative exerting dual inhibition of COX/LOX.7) Comparing the structures of this compound and compound 18b shows that these two have a high extent of similarity; suggesting that they also may share common mechanisms of action and structure-activity relationships. This is also in line with the findings of Barreiro's group to discover some derivatives with furan substituent as potent inhibitors of AA induced platelet aggregation [23]. Of the synthesized derivatives, compound 16f with dual inhibition of ADP/AA-induced platelet aggregation and compounds 18b-f as the most active inhibitors of AA induced platelet aggregation have molecular weights ranging from 194 to 249 and C logP values ranging from 2.23 to 3.18. Therefore the mentioned derivatives have ideal physicochemical characteristics to be considered as starting points for further lead optimization studies in the area of antiplatelet aggregation drug research.

Compound No.	ADP		AA	
	Inhibition (%)[a,b]	IC$_{50}$ (μm)	Inhibition (%)[b]	IC$_{50}$ (μm)
5a	32.1	>100	99.8	84.9
5b	45.8	>100	99.9	18.3
8a	38.4	>100	100.3	21.4
8b	46.6	>100	100.0	79.8
11a	nd	nd	100.0	5.8
11b	60.1	81.92	99.7	7.3
11c	24.9	>100	100.0	63.4
11d	48.9	>100	99.9	15.5
11e	67.6	95.23	99.8	7.8
11f	49.7	>100	100.5	8.3
16a	38.3	>100	53.0	96.2
16b	2.4	>100	99.1	22.1
16c	34.1	>100	99.8	78.4
16d	49.2	>100	100.3	11.3
16e	3.7	>100	99.7	8.8
16f	81.3	18.1	103.0	21.1
18a	70.6	34.2	85.0	51.9
18b	33.4	>100	100.0	3.8
18c	3.1	>100	99.9	7.8
18d	29.7	>100	100.0	2.6
18e	1.9	>100	99.1	18.1
18f	34.8	>100	100.1	2.2
Indomethacin	42.2[c]	>100	100.0	3.0
Aspirin	21.4[c]	>100	99.8	30.3

Table 1: Antiplatelet activity evaluation of 2-/4-alkylamino-pyrimidine S and N-nucleosides.

CONCLUSION

This paper describes a class of novel 2- and 4-alkylamino-pyrimidines and their nucleosides designed by considering the structural features of the previously reported derivatives proved to be bioactive inhibitor compounds. The derivatives were prepared by a one-pot procedure and evaluated for their antiplatelet aggregation activity using AA and ADP as aggregation inducers. The derivatives effectively inhibited platelet aggregation at 100 μm concentration and some of them exhibited inhibitory activities comparable to those of aspirin and indomethacin. Synthesis and study of antiplatelet activity of 2 and 4-substituted pyrimidine nucleosides will provide valuable information about the SAR of this group of compounds.

EXPERIMENTAL

Chemistry

Melting points were determined on the Electrothermal 9100 melting point apparatus (Electrothermal, UK) and are uncorrected. The IR spectra (KBr) were recorded on an FT-IR NEXCES spectrophotometer (Shimadzu, Japan). The ^{1}H NMR spectra were measured with a Jeol ECA 500 MHz (Japan) in DMSO-d$_6$ and chemical shifts were recorded in d ppm relative to TMS. Mass spectra (EI) were run at 70 eV with a Finnigan SSQ 7000 spectrometer. The purity of the compounds was checked on Aluminium plates coated with silica gel (Merck). The elemental analysis for C, H, N and S was performed by a Costech model 4010 and the percentage values agreed with the proposed structures within ± 0.4% of the theoretical values. The Pharmacological evaluations of the products were carried out in Pharmacological Unit Pharmacology Department (NRC, Cairo, Egypt).

Synthesis of 6-[4-(chlorophenyl)methylene]amino- 2,3-dihydropyrimidin-4(1H)-one (2). A mixture of 6- aminothiouracil 1 (10 m mol) and 4-chloro-benzaldehyde was refluxed in amixture of ethanol/piperdine (50 mL) for 5 hrs. (under TLC control), The reaction mixture was allowed to cool to room tempreature and then add water (100 mL), the precipitate was filtered off, dried and crystallized from dioxane (40 mL); as a pale yellow powder in 83% yield; m. p. 275°C -277°C, IR (cm^{-1}, n): 3380 (br, NH), 1675 (CO); ^{1}H NMR: d 3.42 (br, NH, D$_2$O exchangeable), 7.34 (d, J = 7.5 Hz, aryl-H), 7.97 (d, J = 7.5 Hz, aryl-H), 8.02 (s, pyrimidine-H), 8.36 (s, CH), 9.10 (br, NH); Its MS (m/z), 265 (M$^+$, 100%); C$_{11}$H$_8$ClN$_3$OS (265.7).

Synthesis of 6-[(4-chlorophenyl)methylene]amino- 2-methylthio-pyrimidin-4(1H)-one (3). To a warmed ethanolic KOH solution prepared by dissolving (10 mmol) of KOH in 50 mL (ethanol) was added each of compound 2 (10 mmol), the heating was continued for 30 min and the mixture was allowed to cool to room temperature, and the proper methyliodide (12 mmol) was added. The mixture was stirred under reflux for 5 h, then cool to room temperature, poured into cold water (100 mL). The solid product precipitated was filtered off washed with 100 mL water. The product was dried and crystallized from dioxane; as a white powder in 87% yield; m. p. 242°C - 244°C, IR (cm^{-1}, n): 3376 (br, NH), 1682 (CO); ^{1}H NMR: d 2.34 (s, SCH$_3$), 3.43 (br, NH), 7.34 (d, J = 7.5 Hz, aryl-H), 7.97 (d, J = 7.5 Hz, aryl-H), 8.19 (s, pyrimidine-H), 8.42 (s, CH); Its MS (m/z), 279 (M$^+$, 100%); C$_{12}$H$_{10}$ClN$_3$OS (279.7).

Synthesis of 6-{[4-(chlorophenyl)methylene]amino} -2-(methylsulfonyl)-pyrimidin-4(1H)-one (6). The solution of 3 (0.01 mol) in hydrogen peroxide solution (30 ml) (AcOH, H_2O_2; 2:1) was stirred at room temperature for 10 hs. (under TLC control). The solvent was evaporated under reduced pressure at 90°C, and the crude product collected, dried and crystallized from dimethyl formamide; as a white powder, in 76 % yield; m. p. 287°C - 289°C, IR (cm^{-1}, n): 3365 (br, NH), 1680 (CO);^{1}H NMR: d 2.23 (s, SCH$_3$), 3.50 (br, NH, D$_2$O exchangeable), 7.34 (d, J = 7.46 Hz, aryl-H), 7.97 (d, J = 7.48 Hz, aryl-H), 8.11 (s, pyrimidine-H), 8.38 (s, CH); Its MS (m/z), 311 (M^{+}, 65%); $C_{12}H_{10}ClN_3O_3S$ (311.7).

Preparation of the acetylated N-nucleosides of 6-[4- (aryl)methylene]amino-pyrimidin-4(1H)-one (4a,b) and (7a,b); General procedure: To a solution of 3 or 6 (0.01 mol) in aqueous potassium hydroxide (0.01 mol) in distilled water (5 ml) was added a solution of 1-bromo- 2,3,5-tri-O-acetyl-α-D-arabinofuranose/2,3,5-Tri-Obenzoyl-β-D-ribofuranosyl bromide (0.015 mol) in acetone (40 ml). The reaction mixture was stirred at room temperature for 18 - 24 h (under TLC control). The solvent was evaporated under reduced pressure at 40°C, and the crude product was collected and washed with distilled water to remove KBr formed. The product was dried, and crystallized from dimethyl formamide (50 mL).

N-(2',3',5'-tri-O-acetyl-β-D-arabinofuranosyl)-6-[4-   (chlorophenyl) methylene]amino-2-(methylthio)-pyrimidin-4(1H)-one (4a). It was obtained from 3 and 1-bromo- 2,3,5-tri-O-acetyl-α-D-arabinofuranose; as white powder, m. p. 223°C - 225°C; IR (cm^{-1}, v); 1710 (3CO), 1686 (CO);^{1}H.NMR: d 1.92, 1.98, 2.01 (3s, 3CH$_3$CO), 2.33 (s, SCH$_3$), 3.99 (m, H-4'), 4.12 (m, H-5', H-5"), 5.28 (m, H-3'), 5.35 (m, H-2'), 6.67 (d, J = 3.67 Hz, H-1'), 7.30 (d, J = 7.51 Hz, Ar-H), 7.97 (d, J = 7.50 Hz, Ar-H), 8.18 (s, pyrimidine-H), 8.45 (s, CH); ^{13}C. NMR: 21.43, 22.19, 22.24 (3CH$_3$), 28.32 (SCH$_3$), 60.59 (C-5'), 65.27 (C-3'), 66.87 (C-2'), 67.60 (C-4'), 86.60 (C-1'), 124.3-148.5 (10C-Ar), 166.8 (CO), 169.5, 170.9, 172.5 (3CO); Its MS (m/z), 537 (M^{+}, 38%); $C_{23}H_{24}ClN_3O_8S$ (537.9).

N-(2',3',5'-tri-O-benzoyl-β-D-ribofuranosyl)-6-[4-   (chlorophenyl) methylene]amino-2-(methylthio)-pyrimidin-4(1H)-one (4b). It was obtained from 3 and 2,3,5- Tri-O-benzoyl-β-D-ribofuranosyl bromide; as pale yellow powder, m. p. 289°C - 291°C; IR (cm^{-1}, v); 1720 (3CO), 1682 (CO);^{1}H.NMR: d 2.26 (SCH$_3$), 4.79 (m, H-4'), 4.62 (m, H-5', H-5"), 4.99 (d, J = 4.0 Hz, H-2'), 5.87 (t, J = 5.3 Hz, H-3'), 6.58 (d, J = 4.8 Hz, H-1'), 7.13 (m, Ar-H), 8.13 (m, 19H, Ar-H), 8.36 (s, pyrimidine-H), 8.72 (s, CH); ^{13}C. NMR: 21.16 (CH$_3$), 61.78 (C-5'), 65.69 (C-3'), 66.84 (C-2'), 67.38 (C-4'), 85.69 (C-1'), 121.157.6 (25C-Ar), 167.5 (CO), 170.3, 171.5, 173.6 (3CO); Its MS (m/z), 724 (M^{+},

19%); $C_{38}H_{30}N_3O_8S$ (724.1). N-(2',3',5'-tri-O-acetyl-β-D-arabinofuranosyl) sulfone-   6-[4-(chlorophenyl)-methylene]-amino-2-(methylsulfone)-pyrimidin-4(1H)-one (7a). It was obtained from 6 and 1-bromo-2,3,5-tri-O-acetyl-α-D-arabinofuranose; as yellow powder, m. p. 253°C - 255°C; IR (cm^{-1}, v); 1712 (3CO), 1678 (CO); ^{1}H.NMR: d 1.98, 2.02, 2.11 (3s, 3CH$_3$CO), 2.50 (s, SCH$_3$), 4.02 (m, H-4'), 4.14 (m, H-5', H-5''), 5.27 (m, H-3'), 5.33 (m, H-2'), 6.61 (d, J = 3.70 Hz, H-1'), 7.33 (d, J = 7.50 Hz, Ar-H), 7.88 (d, J = 7.50 Hz, Ar-H), 8.12 (s, pyrimidine-H), 8.38 (s, CH); ^{13}C. NMR: 20.39, 21.89, 22.74 (3CH$_3$), 30.17 (SCH$_3$), 61.31 (C-5'), 65.35 (C-3'), 66.79 (C-2'), 66.89 (C-4'), 86.56 (C-1'), 125.1 - 149.2 (10C-Ar), 167.3 (CO), 169.2, 170.4, 172.8 (3CO); Its MS (m/z), 569 (M$^+$, 61%); $C_{23}H_{24}ClN_3O_{10}S$ (569.9).

N-(2',3',5'-tri-O-benzoyl-β-D-ribofuranosyl)sulfone-6-[4-(chlorophenyl)-methylene]-amino-2-(methylsulfone)   pyrimidin-4(1H)-one (7b). It was obtained from 6 and 2,3,5-Tri-O-benzoyl-β-D-ribofuranosyl bromide; as yellow powder, m. p. 304°C - 306°C; IR (cm^{-1}, v); 1728 (3CO), 1672 (CO), 1335 (SO); ^{1}H.NMR: d 2.23 (s, CH$_3$), 3.89 (m, H-4'), 4.09 (m, H-5', H-5''), 5.19 (m, H-3'), 5.40 (m, H-2'), 6.78 (d, J = 3.65 Hz, H-1'), 7.13 - 8.03 (m, 19H, Ar-H), 8.42 (s, pyrimidine-H), 8.76 (s, CH); Its MS (m/z), 756 (M$^+$, 54%); $C_{38}H_{30}N_3O_{10}S$ (756.1).

Synthesis of diacetylated 2-{S-(β-D-glycofuranosyl)- 6-[4-(chlorophenyl) methylene]-aminopyrimidin- 4(1H)-one (5a,b) and (8a,b); General procedure: Acetylated compound 4a,b or 7a,b (1.0 mmol) was dissolved in methanolic ammonia (saturated with NH$_3$ at 0°C, 100 ml). The reaction mixture was stirred overnight and then heated the reaction mixture for 1 h at 120°C - 130°C. The mixture was then cooled and the solvent was evaporated to provide the crude nucleoside. The crude was purified by heating the crude in n-hexane (100 ml, three times) and crystallized from methanol (20 - 30 mL).

N-(β-D-arabinofuranosyl)-6-[4-(chlorophenyl)-   methylene] aminopyrimidin-4(1H)-one (5a). It was obtained from 4a, as white powder, m. p. 239°C - 241°C; IR (cm^{-1}, v); 3500 (brs, OH), 1676 (CO), 1330 (SO); ^{1}H. NMR: d 2.34 (s, SCH$_3$), 3.86 (m, H-5', H-5''), 4.09 (m, H-4'), 4.76 (t, H-2'), 5.08 (t, J = 5.41 Hz, J = 4.89 Hz, OH-C(5'), 5.24 (d, J = 4.48 Hz, OH-C(3'), 5.50 (d, J = 5.73 Hz, OH-C(2'), 5.72 (t, J = 9.76 Hz, H-3'), 6.91 (d, J = 5.66 Hz, H-1'), 7.28 (d, 2H, Ar-H), 8.07 (d, 2H, Ar-H), 8.35 (s, pyrimidine-H), 8.58 (s, CH); ^{13}C.NMR: 26.03 (SCH$_3$), 60.91 (C-5'), 65.29 (C-3'), 67.45 (C-2'), 69.19 (C-4'), 87.67 (C-1'), 121.6 - 147.7 (10C-Ar), 167.5 (CO); Its MS (m/z), 411 (M$^+$, 31%); $C_{17}H_{18}ClN_3O_5S$ (411.8).

N-(β-D-ribofuranosyl)-6-[4-(chlorophenyl)methylene]   aminopyrimidin-4(1H)-one (5b). It was obtained from 4b, as white powder, m. p. 221°C - 223°C; IR (cm^{-1}, v); 3486 (brs, OH), 1682 (CO); ^{1}H.NMR: d 2.38 (s, SCH$_3$),

3.84 (m, H-5', H-5''), 4.04 (m, H-4'), 4.75 (t, H-2'), 5.06 (t, J = 5.40 Hz, J = 4.86 Hz, OH-C(5'), 5.26 (d, J = 4.47 Hz, OH-C(3'), 5.52 (d, J = 5.71 Hz, OH-C(2'), 5.72 (t, J = 9.74 Hz, H-3'), 6.90 (d, J = 5.67 Hz, H-1'), 7.30 (d, 2H, Ar-H), 8.08 (d, 2H, Ar-H), 8.38 (s, pyrimidine-H), 8.64 (s, CH); ^{13}C.NMR: 26.35 (SCH$_3$), 60.78 (C-5'), 65.31 (C-3'), 67.54 (C-2'), 70.11 (C-4'), 88.23 (C-1'), 121.9 - 148.5 (10C-Ar), 167.8 (CO); Its MS (m/z), 411 (M$^+$, 27%); C$_{17}$H$_{18}$ClN$_3$O$_5$S (411.8).

N-(β-D-arabinofuranosyl)sulfone-6-[4-(chlorophenyl)   methylene] aminopyrimidin-4(1H)-one (8a). It was obtained from 7a, as white powder, m. p. 271°C - 273°C; IR (cm^{-1}, v); 3470 (brs, OH), 1680 (CO), 1320 (SO); ^{1}H. NMR: d 2.45 (s, SCH$_3$), 3.90 (m, H-5', H-5''), 4.05 (m, H-4'), 4.82 (t, H-2'), 5.11 (t, J = 5.36 Hz, J = 4.91 Hz, OH-C(5'), 5.21 (d, J = 4.50 Hz, OH-C(3'), 5.46 (d, J = 5.67 Hz, OH-C(2'), 5.68 (t, J = 9.73 Hz, H-3'), 6.79 (d, J = 5.66 Hz, H-1'), 7.23 (d, Ar-H), 8.00 (d, Ar-H), 8.20 (s, pyrimidine-H), 8.65 (s, CH); ^{13}C. NMR: 27.01 (SCH$_3$), 61.21 (C-5'), 65.32 (C-3'), 67.54 (C-2'), 69.22 (C-4'), 86.97 (C-1'), 122.3 - 147.9 (10C-Ar), 166.8 (CO); Its MS (m/z), 443 (M$^+$, 34%); C$_{17}$H$_{18}$ClN$_3$O$_7$S (443.8).

N-(β-D-ribofuranosyl)sulfone-6-[4-(chlorophenyl)   methylene] aminopyrimidin-4(1H)-one (8b). It was obtained from 7b, as white powder, m. p. 263°C - 265°C; IR (cm^{-1}, v); 3485 (brs, OH), 1680 (CO), 1315 (SO); ^{1}H. NMR: d 2.42 (s, SCH$_3$), 3.73 (m, H-5', H-5''), 4.08 (m, H-4'), 4.82 (t, H-2'), 5.15 (t, J = 5.38 Hz, J = 4.87 Hz, OH-C(5'), 5.23 (d, J = 5.46 Hz, OH-C(3'), 5.48 (d, J = 3.67 Hz, OH-C(2'), 5.70 (t, J = 9.69 Hz, H-3'), 6.81 (d, J = 3.59 Hz, H-1'), 7.25 (d, Ar-H), 8.06 (d, Ar-H), 8.42 (s, pyrimidine-H), 8.68 (s, CH); ^{13}C. NMR: 27.16 (SCH$_3$), 61.32 (C-5'), 65.41 (C-3'), 67.69 (C-2'), 69.34 (C-4'), 86.94 (C-1'), 122.6-149.0 (10C-Ar), 167.4 (CO); Its MS (m/z), 443 (M$^+$, 41%); C$_{17}$H$_{18}$ClN$_3$O$_7$S (443.8).

Synthesis   of   2-alkylamino-6-[4-(chlorophenyl)methylene] aminopyrimidin-4(1H)-one (9a-c). General procedure: A mixture of 3 (10 m mol) fused with morphline/ methylpiprazine/and or piprazine (15 m mol) in sand bath at 180°C for 3 h. The reaction mixture was allowed to cool to room temp., and then add ethanol (20 mL), the precipitate was filtered off, dried and crystallized from an appropriate solvent to produce 9a-c.

Synthesis of 6-[4-(chlorophenyl)methylene]amino-2- morpholin-4-ylpyrimidin-4(3H)-one (9a). It was obtained from morpholine (15 m mol) as a apale yellow powder crystallized from dioxane in 68% yield; m. p. 254°C - 256°C, IR (cm^{-1}, n): 1678 (CO); ^{1}H NMR: d 3.23 (t, 2NCH$_2$, J = 4.87 Hz), 3.56 (t, 2OCH$_2$, J = 4.95 Hz), 7.31 (d, J = 7.46 Hz, Ar-H), 7.86 (d, J = 7.48 Hz, Ar-H), 8.07 (s, pyrimidine-H), 8.58 (s, CH); C^{13} NMR: 167.4 (CO), 122.6 - 156.8

(10C Ar), 66.54, 47.09 (4C, O(CH$_2$)$_2$, N(CH$_2$)$_2$); Its MS (m/z), 318 (M$^+$, 69%); C$_{15}$H$_{15}$ClN$_4$O$_2$ (318.7).

Synthesis of 6-[4-(chlorophenyl)methylene]amino-2- piprazin-1-ylpyrimidin-4(3H)-one (9b). It was obtained from piprazine (15 m mol) as a yellow powder crystallized from DMF in 72% yield; m. p. 232°C - 234°C, IR (cm^{-1}, n): 3390 (br, NH), 1669 (CO); 1H NMR: d 2.54 (br, 2NCH2), 3.37 (br, 2NCH$_2$), 7.26 (d, J = 7.48 Hz, Ar-H), 7.89 (d, J = 7.50 Hz, Ar-H), 8.12 (s, pyrimidineH), 8.67 (s, CH); 9.34 (br, NH); C13 NMR: 164.9 (CO), 161.4 (C-2, pyrimidine), 123.2 - 157.1 (10C Ar)), 56.47, 46.18 (4C, N(CH$_2$)$_2$, N(CH$_2$)$_2$); Its MS (m/z), 317 (M$^+$, 29%); C$_{15}$H$_{16}$ClN$_5$O (317.7).

Synthesis of 6-[4-(chlorophenyl)methylene]amino-2- (4-methylpiprazine-1-yl) pyrimidin-4(3H)-one (9c). It was obtained from methylpiprazine (15 m mol) as a pale yellow powder crystallized from DMF in 65% yield; m. p. 273°C - 275°C, IR (cm^{-1}, n): 1667 (CO); ^{1}H NMR: d 2.32 (s, NCH$_3$), 2.50 (br, 2NCH$_2$), 3.29 (br, 2NCH$_2$), 7.28 (d, J = 7.50 Hz, Ar-H), 7.92 (d, J = 7.50 Hz, Ar-H), 8.17 (s, pyrimidine-H), 8.73 (s, CH); Its MS (m/z), 331 (M$^+$, 38%); C$_{16}$H$_{18}$ClN$_5$O (331.3).

Preparation of the acetylated N-nucleosides of 6- [4-(chlorophenyl)methylene]amino-2-(4-alkylamino)-pyrimidin-4(3H)-one (10a-f); General procedure: To a solution of each of 9a-c (0.01 mol) in aqueous potassium hydroxide (0.01 mol) in distilled water (5 ml) was added a solution of 1-bromo-2,3,5-tri-O-acetyl-α-D-arabinofuranose/2,3,5-Tri-O-benzoyl-β-D-ribofuranosyl bromide (0.015 mol) in acetone (40 ml). The reaction mixture was stirred at room temperature for 24 h (under TLC control). The solvent was evaporated under reduced pressure at 40°C, and the crude product was washed with distilled water to remove KBr formed, the product was collected, dried, and crystallized from ethanol (50 - 80 mL).

N-(2',3',5'-tri-O-acetyl-β-D-arabinofuranosyl)-6-[4- (chlorophenyl)methylene]amino-2-(4-morpholin-4-yl) pyrimidin-4(3H)-one (10a). It was obtained from 9a and 2,3,5-tri-O-acetyl-α-D-arabinofuranosyl)-bromide; as white powder, m. p. 187°C - 189°C; IR (cm^{-1}, v); 1734 (3CO), 1683 (CO); ^{1}H. NMR: d 1.93, 1.99, 2.11 (3s, 3CH$_3$CO), 2.54 (m, N(CH$_2$)$_2$), 3.61 (m, O(CH$_2$)$_2$), 4.13 (m, H-4'), 4.18 (m, H-5', H-5"), 5.31 (m, H-3'), 5.39 (m, H-2'), 6.81 (d, J = 3.72 Hz, H-1'), 7.23 (d, Ar-H), 7.79 (d, Ar-H), 8.15 (s, pyrimidine-H), 8.83 (s, CH); ^{13}C.NMR: 22.20, 22.31, 22.57 (3CH$_3$), 45.76 (2C, N(CH$_2$)$_2$), 60.98 (C-5'), 63.28 (2C, O(CH$_2$)$_2$), 66.30 (C-3'), 66.87 (C-2'), 67.41 (C-4'), 86.03 (C-1'), 124.8-148.3 (10C-Ar), 166.8 (CO), 170.1, 171.3, 172.5 (3CO); Its MS (m/z), 576 (M$^+$, 21%); C$_{26}$H$_{29}$ClN$_4$O$_9$ (576.9).

N-(2',3',5'-tri-O-benzoyl-β-D-ribofuranosyl)-6-[4- (chlorophenyl) methylene]amino-2-(4-morpholin-4-yl) pyrimidin-4(3H)-one (10b). It was obtained from 9a and 2,3,5-Tri-O-benzoyl-β-D-ribofuranosyl bromide as pale yellow powder, m. p. 289°C - 291°C; IR (cm^{-1}, v); 1737 (3CO), 1680 (CO); ^{1}H. NMR: d 2.53 (m, N(CH$_2$)$_2$), 3.51 (m, O(CH$_2$)$_2$), 4.07 (m, H-4'), 4.19 (m, H-5', H-5"), 5.23 (m, H-3'), 5.37 (m, H-2'), 6.72 (d, J = 3.65 Hz, H-1'), 7.12 - 8.08 (m, 19H, Ar-H), 8.35 (s, pyrimidine-H), 8.74 (s, CH); ^{13}C. NMR: 43.56 (2C, HN(CH$_2$)$_2$), 45.11 (2C, N(CH$_2$)$_2$), 61.39 (C-5'), 65.04 (2C, O(CH$_2$)$_2$), 66.14 (C-3'), 67.82 (C-2'), 69.30 (C-4'), 85.31 (C-1'), 123.1 - 157.6 (28C-Ar), 166.5 (CO), 170.5, 172.3, 173.8 (3CO); Its MS (m/z), 763 (M$^+$, 19%); C$_{41}$H$_{35}$ClN$_4$O$_9$(763.1).

N-(2',3',5'-tri-O-acetyl-β-D-arabinofuranosyl)-6-[4-   (chlorophenyl) methylene]amino-2-(4-piprazin-1-yl) pyrimidin-4(3H)-one (10c). It was obtained from 9b and 2,3,5-tri-O-acetyl-α-D-arabinofuranosyl)-bromide; as white powder, m. p. 212°C - 214°C; IR (cm^{-1}, v); 3355 (NH), 1728 (3CO), 1672 (CO); ^{1}H.NMR: d 1.98, 2.00, 2.09 (3s, 3CH$_3$CO), 2.45 (m, N(CH$_2$)$_2$), 3.03 (m, HN(CH$_2$)$_2$), 4.08 (m, H-4'), 4.15 (m, H-5', H-5"), 5.31 (m, H-3'), 5.38 (m, H-2'), 6.73 (d, J = 3.67 Hz, H-1'), 7.24 (d, Ar-H), 7.81 (d, Ar-H), 8.25 (s, pyrimidine-H), 8.82 (s, CH), 10.12 (br, NH D$_2$O exchangeable); ^{13}C. NMR: 22.17, 22.25, 22.63 (3CH$_3$), 43.74 (2C, HN(CH$_2$)$_2$), 45.87 (2C, N(CH$_2$)$_2$), 61.55 (C-5'), 66.31 (C-3'), 66.81 (C-2'), 67.35 (C-4'), 85.73 (C-1'), 124.3 - 156.5 (10C-Ar), 166.8 (CO), 169.4, 170.6, 172.9 (3CO); Its MS (m/z), 575 (M$^+$, 28%); C$_{26}$H$_{30}$ClN$_5$O$_8$(575.9).

N-(2',3',5'-tri-O-benzoyl-β-D-ribofuranosyl)-6-[4-   (chlorophenyl) methylene]amino-2-(4-piprazin-1-yl) pyrimidin-4(3H)-one (10d). It was obtained from 9b and 2,3,5-Tri-O-benzoyl-β-D-ribofuranosyl bromide; as yellow powder, m. p. 261°C - 263°C; IR (cm^{-1}, v); 1721 (3CO), 1681 (CO); ^{1}H. NMR: d 2.34 (m, N(CH$_2$)$_2$), 3.21 (m, N(CH$_2$)$_2$), 4.08 (m, H-4'), 4.19 (m, H-5', H-5"), 5.28 (m, H-3'), 5.37 (m, H-2'), 6.78 (d, J = 3.63 Hz, H-1'), 7.11 - 8.02 (m, 19H, Ar-H), 8.40 (s, pyrimidine-H), 8.69 (s, CH); Its MS (m/z), 762 (M$^+$, 37%); C$_{41}$H$_{36}$ClN$_5$O$_8$ (762.2).

N-(2',3',5'-tri-O-acetyl-β-D-arabinofuranosyl)-6-[4-   (chlorophenyl) methylene]amino-2-(4-methylpiprazin-1- yl)pyrimidin-4(3H)-one (10e). It was obtained from 9c and 2,3,5-tri-O-acetyl-α-D-arabinofuranosyl)-bromide; as yellow powder, m. p. 209°C - 211°C; IR (cm^{-1}, v); 1728 (3CO), 1672 (CO); ^{1}H.NMR: d 1.93, 1.99, 2.11 (3s, 3CH$_3$CO), 2.32 (s, CH$_3$), 2.45 (m, N(CH$_2$)$_2$), 3.03 (m, HN(CH$_2$)$_2$), 4.13 (m, H-4'), 4.18 (m, H-5', H-5"), 5.31 (m, H-3'), 5.39 (m, H-2'), 6.81 (d, J = 3.72 Hz, H-1'), 7.27 (d, Ar-H), 8.00 (d, Ar-H), 8.43 (s, pyrimidine-H), 8.89 (s, CH); ^{13}C.NMR: 22.21, 22.30, 22.57 (3CH$_3$), 30.05 (NCH$_3$), 43.76 (2C, N(CH$_2$)$_2$), 45.93 (2C, N(CH$_2$)$_2$), 60.98 (C-5'), 66.30 (C-3'), 66.87 (C-2'), 67.41 (C-4'), 86.03 (C-1'), 119.8-148.3 (10C-

Ar), 166.8 (CO), 170.1, 171.3, 172.9 (3CO); Its MS (m/z), 590 (M$^+$, 21%); $C_{27}H_{32}ClN_5O_8S$ (590.9).

N-(2',3',5'-tri-O-benzoyl-β-D-ribofuranosyl)-6-[4-   (chlorophenyl) methylene]amino-2-(4-methylpiprazin-1-yl)pyrimidin-4(3H)-one   (10f).   It was obtained from 9c and 2,3,5-Tri-O-benzoyl-β-D-ribofuranosyl bromide as yellow powder, m. p. 281°C - 283°C; IR (cm^{-1}, ν); 1719 (3CO), 1668 (CO); 2.28 (s, NCH$_3$), 2.49 (m, N(CH$_2$)$_2$), 3.26 (m, N(CH$_2$)$_2$), 4.06 (m, H-4'), 4.13 (m, H-5', H-5''), 5.30 (m, H-3'), 5.33 (m, H-2'), 6.73 (d, J = 3.66 Hz, H-1'), 7.16-8.11 (m, 19H, Ar-H), 8.35 (s, pyrimidine-H), 8.74 (s, CH); ^{13}C. NMR: 31.10 (NCH$_3$), 43.56 (2C, HN(CH$_2$)$_2$), 45.11 (2C, N(CH$_2$)$_2$), 60.69 (C-5'), 66.21 (C-3'), 66.76 (C-2'), 67.42 (C-4'), 85.58 (C-1'), 123.1 - 157.3 (28C-Ar), 165.9 (CO), 170.3, 171.3, 173.8 (3CO); Its MS (m/z), 776 (M$^+$, 19%); $C_{42}H_{38}ClN_5O_8$(776.2).

Synthesis of diacetylated N-(-β-D-glycofuranosyl)-6- [4-(chlorophenyl) methylene]-amino-2-(4-alkylamino)pyrimidin-4(3H)-one (11a-f); (see 3.1.4, General procedures), and crystallized from ethanol (80 mL).

N-(β-D-arabinofuranosyl)-6-[4-(chlorophenyl)methylene]   amino-2- (4-morpholin-4-yl)-pyrimidin-4(3H)-one (11a). It was obtained from 10a; as yellow powder, m. p. 239°C - 241°C; IR (cm^{-1}, ν);3450 (br, OH), 1686 (CO); ^{1}H.NMR: d 2.39 (m, N(CH$_2$)$_2$), 3.56 (m, O(CH$_2$)$_2$), 3.88 (m, H-5', H-5''), 4.02 (m, H-4'), 4.78 (t, H-2'), 5.09 (t, J = 5.33 Hz, J = 4.87 Hz, OH-C(5'), 5.19 (d, J = 4.47 Hz, OH-C(3'), 5.39 (d, J = 5.63 Hz, OH-C(2'), 5.70 (t, J = 9.67 Hz, H-3'), 6.81 (d, J = 5.68 Hz, H-1'), 7.30 (d, Ar-H), 8.00 (d, Ar-H), 8.36 (s, pyrimidine-H), 8.92 (s, CH); ^{13}C. NMR: 43.71 (2C, HN(CH$_2$)$_2$), 61.52 (C-5'), 63.28 (2C, O(CH$_2$)$_2$), 66.23 (C-3'), 66.78 (C-2'), 67.37 (C-4'), 85.73 (C-1'), 124.6 - 155.8 (10C-Ar), 167.8 (CO); Its MS (m/z), 450 (M$^+$, 28%); $C_{20}H_{23}ClN_4O_6$(450.8).

N-(β-D-ribofuranosyl)-6-[4-(chlorophenyl)methylene]   amino-2-(4- morpholin-4-yl)pyri-midin-4(3H)-one (11b). It was obtained from 10b; as a yellow powder, m. p. 223°C - 225°C; IR (cm^{-1}, ν); 3435 (OH), 1676 (CO); ^{1}H. NMR: d 2.32 (m, N(CH$_2$)$_2$), 3.52 (m, O(CH$_2$)$_2$), 3.90 (m, H-5', H-5''), 4.05 (m, H-4'), 4.81 (t, H-2'), 5.12 (t, J = 5.35 Hz, J = 4.88 Hz, OH-C(5'), 5.23 (d, J = 4.52 Hz, OH-C(3'), 5.42 (d, J = 5.70 Hz, OH-C(2'), 5.70 (t, J = 9.72 Hz, H-3'), 6.86 (d, J = 5.70 Hz, H-1'), 7.28 (d, ArH), 7.97 (d, Ar-H), 8.32 (s, pyrimidine-H), 8.86 (s, CH; Its MS (m/z), 450 (M$^+$, 32%); $C_{20}H_{23}ClN_4O_6$ (450.8).

N-(β-D-arabinofuranosyl)-6-[4-(chlorophenyl)methyl-ene]amino-2-(4-piprazin-1-yl)-pyrimidin-4(3H)-one (11c). It was obtained from 10c; as yellow powder, m. p. 268°C - 270°C; IR (cm^{-1}, ν); 3425 (brs, OH), 3328 (br, NH) 1672 (CO); ^{1}H.NMR: d 2.48 (m, N(CH$_2$)$_2$), 3.19 (m, O(CH$_2$)$_2$), 3.81 (m, H-5', H-5''),

3.98 (m, H-4'), 4.69 (t, H-2'), 5.09 (t, J = 5.41 Hz, J = 4.88 Hz, OH-C(5'), 5.26 (d, J = 4.47 Hz, OH-C(3'), 5.47 (d, J = 5.61 Hz, OHC(2'), 5.70 (t, J = 9.78 Hz, H-3'), 6.79 (d, J = 5.60 Hz, H-1'), 7.30 (d, Ar-H), 8.00 (d, Ar-H), 8.41 (s, pyrimidineH), 8.67 (s, CH), 9.46 (brs, NH); ^{13}C.NMR: 42.83 (2C, HN(CH$_2$)$_2$), 46.39 (2C, HN(CH$_2$)$_2$), 60.88 (C-5'), 64.81 (C-3'), 67.23 (C-2'), 70.16 (C-4'), 87.09 (C-1'), 125.8 - 156.6 (10C-Ar), 165.9 (CO); Its MS (m/z), 449 (M$^+$, 36%); C$_{20}$H$_{24}$ClN$_5$O$_5$ (449.8).

N-(β-D-ribofuranosyl)-6-[4-(chlorophenyl)methylene]   amino-2-(4-piprazin-1-yl)pyrim-idin-4(3H)-one (11d). It was obtained from 10d; as yellow powder, m. p. 252°C - 254°C; IR (cm^{-1}, v); 3480 (brs, OH), 3290 (br, NH), 1668 (CO); ^{1}H.NMR: d 2.53 (m, N(CH$_2$)$_2$), 3.21 (m, O(CH$_2$)$_2$), 3.89 (m, H-5', H-5''), 4.03 (m, H-4'), 4.73 (t, H-2'), 5.12 (t, J = 5.49 Hz, J = 4.90 Hz, OH-C(5'), 5.29 (d, J = 4.51 Hz, OH-C(3'), 5.51 (d, J = 5.59 Hz, OHC(2'), 5.83 (t, J = 9.80 Hz, H-3'), 6.81 (d, J = 5.56 Hz, H-1'), 7.29 (d, Ar-H), 8.02 (d, Ar-H), 8.38 (s, pyrimidineH), 8.80 (s, CH), 9.60 (brs, NH); Its MS (m/z), 449 (M$^+$, 29%); C$_{20}$H$_{24}$ClN$_5$O$_5$ (449.8).

N-(β-D-arabinofuranosyl)-6-[4-(chlorophenyl)methylene]amino-2-(4-methylpiprazin-1-yl)pyrimidin-4(3H)- one (11e). It was obtained from 10e; as yellow powder, m. p. 239°C - 241°C; IR (cm^{-1}, v); 3456 (brs, OH), 1683 (CO); ^{1}H.NMR: d 2.29 (s, CH$_3$), 2.53 (m, N(CH$_2$)$_2$), 3.07 (m, HN(CH$_2$)$_2$), 3.91 (m, H-5', H-5''), 4.05 (m, H-4'), 4.80 (t, H-2'), 5.13 (t, J = 5.40 Hz, J = 4.85 Hz, OHC(5'), 5.25 (d, J = 4.48 Hz, OH-C(3'), 5.50 (d, J = 5.65 Hz, OH-C(2'), 5.73 (t, J = 9.72 Hz, H-3'), 6.83 (d, J = 5.61 Hz, H-1'), 7.21 (d, Ar-H), 7.96 (d, Ar-H), 8.35 (s, pyrimidine-H), 8.70 (s, CH); ^{13}C.NMR: 30.16 (NCH$_3$), 43.51 (2C, N(CH$_2$)$_2$), 45.87 (2C, N(CH$_2$)$_2$), 61.08 (C-5'), 66.51 (C-3'), 66.90 (C-2'), 67.47 (C-4'), 86.12 (C-1'), 126.7 - 154.9 (10C-Ar), 166.4 (CO); Its MS (m/z), 463 (M$^+$, 27%); C$_{21}$H$_{26}$ClN$_5$O$_5$ (463.9).

N-(β-D-ribofuranosyl)-6-[4-(chlorophenyl)methylene]   amino-2-(4-methylpiprazin-1-yl)pyrimidin-4(3H)-one (11f). It was obtained from 10f as yellow powder, m. p. 247°C - 249°C; IR (cm^{-1}, v); 3428 (brs, OH), 1689 (CO); ^{1}H.NMR: d 2.26 (s, CH$_3$), 2.52 (m, N(CH$_2$)$_2$), 3.05 (m, HN(CH$_2$)$_2$), 3.40 (m, H-5', H-5''), 4.12 (m, H-4'), 4.85 (t, H-2'), 5.18 (t, J = 5.47 Hz, J = 4.90 Hz, OH-C(5'), 5.31 (d, J = 4.51 Hz, OH-C(3'), 5.62 (d, J = 5.67 Hz, OHC(2'), 5.76 (t, J = 9.77 Hz, H-3'), 6.81 (d, J = 5.64 Hz, H-1'), 7.24 (d, Ar-H), 7.99 (d, Ar-H), 8.39 (s, pyrimidineH), 8.81 (s, CH); Its MS (m/z), 463 (M$^+$, 31%); C$_{21}$H$_{26}$ClN$_5$O$_5$ (463.9).

Synthesis of 4-[(4-aryl)methylene]amino-6-chloropyrimidine-2-thiol (13). A solution of 2a (0.01 mol) in dry dioxane (40 mL) was treated with 10 mL of phosphorus oxychloride, and the mixture was stirred under reflux for 5 h. The reaction mixture was allowed to cool to 25°C, and poured into cold water

(100 mL), a solid was separated, filtered off, and crystallized from the benzene (50 mL), as yellow powder, m. p. 178°C - 179°C ; IR (cm⁻¹, v); 3245 (br, NH); 1H.NMR: d 7.20 (d, J = 7.50 Hz, Ar-H), 7.99 (d, J = 7.50 Hz, Ar-H), 8.39 (s, pyrimidine-H), 8.81 (s, CH); Its MS (m/z), 284 (M+, 31%), 285 (M⁺ +1, 21%); C11H7Cl2N3S (284.1).

Synthesis of 4-[(4-chlorophenyl)methylene]amino- 6-(4-alkylamino-yl) pyrimidine-2-thiol (14a-c). General procedure: A mixture of each of compound 13 (10 m mol) fused with morphline/methylpiprazine/ and or piprazine (15 m mol) in sand bath at 180°C for 2 h. The reaction mixture was allowed to cool to room temp., and then add ethanol (20 mL), the precipitate was filtered off, dried and crystallized from dimethyl formamide to produce 14a-c.

4-[(4-chlorophenyl)methylene]amino-6-morpholin-4-ylpyrimidine-2-thiol (14a). It was obtained from morpholine (15 mmol) as a brown powder crystallized from dioxane in 71% yield; m. p. 286°C - 288°C, IR (cm⁻¹, n): 3410 (br, NH); 1H NMR: d 3.26 (t, 2NCH₂, J = 4.87 Hz), 3.51 (t, 2OCH2, J = 4.95 Hz), 7.25 (d, J = 7.50 Hz, ArH), 7.96 (d, J = 7.48 Hz, Ar-H), 8.34 (s, pyrimidine-H), 8.63 (s, CH), 9.02 (br, NH); 13C NMR: 168.3 (C=S), 125.3 - 156.6 (10C Ar), 64.74 (4C, O(CH2)2), 47.09 (N(CH2)2); Its MS (m/z), 334 (M⁺, 69%); C₁₅H₁₅ClN₄OS (334.8).

4-[4-(chlorophenyl)methylene]amino-6-piprazin-1-   ylpyrimidine-2-thiol (14b). It was obtained from piprazine (15 mmol) as a yellow powder crystallized from DMF in 68 % yield; m. p. 261°C - 263°C, IR (cm⁻¹, n): 3390 (br, NH); ¹H NMR: d 2.53 (brs, 2NCH₂), 3.36 (brs, 2NCH₂), 7.18 (d, J = 7.47 Hz, Ar-H),7.66 (d, J = 7.50 Hz, Ar-H), 8.29 (s, pyrimidine-H), 8.71 (s, CH), 9.16 (br, NH), 9.65 (br, NH); Its MS (m/z), 333 (M⁺, 29%); C₁₅H₁₆ClN₅S (333.8).

4-[4-(chlorophenyl)methylene]amino-6-(4-methylpiprazine-1-yl)pyrimi-dine-2-thiol (14c). It was obtained from 4-methylpiprazine (15 mmol) as a pale yellow powder crystallized from DMF in 71% yield; m. p. 281°C - 283°C, IR (cm⁻¹, n): 3345 (br, NH); ¹H NMR: d 2.24 (s, NCH₃), 2.48 (m, 2NCH₂), 3.29 (m, 2NCH₂), 7.21 (d, J = 7.47 Hz, Ar-H), 7.75 (d, J = 7.50 Hz, Ar-H), 8.33 (s, pyrimidine-H), 8.87 (s, CH), 9.20 (brs, NH); Its MS (m/z), 347 (M⁺, 38%); C₁₆H₁₈ClN₅S (347.8).

Preparation of the acetylated S-nucleosides of 4- [4-(chlorophenyl) methylene]amino-6-(4-alkylamino)- pyrimidinee (15a-f). (See 3.1.7; General procedure), crystallized from ethanol (40 - 80 mL).

2-(S-2',3',5'-tri-O-acetyl-β-D-arabinofuranosyl)-4-[4-   (chlorophenyl) methylene]-amino-6-morpholin-4-ylpyrimidine (15a). It was obtained from 14a and 2,3,5-tri-Oacetyl-α-D-arabinofuranosyl)-bromide; as white powder,

m. p. 268°C - 270°C; IR (cm^{-1}, ν); 1741 (3CO);^{1}H.NMR: d 1.93, 1.99, 2.02 (3s, 3CH$_3$CO), 3.19 (m, N(CH$_2$)$_2$), 3.53 (m, O(CH$_2$)$_2$), 4.03 (m, H-4'), 4.18 (m, H-5', H-5''), 5.29 (m, H-3'), 5.40 (m, H-2'), 6.81 (d, J = 3.70 Hz, H-1'), 7.28 (d, Ar-H), 7.98 (d, Ar-H), 8.56 (s, pyrimidine-H), 8.95 (s, CH); ^{13}C. NMR: 22.19, 22.24, 22.61 (3CH$_3$), 43.71 (2C, HN(CH$_2$)$_2$), 61.52 (C-5'), 65.93 (2C, O(CH$_2$)$_2$), 66.23 (C-3'), 66.78 (C-2'), 67.37 (C-4'), 85.73 (C-1'), 121.3 - 147.5 (11C-Ar), 168.9, 170.8, 172.5 (3CO); Its MS (m/z), 593 (M$^+$, 37%); C$_{26}$H$_{29}$ClN$_4$O$_8$S (593.0).

2-(S-2',3',5'-tri-O-benzoyl-β-D-ribofuranosyl)-4-[4-   (chlorophenyl) methylene]amino-6-morpholin-4-ylpyrimidine (15b). It was obtained from 14a and 2,3,5-Tri-Obenzoyl-β-D-ribofuranosyl bromide; as pale yellow powder, m. p. 278°C - 280°C; IR (cm^{-1}, ν); 1737 (3CO); ^{1}H.NMR: d 2.39 (m, N(CH$_2$)$_2$), 3.38 (m, O(CH$_2$)$_2$), 4.11 (m, H-4'), 4.18 (m, H-5', H-5''), 5.30 (m, H-3'), 5.37 (m, H-2'), 6.83 (d, J = 3.67 Hz, H-1'),7.09 - 8.00 (m, 19H, Ar-H), 8.29(s, pyrimidine-H), 8.76(s, CH);^{13}C.NMR: 43.56 (2C, N (CH$_2$)$_2$), 61.40 (C-5'), 65.11 (2C, O(CH$_2$)$_2$), 66.23(C-3'), 68.84 (C-2'), 70.19 (C-4'), 84.78 (C-1'), 121.1 - 155.6 (29C-Ar),(CO), 169.9, 170.7, 173.4 (3CO); Its MS (m/z), 779(M$^+$, 23%);C$_{41}$H$_{35}$ClN$_4$O$_8$S (779.2).

2-(S-2',3',5'-tri-O-acetyl-β-D-arabinofuranosyl)-6-[4-   (chlorophenyl) methylene]-amino-6-piprazin-1-ylpyrimidine (15c). It was obtained from 14b and 2,3,5-tri-Oacetyl-α-D-arabinofuranosyl)-bromide; as yellow powder, m. p. 259°C - 261°C; IR (cm^{-1}, ν); 1728 (3CO); ^{1}H.NMR: d 1.96, 1.99, 2.08 (3s, 3CH$_3$CO), 2.57 (m, N(CH$_2$)$_2$), 3.21 (m, N(CH$_2$)$_2$), 4.11 (m, H-4'), 4.17 (m, H-5', H-5''), 5.28 (m, H-3'), 5.37 (m, H-2'), 6.80 (d, J = 3.73 Hz, H-1'), 7.31 (d, Ar-H), 8.02 (d, Ar-H), 8.30 (s, pyrimidine-H), 8.78 (s, CH); ^{13}C.NMR: 22.21, 22.30, 22.57 (3CH$_3$), 44.71 (2C, N(CH$_2$)$_2$), 46.28 (2C, N(CH$_2$)$_2$), 60.87 (C-5'), 66.26 (C-3'), 66.90(C-2'), 67.36 (C-4'), 86.00 (C-1'), 126.8 - 148.3 (11C-Ar),170.1, 171.3, 172.9 (3CO);Its MS (m/z),592(M$^+$,21%); C$_{26}$H$_{30}$ClN$_5$O$_7$S (592.0).

2-(S-2',3',5'-tri-O-benzoyl-β-D-ribofuranosyl)-4-[4-   (chlorophenyl) methylene]amino-6-piprazin-1-ylpyrimidine (15d). It was obtained from 14b and 2,3,5-Tri-Obenzoyl-β-D-ribofuranosyl bromide; as yellow powder, m. p. 263°C - 265°C; IR (cm^{-1}, ν); 1728 (3CO); ^{1}H.NMR: d 2.36 (m, N(CH$_2$)$_2$), 3.21 (m, N(CH$_2$)$_2$), 4.09 (m, H-4'), 4.15 (m, H-5', H-5''), 5.28 (m, H-3'), 5.35 (m, H-2'), 6.78 (d, J = 3.70 Hz, H-1'), 7.07-8.02 (m, 19H, Ar-H), 8.29 (s, pyrimidine-H), 8.76 (s, CH); Its MS (m/z), 778 (M$^+$, 31%); C$_{41}$H$_{36}$ClN$_5$O$_7$S (778.2).

2-(S-2',3',5'-tri-O-acetyl-β-D-arabinofuranosyl)-4-[4-   (chlorophenyl) methylene]-amino-6-(4-methylpiprazin-1-yl)pyrimidine (15e). It was obtained from 14c and 2,3,5- tri-O-acetyl-α-D-arabinofuranosyl)-bromide; as yellow powder, m. p. 255°C - 257°C; IR (cm⁻¹, v); 1721 (3CO);¹H.NMR: d 1.93, 1.99, 2.11 (3s, 3CH₃CO), 2.28 (s, NCH₃), 2.54 (m, N(CH₂)₂), 3.26 (m, 4H, N(CH₂)₂), 4.13 (m, H-4'), 4.18 (m, H-5', H-5"), 5.31 (m, H-3'), 5.39 (m, H-2'), 6.81 (d, J = 3.72 Hz, H-1'), 7.27 (d, 2H, Ar-H), 7.95 (d, Ar-H), 8.34 (s, pyrimidine-H), 8.69 (s, CH); ¹³C.NMR: 22.21, 22.30, 22.57 (3CH₃), 30.12 (N-CH₃), 45.76 (2C, N(CH₂)₂), 46.28 (2C, N(CH₂)₂), 60.98 (C- 5'), 66.30 (C-3'), 66.87 (C-2'), 67.41 (C-4'), 86.03 (C-1'), 123.4 - 151.5 (11C-Ar), 170.1, 171.3, 172.9 (3CO); Its MS (m/z), 606 (M⁺, 31%); C₂₇H₃₂ClN₅O₇S (606.0).

2-(S-2',3',5'-tri-O-benzoyl-β-D-ribofuranosyl)-4-[4-   (chlorophenyl) methylene]amino-6-(4-methylpiprazin-1-yl)pyrimidine (15f). It was obtained from 14c and 2,3,5- Tri-O-benzoyl-β-D-ribofuranosyl bromide as yellow powder, m. p. 291°C - 293°C; IR (cm⁻¹, v); 1728 (3CO);¹H.NMR: d 2.23 (s, CH₃), 2.45 (m, N(CH₂)₂), 3.46 (m, N(CH₂)₂), 4.11 (m, H-4'), 4.21 (m, H-5', H-5"), 5.27 (m, H-3'), 5.34 (m, H-2'), 6.89 (d, J = 3.62 Hz, H-1'), 7.12 - 8.08 (m, 19H, Ar-H), 8.35 (s, pyrimidine-H), 8.79 (s, CH); ¹³C.NMR: 29.90 (NCH₃), 44.90 (2C, N(CH₂)₂), 47.43 (2C, N(CH₂)₂), 60.76 (C-5'), 66.23 (C-3'), 66.76 (C-2'), 68.34 (C-4'), 87.01 (C-1'), 122.5 - 154.3 (29CAr), 170.2, 172.1, 173.7 (3CO); Its MS (m/z), 792 (M⁺, 27%); C₄₂H₃₈ClN₅O₇S (792.2).

Synthesis of diacetylated S-(-β-D-glycofuranosyl)- 6-[4-(chlorophenyl) methylene]-amino-2-(4-alkylamino) pyrimidine (16a-f); (see 3.1.4, General procedures), crystallized from dimethyl formamide (30 - 50 mL).

2-[S-(β-D-arabinofuranosyl)]-4-[4-(chlorophenyl)methylene]amino-6-morpholin-4-ylpyrimidine (16a). It was obtained from 15a; as white powder, m. p. 289°C - 291°C; IR (cm⁻¹, v); 3463 (OH);¹H.NMR: d 2.37 (m, N(CH₂)₂), 3.51 (m, O(CH₂)₂), 3.89 (m, H-5', H-5"), 4.01 (m, H-4'), 4.77 (t, H-2'), 5.10 (t, J = 5.34 Hz, J = 4.86 Hz, OHC(5'), 5.23 (d, J = 4.49 Hz, OH-C(3'), 5.41 (d, J = 5.60 Hz, OH-C(2'), 5.70 (t, J = 9.72 Hz, H-3'), 6.85(d, J = 5.64 Hz, H-1'), 7.25 (d, Ar-H), 8.01 (d, Ar-H), 8.33 (s, pyrimidine-H), 8.80 (s, CH); ¹³C. NMR: 43.70 (2C, N(CH₂)₂), 61.51 (C-5'), 64.29 (2C, O(CH₂)₂), 66.28 (C- 3'), 66.83 (C-2'), 67.40 (C-4'), 85.76 (C-1'), 124.6 - 155.5 (10C-Ar); Its MS (m/z), 466 (M⁺, 32%); C₂₀H₂₃ClN₄O₅S (466.9).

2-[S-(β-D-ribofuranosyl)]-4-[4-(chlorophenyl)methylene]amino-6-morpholin-4-ylpyr-imidin-4(3H)-one (16b). It was obtained from 15b; as pale powder, m. p. 305°C - 307°C; IR (cm⁻¹, v); 3412 (OH); ¹H.NMR: d 2.30 (m, N(CH₂)₂), 3.51 (m, O(CH₂)₂), 3.95 (m, H-5', H-5"), 4.07 (m, H-4'), 4.83 (t, H-2'), 5.18 (t, J = 5.40 Hz, J = 4.82 Hz, OH-C(5'), 5.27 (d, J = 4.55 Hz, OH-C(3'), 5.45 (d, J = 5.65 Hz, OH-C(2'), 5.71 (t, J = 9.73 Hz, H-3'), 6.87 (d, J = 5.70 Hz, H-1'), 7.29 (d, Ar-H), 7.98 (d, Ar-H), 8.32 (s, pyrimidine-H), 8.81 (s, CH); Its MS (m/z), 466 (M⁺, 38%); C₂₀H₂₃ClN₄O₅S (466.9).

2-[S-(β-D-arabinofuranosyl)]-4-[4-(chlorophenyl)methylene]amino-6-piprazin-1-yl-pyrimidine (16c). It was obtained from 15c; as white powder, m. p. 295°C - 297°C; IR (cm⁻¹, v); 3443 (OH), 3368 (NH); ^{1}H.NMR: d 2.43 (m, N(CH$_2$)$_2$), 3.23 (m, N(CH$_2$)$_2$), 3.90 (m, H-5', H-5''), 3.99 (m, H-4'), 4.70 (t, H-2'), 5.14 (t, J = 5.44 Hz, J = 4.89 Hz, OH-C(5'), 5.28 (d, J = 4.54 Hz, OH-C(3'), 5.53 (d, J = 5.60 Hz, OH-C(2'), 5.73 (t, J = 9.83 Hz, H-3'), 6.76 (d, J = 5.57 Hz, H-1'), 7.30 (d, Ar-H), 8.00 (d, Ar-H), 8.38 (s, pyrimidine-H), 8.75 (s, CH), 9.46 (brs, NH); ^{13}C. NMR: 43.50 (2C, HN(CH$_2$)$_2$), 46.23 (2C, HN(CH$_2$)$_2$), 60.68 (C-5'), 64.79 (C-3'), 67.32 (C-2'), 70.23 (C-4'), 87.13 (C-1'), 125.8 -156.6 (11C-Ar); Its MS (m/z), 465 (M⁺, 36%); C$_{20}$H$_{24}$ClN$_5$O$_4$S (465.9).

2-[S-(β-D-ribofuranosyl)]-4-[4-(chlorophenyl)methylene]amino-6-piprazin-1-yl-pyri-midine (16d). It was obtained from 15d; as yellow powder, m. p. 283°C - 285°C; IR (cm⁻¹, v); 3456 (OH), 3290 (NH); ^{1}H.NMR: d 2.46 (m, N(CH$_2$)$_2$), 3.20 (m, N(CH$_2$)$_2$), 3.88 (m, H-5', H-5''), 4.05 (m, H-4'), 4.74 (t, H-2'), 5.15 (t, J = 5.46 Hz, J = 4.87 Hz, OH-C(5'), 5.32 (d, J = 4.50 Hz, OH-C(3'), 5.54 (d, J = 5.61 Hz, OH-C(2'), 5.87 (t, J = 9.83 Hz, H-3'), 6.86 (d, J = 5.60 Hz, H-1'), 7.23 (d, Ar-H), 8.00 (d, Ar-H), 8.29 (s,pyrimidine-H), 8.79(s, CH), 9.25 (br, NH); Its MS (m/z), 465 (M⁺, 33%); C$_{20}$H$_{24}$ClN$_5$O$_4$S (465.9).

2-[S-(β-D-arabinofuranosyl)]-4-[4-(chlorophenyl)methylene]amino-6-(4-methylpip-razin-1-yl)pyrimidine (16e). It was obtained from 15e; as yellow powder, m. p. 278°C - 280°C; IR (cm⁻¹, v); 3437 (OH); ^{1}H.NMR: d 2.31 (s, NCH$_3$), 2.46 (m, N(CH$_2$)$_2$), 3.12 (m, N(CH$_2$)$_2$), 3.97 (m, H-5', H-5''), 4.09 (m, H-4'), 4.78 (t, H-2'), 5.15 (t, J = 5.43 Hz, J = 4.89 Hz, OH-C(5'), 5.28 (d, J = 4.50 Hz, OH-C(3'), 5.48 (d, J = 5.70 Hz, OH-C(2'), 5.81 (t, J = 9.72 Hz, H-3'), 6.89 (d, J = 5.64 Hz, H-1'), 7.27 (d, ArH), 7.98 (d, Ar-H), 8.32 (s, pyrimidine-H), 8.78 (s, CH); ^{13}C.NMR: 30.21 (NCH$_3$), 43.65 (2C, N(CH$_2$)$_2$), 45.67 (2C, N(CH$_2$)$_2$), 61.43 (C-5'), 66.49 (C-3'), 66.94 (C-2'), 67.51 (C-4'), 86.19 (C-1'), 125.7-154.9 (11C-Ar); Its MS (m/z), 479 (M⁺, 27%); C$_{21}$H$_{26}$ClN$_5$O$_4$S (479.9).

2-[S-(β-D-ribofuranosyl)]-4-[4-(chlorophenyl)methylene]amino-6-(4-methylpiprazin-1-yl)pyrimidine (16f). It was obtained from 15f as yellow powder, m. p. 321°C - 323°C; IR (cm⁻¹, v); 3450 (OH); ^{1}H.NMR: d 2.36 (s, NCH$_3$), 2.48 (m, N(CH$_2$)$_2$), 3.09 (m, N(CH$_2$)$_2$), 3.65 (m, H-5', H-5''), 4.11 (m, H-4'), 4.82 (t, H-2'), 5.19 (t, J = 5.50 Hz, J = 4.92 Hz, OH-C(5'), 5.34 (d, J = 4.50 Hz, OH-C(3'), 5.66 (d, J = 5.69 Hz, OH-C(2'), 5.78 (t, J = 9.73 Hz, H-3'), 6.86 (d, J = 5.61 Hz, H-1'), 7.23 (d, J = 7.52 Hz, Ar-H), 7.98 (d, J = 7.51 Hz, Ar-H), 8.34 (s, pyrimidine-H), 8.74 (s, CH); Its MS (m/z), 479 (M⁺, 27%); C$_{21}$H$_{26}$ClN$_5$O$_4$S (479.9).

Preparation of the acetylated S-nucleoside sulfone of 4-[4-(chlorophenyl)-methylene]amino-6-(4- alkylamino) pyrimidinee (17a-f). General procedure:

The solution of 15a-f (0.01 mol) in hydrogen peroxide solution (20 ml) (AcOH, H_2O_2; 2:1) was stirred at room temperature for 18 - 24 hs. (under TLC control). The solvent was evaporated under reduced pressure at 40°C, and the crude product was filtered off. The product was dried, and crystallized from the ethanol (50 - 80 mL).

2-(S-2',3',5'-tri-O-acetyl-β-D-arabinofuranosyl)-4-[4-   (chlorophenyl) methylene]-amino-6-morpholin-4-ylpyrimidine (17a). It was obtained from 14a and 2,3,5-tri-Oacetyl-α-D-arabinofuranosyl)-bromide; as white powder, m. p. 211°C - 213°C; IR (cm^{-1}, v); 1731 (3CO), 1320 (SO); ^{1}H.NMR: d 1.98, 2.06, 2.14 (3s, 3CH$_3$CO), 2.33 (s, CH$_3$), 2.49 (m, 4H, N(CH$_2$)$_2$), 3.35 (m, 4H, O(CH$_2$)$_2$), 4.11 (m, H-4'), 4.21 (m, H-5', H-5''), 5.35 (m, H-3'), 5.48 (m, H-2'), 6.81 (d, J = 3.69 Hz, H-1'), 7.31 (d, 2H, Ar-H), 8.03 (d, 2H, Ar-H), 8.32 (s, pyrimidine-H), 8.70 (s, CH) 10.25; ^{13}C. NMR: 22.19, 22.36, 22.89 (3CH$_3$), 43.72 (2C, HN(CH$_2$)$_2$), 61.53 (C-5'), 64.65 (2C, O(CH$_2$)$_2$), 66.19 (C-3'), 66.83 (C-2'), 67.40 (C-4'), 85.76 (C-1'), 123.3-153.5 (11C-Ar), 169.8, 170.6, 173.5 (3CO); Its MS (m/z), 625 (M$^+$, 42%); C$_{26}$H$_{29}$ClN$_4$O$_{10}$S (625.0).

2-(S-2',3',5'-tri-O-benzoyl-β-D-ribofuranosyl)-4-[4-   (chlorophenyl) methylene]amino-6-morpholin-4-ylpyrimidine (17b). It was obtained from 14a and 2,3,5-Tri-Obenzoyl-β-D-ribofuranosyl bromide; as yellow powde, m. p. 307°C - 309°C; IR (cm^{-1}, v); 1729 (3CO), 1336 (SO); ^{1}H.NMR: d 2.50 (m, 4H, N(CH$_2$)$_2$), 3.36 (m, 4H, O(CH$_2$)$_2$), 4.08 (m, H-4'), 4.19 (m, H-5', H-5''), 5.31 (m, H-3'), 5.47 (m, H-2'), 6.77 (d, J = 3.70 Hz, H-1'), 7.12 - 8.00 (m, 19H, Ar-H), 8.32 (s, pyrimidine-H), 8.90 (s, CH); ^{13}C. NMR: 43.58 (2C, HN(CH$_2$)$_2$), 61.37 (C-5'), 65.12 (2C, O(CH$_2$)$_2$), 66.19 (C-3'), 66.84 (C-2'), 67.36 (C-4'), 85.66 (C-1'), 123.5-156.9 (29C-Ar), 169.7, 171.0, 174.1 (3CO); Its MS (m/z), 811 (M$^+$, 19%); C$_{41}$H$_{35}$N$_4$O$_{10}$S (811.2).

2-(S-2',3',5'-tri-O-acetyl-β-D-arabinofuranosyl)-6-[4-   (chlorophenyl) methylene]-amino-6-piprazin-1-ylpyrimidine (17c). It was obtained from 14b and 2,3,5-tri-Oacetyl-α-D-arabinofuranosyl)-bromide; as yellow powder, m. p. 234°C - 236°C; IR (cm^{-1}, v); 3240 (NH),1724 (3CO), 1330 (SO); ^{1}H.NMR: d 1.95, 1.99, 2.14 (3s, 3CH$_3$CO), 2.49 (m, N(CH$_2$)$_2$), 3.43 (m, HN(CH$_2$)$_2$), 4.15 (m, H-4'), 4.22 (m, H-5', H-5''), 5.33 (m, H-3'), 5.42 (m, H-2'), 6.84 (d, J = 3.69 Hz, H-1'), 7.28 (d, 2H, Ar-H), 7.98 (d, 2H, Ar-H), 8.40 (s, pyrimidine-H), 8.81 (s, CH), 9.18 (br, NH); Its MS (m/z), 624 (M$^+$, 33%); C$_{26}$H$_{30}$ClN$_5$O$_9$S (624.0).

2-(S-2',3',5'-tri-O-benzoyl-β-D-ribofuranosyl)-4-[4-   (chlorophenyl) methylene]amino-6-piprazin-1-ylpyrimidine (17d). It was obtained from 14b and 2,3,5-Tri-Obenzoyl-β-D-ribofuranosyl bromide; as yellow powder, m. p. 281°C - 283°C; IR (cm^{-1}, v); 3280 (NH), 1719 (3CO), 1335(SO); ^{1}H.NMR: d 2.50 (m, N(CH$_2$)$_2$), 3.47 (m, N(CH$_2$)$_2$, 4.09 (m, H-4'), 4.17 (m, H-5', H-5''),

5.36 (m, H-3'), 5.45 (m, H-2'), 6.78 (d, J = 3.73 Hz, H-1'), 7.15 - 8.04 (m, 19H, Ar-H), 8.34 (s, pyrimidine-H), 8.76 (s, CH), 9.11 (br, NH); Its MS (m/z), 810(M$^+$, 27%); C$_{41}$H$_{36}$ClN$_5$O$_9$S (810.2).

2-(S-2',3',5'-tri-O-acetyl-β-D-arabinofuranosyl)-4-[4-   (chlorophenyl) methylene]-amino-6-(4-methylpiprazin-1-yl)pyrimidine (17e). It was obtained from 14c and 2,3, 5-tri-O-acetyl-α-D-arabinofuranosyl)-bromide; as yellow powder, m. p. 229°C - 231°C; IR (cm^{-1}, v); 1728 (3CO), 1321 (SO); ^{1}H.NMR: d 1.96, 2.03, 2.16 (3s, 3CH$_3$CO), 2.37 (s, NCH$_3$), 2.55 (m, N(CH$_2$)$_2$), 3.39 (m, N(CH$_2$)$_2$), 4.16 (m, H-4'), 4.21 (m, H-5', H-5''), 5.32 (m, H-3'), 5.43 (m, H-2'), 6.83 (d, J = 3.72 Hz, H-1'), 7.31 (d, 2H, Ar-H), 8.00 (d, 2H, Ar-H), 8.38 (s, pyrimidine-H), 8.86 (s, CH);^{13}C.NMR: 29.97 (NCH$_3$), 45.65 (2C, N(CH$_2$)$_2$), 46.79 (2C, N(CH$_2$)$_2$), 61.09 (C-5'), 66.56 (C-3'), 67.07 (C-2'), 68.13 (C-4'), 86.21 (C-1'), 122.3 - 153.5 (11CAr), 170.9, 171.7, 173.2 (3CO); Its MS (m/z), 638 (M$^+$, 21%); C$_{27}$H$_{32}$ClN$_5$O$_9$S (638.0).

2-(S-2',3',5'-tri-O-benzoyl-β-D-ribofuranosyl)-4-[4-   (chlorophenyl) methylene]amino-6-(4-methylpiprazin-1-yl)pyrimidine (17f). It was obtained from 14c and 2,3,5- Tri-O-benzoyl-β-D-ribofuranosyl bromide as yellow powder, m. p. 301°C - 303°C; IR (cm^{-1}, v); 1730 (3CO), 1335 (SO); ^{1}H.NMR: d 2.31 (s, NCH$_3$), 2.50 (m, N(CH$_2$)$_2$), 3.53 (m, N(CH$_2$)$_2$), 4.14 (m, H-4'), 4.19 (m, H-5', H-5''), 5.32 (m, H-3'), 5.41 (m, H-2'), 6.85 (d, J = 3.70 Hz, H-1'), 7.10 - 8.01 (m, 19H, Ar-H), 8.31 (s, pyrimidine-H), 8.70 (s, CH); ^{13}C.NMR: 30.20 (NCH$_3$), 43.29 (2C, N(CH$_2$)$_2$), 45.73 (2C, N(CH$_2$)$_2$), 61.08 (C-5'), 66.19 (C-3'), 66.90 (C-2'), 67.43 (C-4'), 86.01 (C-1'), 125.3 - 154.8 (29C-Ar), 170.6, 171.4, 173.2 (3CO); Its MS (m/z), 824 (M$^+$, 18%); C$_{42}$H$_{38}$ClN$_5$O$_9$S (824.2).

Synthesis of diacetylated S-(-β-D-glycofuranosyl)-6- [4-(chlorophenyl) methylene]-amino-2-(4-alkylamino)pyrimidine (18a-f); (see 3.1.4, General procedures) and crystallized from ethanol (50 - 70 mL) .

2-[S-(β-D-arabinofuranosyl)]-4-[4-(chlorophenyl)methylene]amino-6-morpholin-4-ylpyrimidine (18a). It was obtained from 17a; as white powder, m. p. 241°C - 243°C; IR (cm^{-1}, v); 3458 (OH), 3286 (NH), 1686 (CO), 1330 (SO); ^{1}H.NMR: d 2.32 (m, N(CH$_2$)$_2$), 3.45 (m, O(CH$_2$)$_2$), 3.94 (m, H-5', H-5''), 4.03 (m, H-4'), 4.79 (t, H-2'), 5.13 (t, J = 5.34 Hz, J = 4.88 Hz, OH-C(5'), 5.29 (d, J = 4.52 Hz, OH-C(3'), 5.47 (d, J = 5.63 Hz, OH-C(2'), 5.73 (t, J = 9.76 Hz, H-3'), 6.84 (d, J = 5.62 Hz, H-1'), 7.26 (d, Ar-H), 8.00 (d, Ar-H), 8.31 (s, pyrimidine-H), 8.76 (s, CH); ^{13}C. NMR: 43.72 (2C, N(CH$_2$)$_2$), 61.50 (C-5'), 64.19 (2C, O(CH$_2$)$_2$), 66.31 (C-3'), 66.83 (C-2'), 67.42 (C-4'), 85.78 (C-1'), 124.9 - 155.7 (11C-Ar); Its MS (m/z), 498 (M$^+$, 28%); C$_{20}$H$_{23}$ClN$_4$O$_7$S (498.9).

2-[S-(β-D-ribofuranosyl)]-4-[4-(chlorophenyl)methylene]amino-6-morpholin-4-yl-pyrimidin-4(3H)-one (18b). It was obtained from 17b; as pale yellow powder, m. p. 276°C - 278°C; IR (cm^{-1}, v); 3465 (OH), 3270 (NH) 1341 (SO); ^{1}H.NMR: d 2.35 (m, N(CH$_2$)$_2$), 3.47 (m, O(CH$_2$)$_2$), 3.96 (m, H-5', H-5"), 4.05 (m, H-4'), 4.87 (t, H-2'), 5.18 (t, J = 5.40 Hz, J = 4.81 Hz, OH-C(5')), 5.29 (d, J = 4.56 Hz, OH-C(3')), 5.49 (d, J = 5.63 Hz, OHC(2')), 5.74 (t, J = 9.78 Hz, H-3'), 6.86 (d, J = 5.72 Hz, H-1'), 7.35 (d, Ar-H), 8.11 (d, Ar-H), 8.36 (s, pyrimidineH), 8.80 (s, CH); Its MS (m/z), 498 (M$^+$, 19%); C$_{20}$H$_{23}$ClN$_4$O$_7$S (498.9).

2-[S-(β-D-arabinofuranosyl)]-4-[4-(chlorophenyl)methylene]amino-6-piprazin-1-yl-pyrimidine (18c). It was obtained from 17c; as yellow powder, m. p. 251°C - 253°C; IR (cm^{-1}, v); 3428 (br, OH), 3278 (NH), 1335 (SO); ^{1}H. NMR: d 2.39 (m, N(CH$_2$)$_2$), 3.28 (m, N(CH$_2$)$_2$), 3.90 (m, H-5', H-5"), 3.99 (m, H-4'), 4.67 (t, H-2'), 5.13 (t, J = 5.45 Hz, J = 4.88 Hz, OH-C(5')), 5.26 (d, J = 4.52 Hz, OH-C(3')), 5.59 (d, J = 5.58 Hz, OH-C(2')), 5.76 (t, J = 9.84 Hz, H-3'), 6.78 (d, J = 5.59 Hz, H-1'), 7.30 (d, Ar-H), 8.03 (d, Ar-H), 8.39 (s, pyrimidine-H), 8.79 (s, CH), 9.21 (br, NH); ^{13}C.NMR: 43.51 (2C, HN(CH$_2$)$_2$), 46.25 (2C, HN(CH$_2$)$_2$), 60.68 (C-5'), 64.77 (C-3'), 67.35 (C-2'), 70.26 (C-4'), 87.19 (C-1'), 125.3 - 156.9 (11CAr); Its MS (m/z), 497 (M$^+$, 30%); C$_{20}$H$_{24}$ClN$_5$O$_6$S (497.9).

2-[S-(β-D-ribofuranosyl)]-4-[4-(chlorophenyl)methylene]amino-6-piprazin-1-yl-pyrimidine (18d). It was obtained from 17d; as yellow powder, m. p. 269°C - 271°C; IR (cm^{-1}, v); 3435 (OH), 3268 (NH), 1335 (SO); ^{1}H.NMR: d 2.43 (m, N(CH$_2$)$_2$), 3.21 (m, N(CH$_2$)$_2$), 3.85 (m, H-5', H-5"), 4.03 (m, H-4'), 4.74 (t, H-2'), 5.16 (t, J = 5.48 Hz, J = 4.85 Hz, OH-C(5')), 5.30(d, J = 4.51 Hz, OH-C (3')), 5.56 (d, J = 5.60 Hz, OH-C(2')), 5.86 (t, J = 9.81 Hz, H-3'), 6.89 (d, J = 5.59 Hz, H-1'),7.26 (d, ArH), 8.02 (d, Ar-H), 8.33 (s, pyrimidine-H), 8.81 (s, CH), 9.22 (br, NH); Its MS(m/z), 497(M$^+$, 24%); C$_{20}$H$_{24}$ClN$_5$O$_6$S (497.9).

2-[S-(β-D-arabinofuranosyl)]-4-[4-(chlorophenyl)methylene]amino-6-(4-methylpipr-azin-1-yl)pyrimidine (18e). It was obtained from 17e; as yellow powder, m. p. 247°C - 249°C; IR (cm^{-1}, v); 3420 (OH), 1335 (SO); ^{1}H.NMR: d 2.35 (s, NCH$_3$), 2.49 (m, N(CH$_2$)$_2$), 3.17 (m, N(CH$_2$)$_2$), 3.96 (m, H-5', H-5"), 4.07 (m, H-4'), 4.78 (t, H-2'), 5.14 (t, J = 5.43 Hz, J = 4.91 Hz, OH-C(5')), 5.29 (d, J = 4.48 Hz, OH-C(3')), 5.51 (d, J = 5.67 Hz, OH-C(2')), 5.80 (t, J = 9.73 Hz, H-3'), 6.87 (d, J = 5.62 Hz, H-1'), 7.25 (d, Ar-H), 7.99 (d, Ar-H), 8.36 (s, pyrimidine-H), 8.74 (s, CH);^{13}C.NMR: 30.20 (NCH$_3$), 43.68 (2C, N(CH$_2$)$_2$), 45.64 (2C, N(CH$_2$)$_2$), 61.49 (C-5'), 66.53 (C-3'), 66.96 (C-2'), 67.56 (C-4'), 86.21 (C-1'), 125.7 - 154.9 (11CAr); Its MS (m/z), 511 (M$^+$, 30%); C$_{21}$H$_{26}$ClN$_5$O$_6$S (511.9).

2-[S-(β-D-ribofuranosyl)]-4-[4-(chlorophenyl)methylene]amino-6-(4-methylpipra-zin-1-yl)pyrimidine (18f). It was obtained from 17f as yellow powder, m. p. 287°C - 289°C; IR (cm^{-1}, v); 3400 (OH), 1335 (SO); ^{1}H.NMR: d

2.33 (s, NCH$_3$), 2.45 (m, N(CH$_2$)$_2$), 3.19 (m, N(CH$_2$)$_2$), 3.89 (m, H-5', H-5"), 4.10 (m, H-4'), 4.82 (t, H-2'), 5.17 (t, J = 5.51 Hz, J = 4.93 Hz, OH-C(5'), 5.29 (d, J = 4.52 Hz, OH-C(3'), 5.63 (d, J = 5.71 Hz, OH-C(2'), 5.80 (t, J = 9.74 Hz, H-3'), 6.88 (d, J = 5.60 Hz, H-1'), 7.25 (d, J = 7.50 Hz, Ar-H), 8.01 (d, J = 7.51 Hz, Ar-H), 8.40 (s, pyrimidine-H), 8.75 (s, CH); Its MS (m/z), 511 (M$^+$, 28%); C$_{21}$H$_{26}$ClN$_5$O$_6$S (511.9).

In Vitro Evaluation of Antiplatelet Aggregation Activity

The blood samples were obtained from healthy volunteers with negative history of smoking or taking any medications since 14d prior to blood collection. To blood samples was added trisodium citrate dihydrate 3.8% (1 part citrate, 9 part blood) and centrifuged at 1000 rpm for 8 min to obtain PRP. The remaining was centrifuged at 3000 rpm for 15 min and PPP was collected from the above layer which was used as the test blank. The platelet count was adjusted to 250,000 plts/mL by diluting PRP with appropriate amount of PPP once needed. Of the test compounds previously dissolved in dimethyl sulfoxide (DMSO), 1 µL was added to 200 µL of PRP and incubated at 37°C for 5 min. To induce platelet aggregation, a solution of ADP (5 µm) or arachidonic acid (1.25 mg/mL) was added to the samples and aggregation was measured using APACT 4004 aggregometer for a 5-minute period. DMSO (0.5% v/v) was used as negative control and indomethacin and aspirin as standard drugs. Platelet aggregation inhibition (%) was calculated by the formula

$$\text{inhibition } \% = \left[1 - \left(D/S\right)\right] \times 100$$

where: D, platelet aggregation in the presence of test compounds; S, platelet aggregation in the presence of DMSO). Compounds were tested at the initial concentration of 100 µm and IC50 was calculated from log (concentration)-inhibition (%) diagram for those that inhibited the platelet aggregation by 50% or more. IC50 in this test was defined as the concentration at which the test compound inhibits platelet aggregation by 50%.

REFERENCES

1.   J. Clark, M. S. Shahhet, D. Korakas and G. Varvounis, "Synthesis of Thieno[2,3-d] Pyrimidines from 4,6-Dichloropyrimidine-5-Carbaldehydes" Journal of Heterocyclic Chemistry, Vol. 30, No. 4, 1993, pp. 1065-1072.http://dx.doi.org/10.1002/jhet.5570300439

2.   B. Tozkoparan, M. Ertan, P. Kelicen and R. Demirdamar, "Synthesis and Anti-Inflammatory Activities of Some Thiazolo[3,2-a]pyrimidine Derivatives" Farmaco, Vol. 54, No. 9, 1999, pp. 588-593. http://dx.doi.

org/10.1016/S0014-827X(99)00068-3

3.   K. Ogowva, I. Y. Yamawaki, Y. I. Matsusita, N. Nomura, P. F. Kador and J. H. Kinoshita, "Syntheses of Substituted 2,4-Dioxo-Thienopyrimidin-1-Acetic Acids and Their Evaluation as Aldose Reductase Inhibitors," European Journal of Medicinal Chemistry, Vol. 28, No. 10, 1993, pp. 769-781. http://dx.doi.org/10.1016/0223-5234(93)90112-R

4.   M. Santagati, M. Modica, A. Santagati, F. Russo, S. Spampinato, "Synthesis of Aminothienopyrimidine and Thienotriazolopyrimidine Derivatives as Potential Anticonvulsant, Agents" Pharmazie, Vol. 51, No. 1, 1996, pp. 7- 11.

5.   A. B. A. El-Gazzar and H. N. Hafez, "Synthesis of 4- Substituted Pyrido[2,3-d]pyrimidin-4(1H)-one as Analgesic and Anti-Inflammatory Agents," Bioorganic & Medicinal Chemistry Letters, Vol. 19, No. 13, 2009, pp. 3392- 3397.http://dx.doi.org/10.1016/j.bmcl.2009.05.044

6.   V. K. Ahluwalia, M. Chopra and R. Chandra, "Convenient Synthesis of Novel Pyrimidine Analogues of O-Hydroxy Chalcones and Pyrano[2,3-d] pyrimidines and Their Biological Activities," Journal of Chemical Research, No. 4, 2000, pp. 162-163.

7.   L. V. G. Nargund, V. V. Badiger and S. M. Yarnal, "Synthesis and Antibacterial Activity of Substituted 4-Aryloxypyrimido[5,4-c] cinnolines," European Journal of Medicinal Chemistry, Vol. 29, No. 3, 1994, pp. 245-247. http://dx.doi.org/10.1016/0223-5234(94)90043-4

8.   M. van Laar, E. Volerts and M. Verbaten, "Subchronic Effect of the GABA-Agonist Lorazepam and the 5-HTsub(2A/2C) Antagonist Ritanserin on Driving Performance, Slow Wave Sleep and Daytime Sleepiness in the Healthy Volunteers," Psychopharmacology, Vol. 154, No. 2, 2001, pp. 189-197. http://dx.doi.org/10.1007/s002130000633

9.   K. Danel, E. B. Pedersen and C. Nielsen, "Synthesis and Anti-HIV-1 Activity of Novel 2,3-Dihydro-7H-Thiazolo[3, 2-a]pyrimidin-7-ones," Journal of Medicinal Chemistry, Vol. 41, No. 2, 1998, pp. 191-198. http://dx.doi.org/10.1021/jm970443m

10.   H. C. Zhang, B. E. Maryanoff, H. Ye and C. Chen, "PCT International Applied WO2008054795, 2008," Chemical Abstracts, 2008, Vol. 148, Article ID: 517743.

11.   S. W. Kortum, R. M. Lachance, B. A. Schweitzer, G. Yalamanchili, H. Rahman, M. D. Ennis, R. M. Huff and R. E. Tenbrink, "Thienopyrimidine-Based $P2Y_{12}$ Platelet Aggregation Inhibitors," Bioorganic & Medicinal Chemistry Letters, Vol. 19, No. 20, 2009, pp. 5919-5923. http://dx.doi.org/10.1016/j.bmcl.2009.08.059

12. H. S. Kim, D. Barak, T. K. Harden, J. Boyer and K. A. Jacobson, "Acyclic and Cyclopropyl Analogues of Adenosine Bisphosphate Antagonists of the P2Y1 Receptor: Structure-Activity Relationships and Receptor Docking," Journal of Medicinal Chemistry, Vol. 44, No. 19, 2001, pp. 3092-3108. http://dx.doi.org/10.1021/jm010082h

13. B. Xu, A. Stephens, G. Kirschenheuter, A. F. Greslin, X. Cheng, J. Sennelo, M. Cattaneo, M. L. Zighetti, A. Chen, S. A. Kim, H. S. Kim, N. Bischofberger, G. Cook and K. A. Jacobson, "Acyclic Analogues of Adenosine Bisphosphates as P2Y Receptor Antagonists: Phosphate Substitution Leads to Multiple Pathways of Inhibition of Platelet Aggregation," Journal of Medicinal Chemistry, Vol. 45, No. 26, 2002, pp. 5694-5709.http://dx.doi.org/10.1021/jm020173u

14. P. Crepaldi, B. Crepaldi, B. Cacciari, M. C. Bonache, G. Spalluto, K. Varani, P. A. Borea, I. von Kuegelgen, K. Hoffmann, M. Pugliano, C. Razzari and M. Cattaneo, "6- Amino-2-Mercapto-3H-pyrimidin-4-one Derivatives as New Candidates for the Antagonism at the $P2Y_{12}$ Receptors," Bioorganic & Medicinal Chemistry, Vol. 17, No. 13, 2009, pp. 4612-4621. http://dx.doi.org/10.1016/j.bmc.2009.04.061

15. A. H. Ingall, J. Dixon, A. Bailey, M. E. Coombs, D. Cox, J. I. McInally, S. F. Hunt, N. D. Kindon, B. J. Teobald, P. A. Willis, R. G. Humphries, P. Leff, J. A. Clegg, J. A. Smith and W. Tomlinson, "Antagonists of the Platelet P_{2T} Receptor: A Novel Approach to Anti-Thrombotic Therapy," Journal of Medicinal Chemistry, Vol. 42, No. 2, 1999, pp. 213-220.http://dx.doi.org/10.1021/jm981072s

16. H. N. Hafez, H. A. R. Hussein and A. B. A. El-Gazzar, "Synthesis of Substituted Thieno[2,3-d]pyrimidine-2,4- dithiones and Their S-Glycoside Analogues as Potential Antiviral and Antibacterial Agents," European Journal of Medicinal Chemistry, Vol. 45, No. 9, 2010, pp. 4026- 4034. http://dx.doi.org/10.1016/j.ejmech.2010.05.060

17. A. S. Abbas, H. N. Hafez and A. B. A. El-Gazzar, "Synthesis, in Vitro Antimicrobial and in Vivo Antitumor Evaluation of Novel Pyrimidoquinolines and Its Nucleoside Derivatives," European Journal of Medicinal Chemistry, Vol. 46, No. 1, 2011, pp. 21-30.http://dx.doi.org/10.1016/j.ejmech.2010.09.071

18. D. Gravier, J. P. Dupin, F. Casadebaig, G. Hou, M. Boisseau and H. Bernard, "Synthesis and in Vitro Study of Platelet Antiaggregant Activity of Some 4-Quinazolinone Derivatives," Pharmazie, Vol. 47, 1992, pp. 91-94.

19. J. P. Dupin, R. J. Gryglewski, D. Gravier, G. Hou, F. Casadebaig, J.

Swies and S. Chlopicki, "Synthesis and Thrombolytic Activity of New Thienopyrimidinone Derivatives," Journal of Physiology and Pharmacology, Vo. 53, No. 4, 2002, pp. 625-634.

20. A. Vema, S. K. Panigrahi, G. Rambabu, B. Gopalakrishnan, J. A. Sarma and G. R. Desiraju, "Design of EGFR Kinase Inhibitors: A Ligand-Based Approach and Its Confirmation with Structure-Based Studies," Bioorganic & Medicinal Chemistry, Vol. 11, No. 21, 2003, pp. 4643-4653. http://dx.doi.org/10.1016/S0968-0896(03)00482-6

21. Y. M. Zhang, S. Cockerill, S. B. Guntrip, D. Rusnak, K. Smith, D. Vanderwall, E. Wood and K. Lackey, "Synthesis and SAR of Potent EGFR/erbB2 Dual Inhibitors," Bioorganic & Medicinal Chemistry Letters, Vol. 14, No. 1, 2004, pp. 111-114.http://dx.doi.org/10.1016/j.bmcl.2003.10.010

22. I. A. da Silveira, L. G. Paulo, A. L. da Miranda, S. O. Rocha, A. C. Freitas and E. J. Barreiro, "New Pyrazolylhydrazone Derivatives as Inhibitors of Platelet Aggregation," Journal of Pharmacy and Pharmacology, Vol. 45, No. 7, 1993, pp. 646-649.http://dx.doi.org/10.1111/j.2042-7158.1993.tb05670.x

23. C. Bertez, M. Miquel, C. Coquelet, D. Sincholle and C. Bonne, "Dual Inhibition of Cyclo-Oxygenase and Lipoxygenase by 2-Acetylthiophene 2-Thiazolye Hydrazone (CBS-1108) and Effect on Leukocyte Migration in Vivo," Biochemical Pharmacology, Vol. 33, No. 11, 1984, pp. 1757-1762. http://dx.doi.org/10.1016/0006-2952(84)90346-0

Chapter 11

DEVELOPMENT OF IMPEDIMETRIC BIOSENSOR FOR TOTAL CHOLESTEROL ESTIMATION BASED ON POLYPYRROLE AND PLATINUM NANOPARTICLE MULTI LAYER NANOCOMPOSITE

K. Singh[1], Ruchika Chauhan[2], Pratima R. Solanki[2], Tinku Basu[2]

[1]Amity School of Engineering and Technology, Amity University, Noida, India
[2]Amity Institute of Nano Technology, Amity University, Noida, India

ABSTRACT

A novel impedimetric biosensor was fabricated for total cholesterol sensing based on platinum nanoparticle and polypyrrole multilayer nanocomposite electrode. The Pt nanoparticles (PtNP) electrochemically deposited between two polypyrrole layers on indium tin oxide (ITO) glass plates (PtNP/PPY/ITO) have offered high-electroactive surface area and favourable microenvironment for immobilization of cholesterol esterase (ChEt) and cholesterol oxidase (ChOx) resulting in enhanced electron transfer between the enzyme system and the electrode. Impedimetric response studies of the ChEt-ChOx/PtNP/ITO nanobioelectrode exhibit improved linearity (2.5×10^{-4} to 6.5×10^{-3}M/l), low detection limit (2.5×10^{-4} M/l), fast response time (25 s), high sensitivity (196 Ω/mM/cm^{-2}) and a low value of the Michaelis-Menten constant (Km, 0.2 M/l) with a regression coefficient of 0.997.

INTRODUCTION

Innovations in the field of electrochemical biosensors' matrix are of much importance nowadays for the development of highly sensitive, selective, reliable and low cost biosensors for the clinical diagnosis [1]. The metal nanoparticles have been widely exploited due to their ability as electrode modification materials to enhance the efficiency of electrochemical biosensor

[2-4]. This can be attributed to the ability to design novel sensing systems and enhance the performance of bioanalytical assays[1]. Pt is a well-known catalyst that has a high catalytic activity for hydrogen peroxide electro oxidation [5-7]. Pt nanoparticles (PtNP) on the electronically conducting polypyrrole can enhance conduction and charge transport properties compared to the conventionally synthesized polypyrrole film [8-10]. PtNP possess high surface area, nontoxicity, good biocompatibility and chemical stability, and also show fast electron communication features which provide a desirable microenvironment to enzyme for the direct electron transfer between the enzyme's active sites and the electrode. The PtNP not only retain the bioactivity of the immobilized enzyme but also enhance the sensing characteristics such as sensitivity, selectivity and low detection limit of the fabricated amperometric enzymatic biosensors [11]. The PtNP have the advantages of easy electrochemical synthesis and reproducible preparation, long-term stability and ability to incorporate enzyme successfully via covalent bonding by using carbodiimide chemistry i.e. N-hydroxysuccinimide (NHS) and N-ethyl-N-(3-dimethylaminopropyl carbodiimide) (EDC). Chen et al. have developed a glucose biosensor based on conducting chitosan decorated with PtNP with a response time of 5 s [12].

Human serum contains a mixture of free and esterified cholesterol, both of which are measured to determine the total cholesterol level [13]. The determination of cholesterol level is of importance in clinical diagnosis [14] because of diseases such as coronary heart disease, myocardial infarction and arteriosclerosis [15,16]. It is important to note that 70% of the total blood cholesterol exists in ester form and 30% in free form. But the research work on total cholesterol estimation is limited. Most research work till date is focused on the estimation of free cholesterol [17-20]. Singh et al. have utilized conducting polymers (CP) for the total cholesterol determination [21-23]. However, these biosensors have been found to have low sensitivity (7.5×10^{-4} nA/mg/dl) and a high response time, like 240 s. Li et al. have fabricated screen-printed MWCNT polycarbonate electrode to investigate total cholesterol and found that the CNT-modified biosensor offers a reliable calibration profile [24]. Solanki et al. have immobilized cholesterol esterase (ChEt) and ChOx via glutaraldehyde as a cross-linker onto sol-gel-derived silica/chitosan/MWCNT-based nanobiocomposite electrode deposited onto ITO for the determination of esterified cholesterol [25]. This bioelectrode shows a response time of 10 s, sensitivity of 3.8 µA/mM, and shelf life of about 10 weeks for the estimation of cholesterol oleate. The gold nanowire-modified microfluidic-based amperometric total cholesterol bioelectrode has linearity of 1 to 6 mM/l [26]. Arya et al. have reported self-assembled monolayer-based cholesterol sensors and found linearity of 10 to 500 mg/dl [27,28]. Basu et al. have fabricated a nanocomposite electrode comprising of polypyrrole

(PPY) and carboxy functionalized multiwalled carbon nanotubes (MWCNT) to estimate total cholesterol in the range of 4×10^{-4} to 6.5×10^{-3} M/l and they have used to determine cholesterol in blood serum samples [29]. Basu et al. have also developed a reusable total cholesterol electrode based on nano structured conducting polyaniline [30].

Electrochemical impedance spectroscopy (EIS) has recently been important as a nondestructive, sensitive and efficient means for characterization of electrical properties of materials in biological interfaces [31,32]. In this study, an attempt has been made to develop PtNP and PPY based electrodeposited tri-layer nanocomposite film and the same has been evaluated as a transducer matrix for total cholesterol estimation using impedance spectroscopy. The change in impedance of nanobiocomposite electrodes with the change of different concentration of cholesterol oleate is used as a measure of biosensor performance. To the best of our knowledge, electrochemical impedance spectroscopy, as a measure of total cholesterol estimation, has not been exploited till today. The novelty of this study lies on the ease of fabrication of PtNP based PPY multilayer nanocomposite electrode which provides enzyme friendly environment, non-destructive measurement technique for total cholesterol, high sensitivity, reproducibility and shelf life of the biosensor for total cholesterol estimation.

EXPERIMENTAL

Reagent

Hexa Chloroplatinic Acid (H_2PtCl_6) (Pt, 40%), N-hydroxysuccinimide (NHS), N-ethyl-N-(3-dimethylaminopropyl carbodiimide) (EDC), Cholesterol oleate (98%), cholesterol esterase (EC 3.1.1.13, pseudomonas species) with specific activity as 165 U/mg and cholesterol oxidase (EC 1.1.36 Pseudomonas fluorescens) with specific activity of 24 U/mg have been purchased from SigmaAldrich (USA). Pyrrole (PY) (M.W: 67.09) from spectrochem was double distilled prior to polymerization. All other chemicals are of analytical grade and have been used without further purification and the solutions are prepared in deionized water. Cholesterol oleate (400 mg/dl) is first dissolved in 1% polidocanol (Brij) as a surfactant by heating and gentle stirring resulting in clear and colourless suspension and final volume is made by addition of 0.9% NaCl solution.

Instrumentation

The Fourier transform infrared spectra for PPY/ITO, PtNP/PPY/ITO electrodes and ChEt-ChOx/PtNP/PPY/ ITO nanobioelectrodes have been recorded on Perkin Elmer Spectrophotometer BX-1 in the frequency range, 405 - 4000 cm^{-1}. Surface morphologies of PPY/ITO, PtNP/PPY/ITO electrodes and ChOx-ChEt/ PtNP/PPY/ ITO nanobioelectrode have been investigated by scanning electron microscope (LEO Model). Electrochemical impedance spectroscopy (EIS), cyclic voltammetry (CV) and differential pulse voltammetry study (DPV) have been conducted in phosphate buffer (50 mM, 0.9% NaCl) containing 5 mM $[Fe(CN)_6]^{3-/4-}$ in a three-electrodes cell consisting of Ag/AgCl as reference, platinum (Pt) as counter electrode and ITO as a working electrode (0.25 cm^2) using Potentiostat/Glavanostat (Princeton Applied Research, model No. 273).

Preparation of ChEt-ChOx/PtNP/PPY/ITO Nano-Bioelectrode

The PtNP/PPY/ITO nano composite electrode was fabricated in three steps. In the first step, pyrrole was electropolymerised onto ITO coated glass plates (PPY/ITO) in a three-electrode cell containing 0.1 M pyrrole and 0.1 M PTS (p-toluene sulfonic acid) solution in deionized water by chronopotentiometry using a Potentiostat/Galvanostat. In second step, the PPY/ITO electrode was dipped in 0.8 mM hexa chloroplatinic acid solution. The Pt nanoparticle (PtNP) was deposited on the PPY/ITO electrode using chronoamperommetry technique. In final step, the PtNP decorated PPY/ITO electrode was again dipped in 0.1 M pyrrole solution in order to deposit another thin layer of PPY using chronopotentiometry technique to fabricate a try layer nano composite of PPY/PtNP/PPY/ ITO electrode which is defined as PtNP/PPY/ITO in the manuscript. The fabrication process of nano bioelectrode is shown in Scheme 1. The deposition potential and concentration of hexa chloroplatonic acid solution, and deposition time of each layer were varied to achieve the optimum electrochemical properties of PtNP/PPY/ITO nano composite electrode. The PtNP/PPY/ITO nano composite electrode was cleaned with deionized water. The enzymes (ChOx and ChEt (1:1)) were covalently immobilized onto the PtNP/PPY/ITO nano composite electrode using EDC as the coupling agent and NHS as the activator. For enzyme immobilization, the optimal binding was achieved when 30 μl of solution of 5 μg of enzymes (1:1) mixed with 0.4 M EDC and 0.1 M NHS was dropped on 1 cm^2 of PtNP/PPY/ITO nano composite electrode and kept in a humid chamber for 3 h. Thus, fabricated ChEt-ChOx/ PtNP/PPY/ITO (Scheme 1) was then washed with phosphate buffer saline (PBS) solution (50 mM, 0.9% NaCl, pH 7.0) to remove any unbound enzyme and was stored at 4°C when not in use.

RESULTS AND DISCUSSION

Optimization of the Fabrication Process of PtNP/PPY/ITO Nano Composite Electrode

The PtNP/PPY/ITO nano composite film has been fabricated on ITO electrode using both chronoamperometric and chronopotentiometric techniques. The electronic conductivity of the nano composite film has been optimized by varying concentration of H_2PtCl_6, deposition potential of H_2PtCl_6 and multilayer deposition time. After the deposition of a thin layer of polypyrrole on ITO, PtNP has been deposited using chronoamperometric technique on PPY/ITO electrode. The optimum electronic conductivity of the nano composite film (PtNP/ PPY/ITO) and minimum size distribution have been achieved at a concentration of 0.8 mM/l of H_2PtCl_6. The **Figure 1**A represents cyclic voltamommetry of PtNP/ PPY/ITO electrodes, fabricated by varying electrodeposition potential of H_2PtCl_6, in phosphate buffer (50 mM, pH 7.0) containing 5 mM $[Fe(CN)_6]^{3-/4-}$ in the range of −300 mV to 1 V. The **Figure 1**A(I) demonstrates the variation of anodic peak current with deposition potential of H_2PtCl_6 and optimum result has been obtained at −300 mV. Therefore, PtNP has been deposited at −300 mV for entire work.

The deposition time of each layer has been varied in order to achieve the optimum property of try layer nano composite film. The **Figure 1**B represents the cyclic voltammogram of PtNP/PPY/ITO at various deposition time. The **Figure 1**B: Inset shows the plot of anodic peak current vs deposition time. It has been noticed that maximum peak current has been obtained when deposition time is 30, 40 and 60 s for 1st layer of PPY, PtNP and 2nd layer of PPY respectively. However, inner layer of PPY is to improve the contact between the PtNP and the ITO electrode, so a thin inner layer of PPY is required.

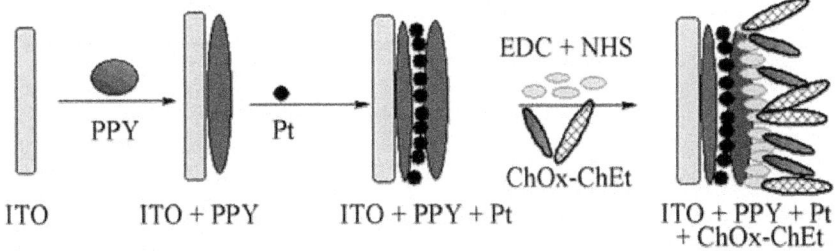

Scheme 1: Schematic diagram for fabrication of bioelectrode ChOx-ChEt/PtNP/PPY/ ITO.

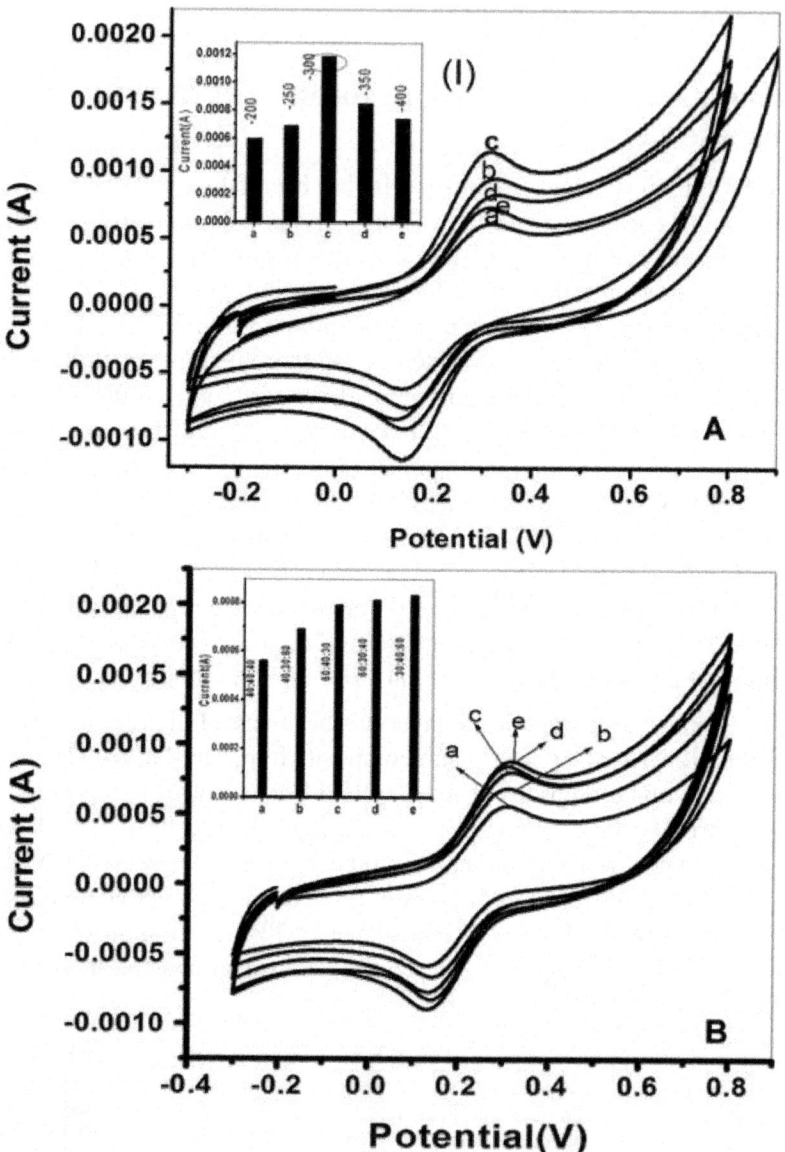

Figure 1: A. Cyclic voltammogram of PtNP/PPY/ITO electrodes fabricated by vary-ing deposition potential of H₂PtCl₆. (a) −200 mV; (b) −250 mV; (c) −300 mV; (d) −350 mV; (e) −400 mV; Inset 1(I): Plot of variation of anodic peak current vs deposition potential of H₂PtCl₆. B. Cyclic voltammogram of PtNP/PPY/ITO electrode fabricated by varying deposition time for each layer (a) 40:40:40; (b) 40:30:60; (c) 60:40:30; (d) 60:30:40; (e) 30:40:60; Inset a: plot of anodic peak current vs deposition time for vari-ous nano composite electrodes fabricated by varying the deposition time of each layer characterization of PtNP/PPY/ITO and ChOx-ChEt/ PtNP/PPY/ITO electrode.

Characterization of PtNP/PPY/ITO and ChOx-ChEt/PtNP/PPY/ITO Electrode

The morphology of PPY/ITO, PtNP/PPY/ITO and ChOxChEt/PtNP/PPY/ITO was studied by scanning electron microscope (SEM). The SEM images are shown in Figures 2A-C. The **Figure 2**A shows that the bulk polymer tends to aggregate in large particles in the globular morphology which becomes more evident with spherical ball shapes. The PtNP, grown in between the PPY thin layers structure during electrodeposition, has been observed in the SEM image of PtNP/PPY/ITO (**Figure 2**B). The SEM micrograph shows that PtNP grown on PPY/ITO electrode (**Figure 2**B) exhibits a granular form like a ball structure along with few cylindrical rod-like structure arrayed in a disordered way. The size of the spherical nanoparticles varies from ca. 50 to 80 nm with average particle size of ca.70 nm. Moreover, PtNP are seen to agglomerate due to high surface energy and strong inter atomic interactions. However, after the immobilization of ChEt and ChOx, the surface morphology of PtNP/PPY/ ITO film changes into well-arranged regular morphology and surface roughness decreases revealing that the enzymes are adsorbed onto PtNP/PPY/ ITO (**Figure 2**C) via covalent bonding and electrostatic interactions. The SEM images of ChEt-ChOx/PtNP/PPY/ITO nanobio-electrode exhibit uniform distribution of the enzymes indicating that PtNP/PPY/ITO film provides a desired microenvironment for strong adsorption of enzymes.

Figure 3 demonstrates FTIR spectra of (a) PPY/ITO, (b) PtNP/PPY/ ITO, (c) ChEt-ChOx/PtNP/PPY/ITO electrodes. The **Figure 3**(a) shows the characteristics absorption bands of polypyrrole. The peaks at 1580 cm^{-1} and 1470 cm^{-1} are due to the fundamental vibration of the pyrrole ring [33]. The peaks at 1105 cm^{-1}, 1055 cm^{-1} and 1320 cm^{-1} are attributed to the =C-H in plane vibration [33]. The broad band at 1178 cm^{-1} may be assigned to N-C stretching band (Nicho and Hu, 2000). All the characteristics peaks of PPY/ ITO electrode are visible in the FTIR spectrum of PtNP/PPY/ITO (**Figure 3**(b)). The peak at 3416 cm^{-1} is attributed to N-H stretch vibration. The band at 1320 cm^{-1}corresponds to =C-H band in plane vibration. The FTIR spectrum of ChEt-ChOx/ PtNP/PPY/ITO nanobioelectrode (**Figure 3**(c)) also reflects the characteristic peaks of PPY/ITO electrode [34]. The ChOx and ChEt binding is indicated by the appearance of additional absorption bands at 1646 and 1560 cm^{-1} assigned to the carbonyl stretch (amide I band) and -N-H bending (amide II band), respectively [35]. Besides that, a broadband seen around 3457 cm^{-1} is attributed to amide bond present in nano bioelectrode.

Figure 4 exhibits the plot of DPV experiments, conducted in phosphate buffer (50 mM, pH 7.0) containing 5 mM [Fe(CN)$_6$]$^{3-/4-}$ in the range 0.02 to 0.8 V (**Figure 4**). The value of maximum response current obtained as 4.39×10^{-5} A for ITO (curve a) increases to 1.88×10^{-4} A for PPY/ITO (curve b). After addition of Pt-NP on PPY/ITO electrode, magnitude of current increases to 2.10×10^{-4} A (curve c).

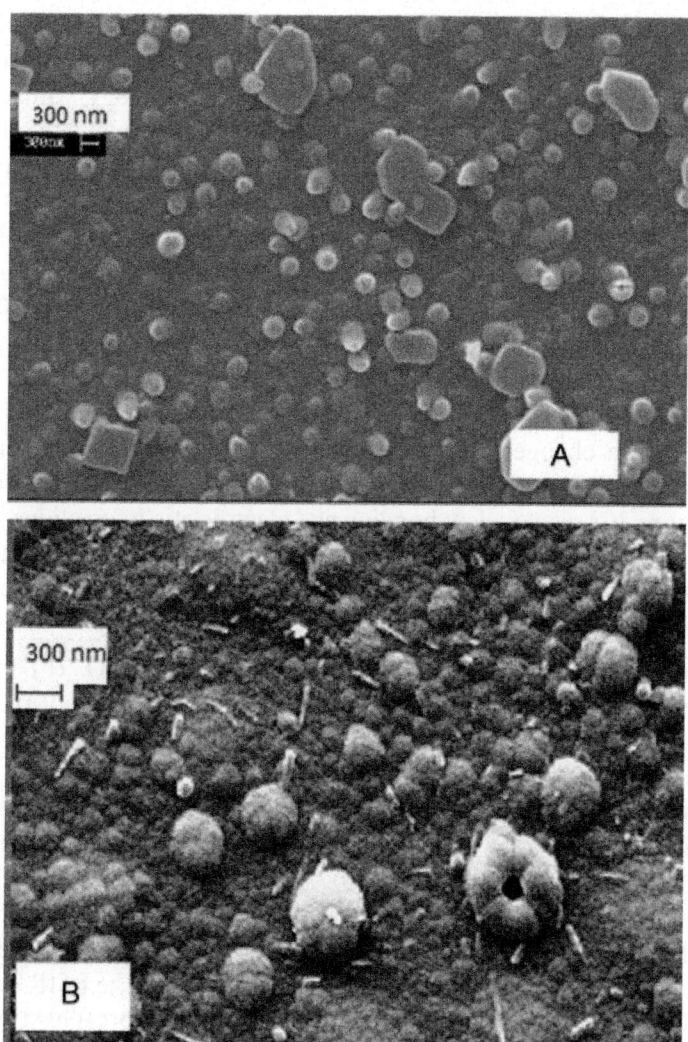

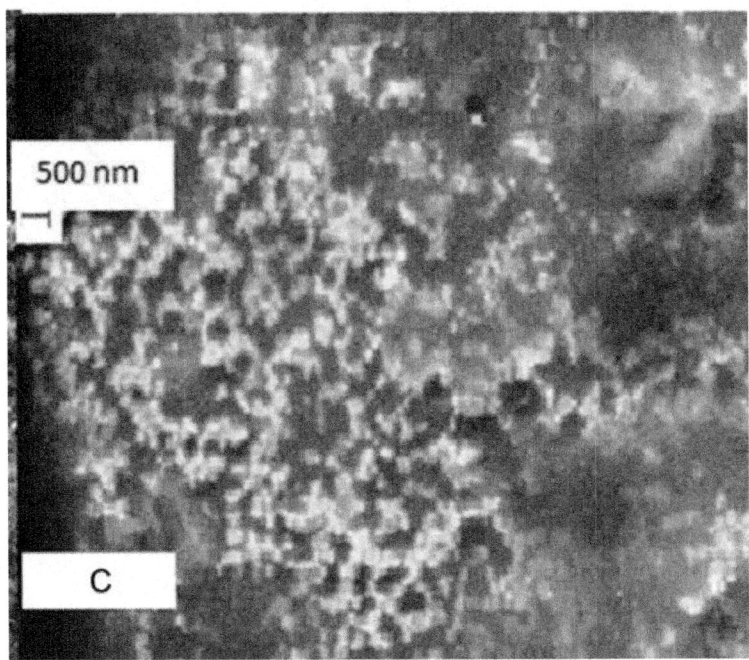

Figure 2: SEM images of the modified electrodes. A. PPY/ ITO; B. PtNP/PPY/ITO; C. ChOx-ChEt/PtNP/PPY/ITO.

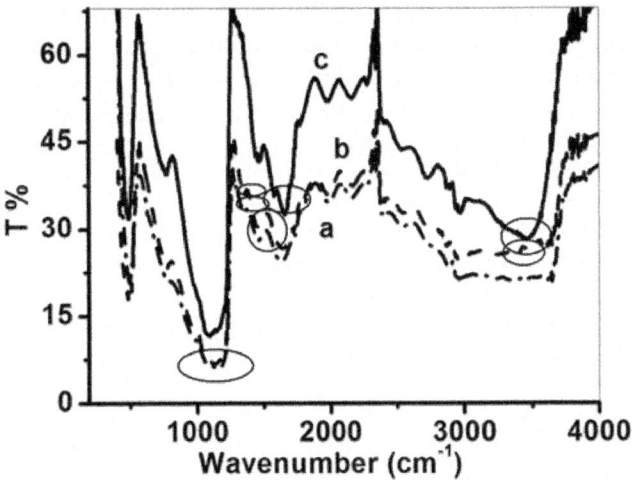

Figure 3: FTIR spectra of (a) PPY/ITO, (b) PtNP/PPY/ITO and (c) ChOx-ChEt/PtNP/ PPY/ITO.

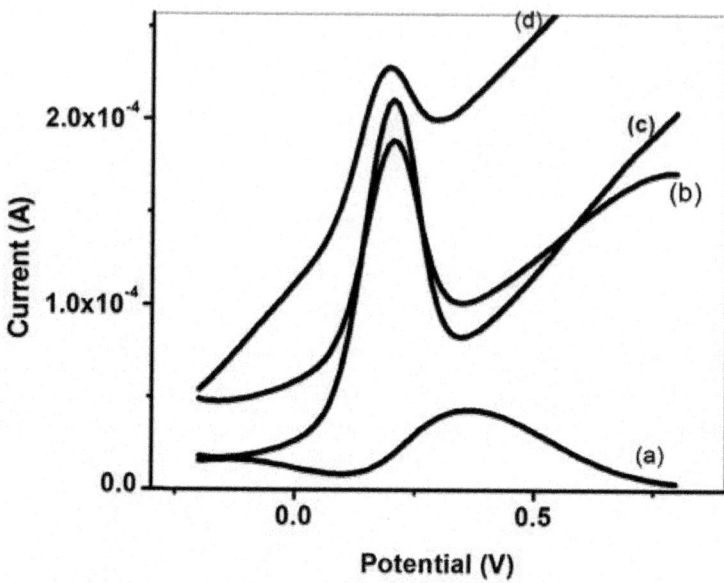

Figure 4: Differential Pulse voltammogram of (a) bare ITO, (b) PPY/ITO, (c) PtNP/PPY/ITO and (d) ChOx-ChEt/PtNP /PPY/ITO.

Interestingly, the magnitude of peak current again increases to 2.29×10^{-4} A to for ChOx-ChEt/ PtNP/PPY/ITO nanobioelectrode indicating PtNP/PPY/ITO nanocomposite provides a favorable microenvironment for immobilization of enzymes wherein the immobilized enzymes have better conformation and they retain their natural activity. These results are further confirmed by cyclic voltammetric and impedimetric study. In general, bioelectrode offers less current as compared to pristine electrode due to insulating nature of the bioelectrode.

The **Figure 5** shows cyclic voltammograms of (a) bare ITO (b) PPY/ITO(c) PtNP/PPY/ITO nano composite electrode and (d) ChOx-ChEt/ PtNP / PPY/ITO nano bioelectrode. The response current obtained for the PPY/ ITO electrode (curve b) is higher than that of the bare ITO electrode (curve a) and the response current of PtNP/PPY/ITO electrode (5.12×10^{-4} A) is 1.5 times higher than that of PPY/ITO electrode (3.32×10^{-4} A). This clearly indicates that PtNP have increased electroactive surface area of PPY/ITO electrode and the use of PtNP/PPY/ITO electrode leads to accelerated electron transfer between medium and electrode. Moreover, the peak response current further increases after immobilization of ChOx-ChEt onto PtNP/PPY/ITO electrode (Figure 5(d)) revealing that PtNP provides a favorable conformational environment for ChOx-ChEt loading resulting in improved electron transport between ChOx-ChEt and electrode. The results are also supported by the DPV results as shown in **Figure 4**.

Figures 6A and B show the CV of PtNP/PPY/ITO electrode and ChOx-ChEt/PtNP/PPY/ITO bioelectrode as a function of scan rate (20 - 160 mV/s). The relationships between the peak current and corresponding potential with scan rate are shown in **Table 1" target="_self"> Table 1**. It can be seen that the magnitudes of cathodic peak and anodic peak current increase with increasing scan rate (Figures 6A(a) and B(a)). This linear dependence can be expressed by the Equations (1) to (4) in **Table 1**.

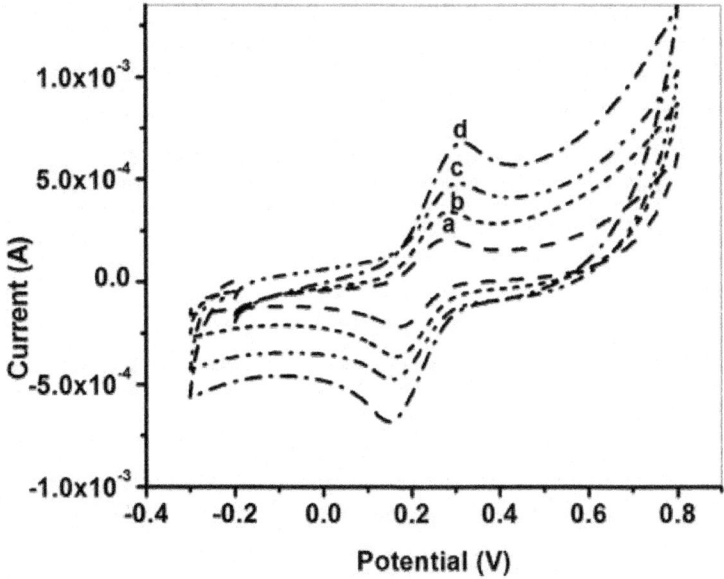

Figure 5: Cyclic voltammogram of (a) bare ITO, (b) PPY/ ITO, (c) PtNP/PPY/ITO nano electrode and (d) ChOxChEt/ PtNP/PPY/ITO nano bioelectrode.

Both the anodic and cathodic peak current exhibit a linear relationship with the scan rate, suggesting that electro-chemical reaction is a diffusion-controlled process for both nano composite and nano bioelectrode. The peak potential also increases linearly with log of scan rate (Figures 6A(b) and B(b)) indicating a diffusion controlled process. The values of slope obtained from the plots of the peak current against the scan rate for ChOx-ChEt/ PtNP/PPY/ITO nanobioelectrode are higher than the slope of the PtNP/PPY/ITO electrode indicating a favorable microenvironment for enzymes with nanobioelectrode. **Table 1" target="_self"> Table 1**(Equations (1)- (8)) shows linear equations, slope, Regression Coefficient, and Standard deviation for the electrochemistry of the electrodes. The characteristics features of the nanobioelectrode as shown in **Table 1" target="_self"> Table 1** reflect that good electron transport is retained in the ChOx-ChEt/PtNP/PPY/ ITO nano bioelectrode.

Electrochemical Impedance Spectroscopic Studies of PtNP/PPY/ ITO Electrode and ChOx-ChEt/PtNP/PPY/ITO

During the last decade, Electrochemcial Impedance Spectroscopy (EIS) has been widely used for probing various types of biomolecular interactions based immunosensors, DNA hybridization, rapid biomolecular screening, cell culture monitoring and relevant literature has been comprehensively reviewed [31,36-38]. EIS is a measure of the response of an electrochemical cell subjected to a small amplitude sinusoidal voltage signal as a function of frequency [39,40]. The resulting sinusoidal current expressed with respect to the perturbing (voltage) wave, and the ratio $V(t)/I(t)$ is defined as the impedance (Z). Impedance is expressed as a complex number, where the real component measures the ohmic resistances and the imaginary one is the capacitive reactance. The Nyquist and Bode plots are the most popular formats for evaluating electrochemical impedance data. of scan rate. B. Cyclic voltammograms of ChOx-ChEt/ PtNP/ PPY/ITO nano bioelectrode as a function of scan rate (20 to 160 mV/s). Inset: (a) plot of variation of current and scan rate and (b) variation of potential and log of scan rate.

In the Nyquist format, the imaginary impedance component (Z out-of-phase) is plotted against the real impedance component (Z in-phase) at each excitation frequency, whereas in the latter format, both the logarithm of the absolute impedance, |Z| and the phase shift, are plotted against the logarithm of the excitation frequency [40][2].

Table 1: Characteristics of PtNP/PPY/ITO nanocomposite electrode and ChOx-ChEt/ PtNP/PPY/ITO nanobioelectrode

Electrode	Slope	Regression Coefficient	Standard Deviation	Linear Equation	
PtNP/PPY/ITO	9.3	0.99	6.22E−05	$I_p = 244 + 9.32 \times$ scan rate	(1)
	−8.6	0.98	1.03E−04	$I_c = -259 - 8.6 \times$ scan rate I	(2)
	0.1	0.98	0.004	$E_p = 0.14 + 0.1 \times \log|(\text{scan rate})|$	(3)
	−0.09	0.98	0.004	$E_c = 0.298 - 0.09 \times \log(\text{scan rate})$	(4)
ChOx-ChEt/PtNP/PPY/ITO	11	0.98	9.8E−05	$I_p (A) = 248\ \mu A + 11\ \mu A \times$ scan rate	(5)
	−9.02	0.97	1.08E−04	$I_c = -367 - 9.02 \times$ scan rate	(6)
	0.1	0.98	0.004	$E_p = 0.14 + 0.1 \times \log(\text{scan rate})$	(7)
	−0.1	0.98	0.005	$E_c = 0.310 - 0.1 \times \log(\text{scan rate})$	(8)

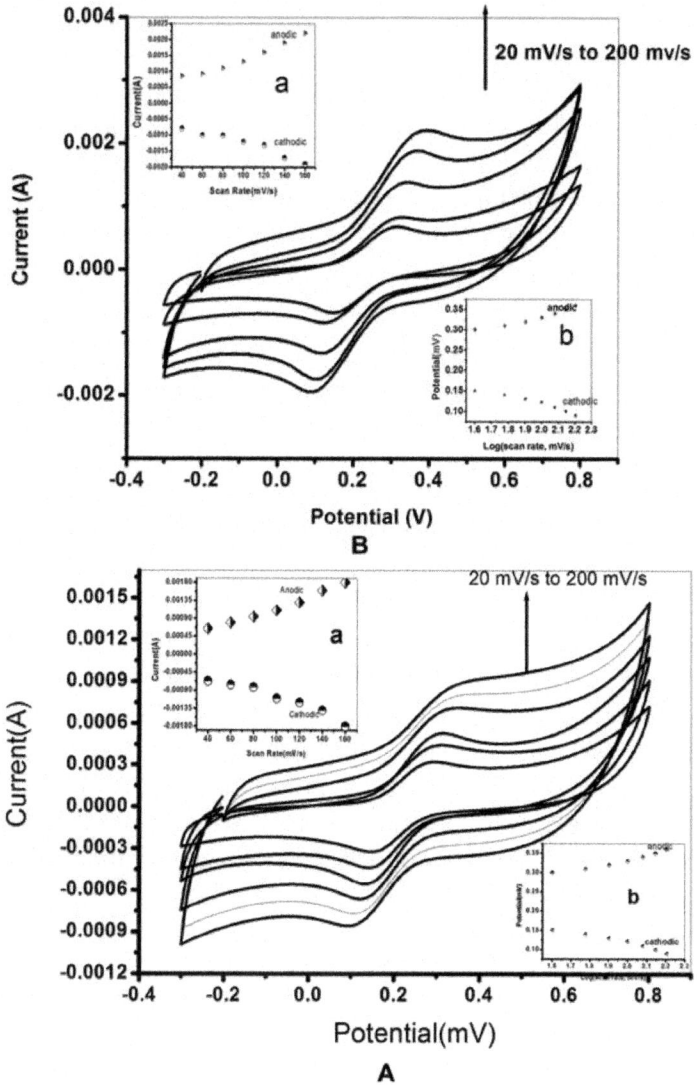

Figure 6: A. Cyclic voltammograms of (a) PtNP/PPY/ITO nano electrode as a function of scan rate (20 to 200 mV/s). Inset: (a) plot of variation of peak current (anodic and cathodic) and scan rate and (b) variation of potential and log

The complex impedance can be presented as the sum of the real (Z) and imaginary (Z) components that mainly originate from the resistance and capacitance of the cell and can be calculated using Equation (9) (for parallel circuit).

$$Z = Z' + jZ = R_s + \frac{R_p}{\left(1 + w^2 C^2 R_p^2\right)} + \frac{C_{dl} R_p^2}{\left(1 + w^2 C^2 R_p^2\right)}$$

(9)

where, R_s = Ohmic resistance of the electrolyte solution; R_p = Polarization resistance [R_p at 0 potential describes as electron charge transfer resistance (R_{CT})], w = Radial frequency and C_{dl} = Double layer capacitance.

C_{dl} calculated by following Equation (10).

$$C_{dl} = \pi f_{max} R_{CT}$$

(10)

C_{dl} and R_{CT}, depend on the dielectric and insulating features at the electrode/ electrolyte interface. The semicircle diameter of EIS spectra gives value of R_{CT} that reveals charge-transfer resistance of redox probe at the electrode interface. However, R_s represents bulk properties of the electrolyte solution and diffusion of applied redox probe are not affected by biochemical reaction occurring at the electrode interface[3]. The Equations (9) and (10) represent relationship between the total impedance and other electrochemical parameters such as R_{CT} and C_{dl}. The interfacial properties of the electrode in the electrolyte depend on charge transfer resistance (R_{CT}) and capacitance (C_{dl}).

The **Figure 7** shows the impedance spectra of (a) bare ITO (b) PPY/ITO (c) PtNP/PPY/ITO and (d) ChOxChEt/PtNP/PPY/ITO. It has been observed that charge transfer resistance (R_{CT}) value for PPY/ITO electrode (821.3 Ω, **Figure** 7, curve b) at 0 V bias potential is smaller than that of the bare ITO electrode (1044.8 Ω). Similarly, the value of charge transfer resistance (586.8 Ω, R_{CT}) of PtNP/PPY/ITO electrode is lower (**Figure 7**, curve c) than PPY/ ITO electrode, revealing that PtNP increases the electroactive surface area of electrode resulting in easier electron transfer from the medium to electrode due to increased diffusion of redox species [Fe(III)/Fe(IV)] onto surface charged nanocomposite PtNP/PPY/ITO film. Interestingly, R_{CT} value further decreases to 353.92 Ω (**Figure 7**, curve d) after the immobilization of ChOx-ChEt onto PtNP/PPY/ITO revealing that PtNP provides a desirable microenvironment for the immobilization of ChOx-ChEt [41]. It can be assumed that ChOx-ChEt rearrange their structure and presents better conformation onto PtNP/PPY/ITO electrode. Thus ChOx-ChEt/PtNP/PPY/ITO nano bioelectrode provides easier electron transfer due to increased active sites for electrical contact between electrode and the redox label in solution. In practice, an ideal semicircle characteristics is generally not observed in impedance spectra. It is normally an inclined semicircle with its centre depressed below the real axis by a finite angle referred to as the depression angle. The depression angle (θ) by which such a semicircular is displaced below the real axes, is related to the width of the relaxation time distribution and is an important parameter. It can be seen that arc of the semicircle with least q value, that is less distorted from centre of the real axis can be calculated using Equation (11).

Depression angle

$$\theta = \left(\frac{(1-n)\pi}{2} \right)$$

(11)

where, n is the fractional exponent and calculated form the standard commercial software available with the instrument. From the above discussion, it is clear that it is necessary to find out the optimum condition at which the minimum depression angle is obtained for both type of electrodes and the cholesterol nano biosensor should be evaluated at that optimum condition. The EIS spectra of PtNP/PPY/ITO electrode and ChEt-ChOx/PtNP/PPY/ ITO bioelectrodes are studied as a function of potential (0 - 0.20 V) to locate an ideal condition.

The Figures 8A and B, shows the EIS spectra of PtNP/PPY/ITO electrode and ChOx-ChEt/PtNP/PPY/ ITO bioelectrode, respectively as a function of potential (0.0 to 0.2 V).

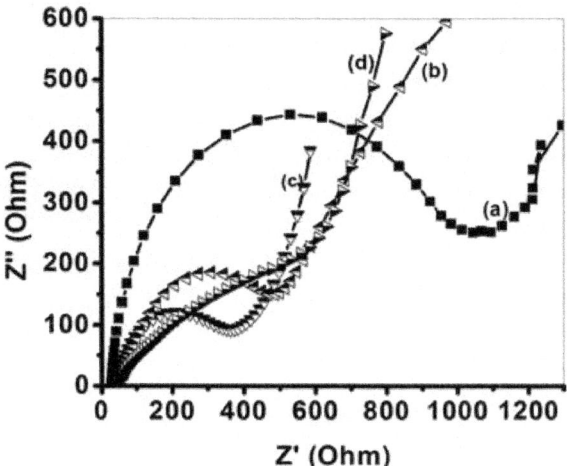

Figure 7: Electrochemical impedience spectra of (a) bare ITO, (b) PPY/ITO, (c) PtNP/PPY/ITO and (d) ChOx-ChEt/ PtNP/PPY/ITO.

The R_{CT} values both for the PtNP/PPY/ ITO electrode and ChOx-ChEt/ PtNP/PPY/ITO nano bio-electrode have been found to decrease linearly with increase in bias potential (Figures 8A(b) and B(b)) and they follow Equations (12) and (13).

$$R_{CT} \left[\text{PtNP/PPY/ITO} \right] = 622 - 2372 \times E$$

with regression coefficient of 0.99.

(12)

$$R_{CT} \left[\text{ChOx-ChEt/PtNP/PPY/ITO} \right] = 349 - 1354 \times E$$

with regression coefficient of 0.99.

(13)

This reveals facile electron transfer kinetics wherein resistance controls the electron transfer kinetics of the redox probe at the electrode. This electron transfer resistance (R_{CT}) can be translated into exchange current under the equilibrium (I_0) and then heterogeneous electrontransfer rate constant (k_{CT}) according to the Equation (14) [30]. The value of k_{CT} for ChOx-ChEt/PtNP/ PPY/ITO bioelectrode is higher (7.6×10^{-5} cm/s) than that of the PtNP/PPY/ ITO electrode (4.3×10^{-5} cm/s).

$$R_{CT} = \frac{RT}{nFI_0}, \; I_0 = nFAk_{CT}\left[S\right], \; k_{CT} = \frac{RT}{n^2 F^2 AR_{CT}\left[S\right]}$$

(14)

where, R = Gas constant, T = Absolute temperature (K), F = Faraday constant, A = Electrode area (cm²), S = Bulk concentration of redox probe (mol·cm³) and n = Number of transferred electrons per molecule of the redox probe.

Figure 9 shows the variation of depression angle (θ) obtained for PtNP/ PPY/ITO electrode and ChOx-ChEt/ PtNP/PPY/ITO bioelectrode as a function of applied potential. It is observed that values of depression angle (θ) for PtNP/ PPY/ITO electrode and ChOx-ChEt/PtNP/PPY/ ITO bioelectrode decrease on increasing the potential from 0 to 0.2 V. However, 0.06 V is the minimum potential at which the values of (θ) obtained for PtNP/PPY/ ITO electrode (0.4°) and ChOx-ChEt/PtNP/PPY/ITO bioelectrode (0.3°) are close to the values of depression angle (θ) at higher potential (0.2 V). This suggests that 0.06 V can be chosen as the working potential for electro-chemical sensing wherein the semicircle of the experimental data is near the centre of real axes with complex planes.

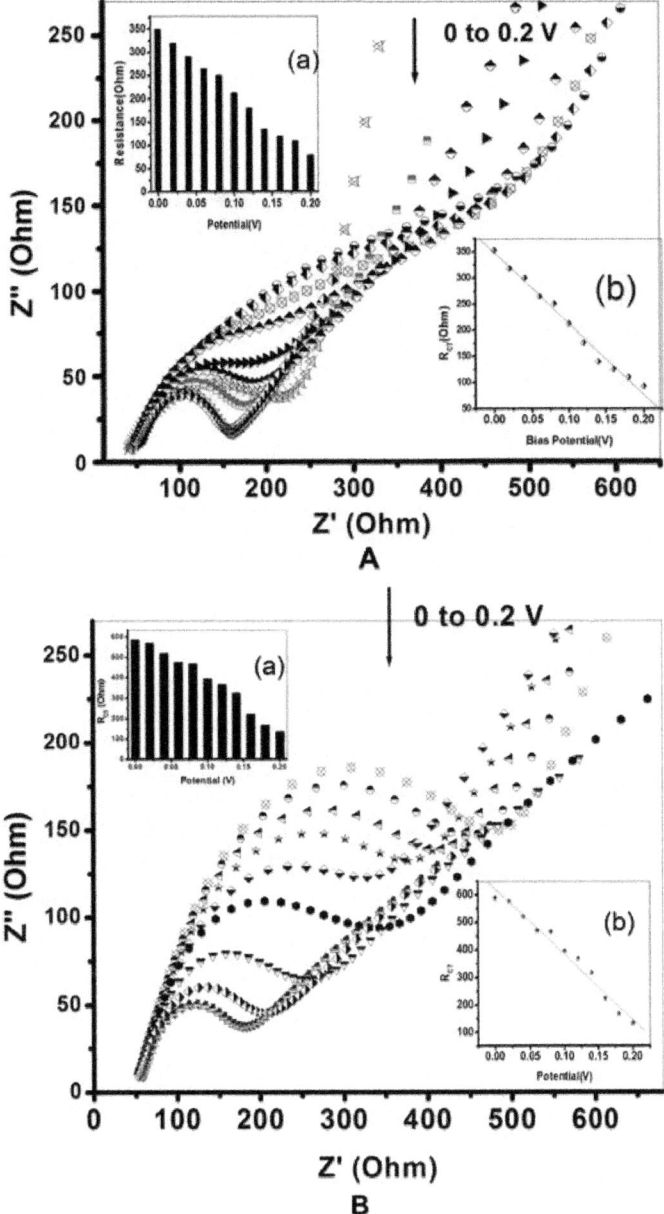

Figure 8: A. Electrochemical Impedance spectra [EIS] of PtNP/PPY/ITO nano-electrode as function of potential (0 - 0.2 v). Inserts: (a) Plot of charge transfer resistance (R_{CT}) and applied potential; (b) Linear relationship between R_{CT} and potential. B. Electrochemical Impedance spectra [EIS] of Chox-ChEt/PtNP/PPY/ITO nano bioelectrode as function of potential (0 - 0.2 v). Inserts: (a) Plot of charge transfer resistance (R_{CT}) and potential; (b) Linear relationship between R_{CT} and Potential.

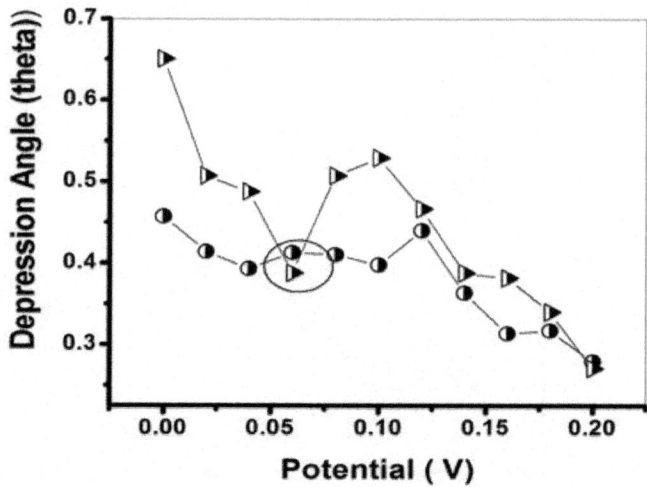

Figure 9: Plot between depression angle and applied potential for PtNP-PPY/ITO and ChOx-ChEt/PtNP-PPY/ITO electrode.

It is necessary to mention here that impedance behavior of an electrode depends on the circuit, it follows. In order to find out the best fitted circuit, impedance spectra of PtNP/PPY/ITO has been compared with computer simulated spectra using six different electronic circuits (EC) based on Randles and Ershler theoretical model. **Figure 10** shows that the Faradic impedance spectra of PtNP/PPY/ITO electrode, fitted with computer simulated spectra using six different electronic circuits (EC). The best fit between the simulated electronic circuit and experimental spectra is obtained for electronic circuit EC_4 i.e., EC_4 [R(Q[R(Q[RT])])]. And the minimum error has been observed in EC4 as compared to other circuits.

Figure 11 shows EIS spectra simulated by EC_4 [R(Q[R(Q[RT])])] for PtNP/PPY/ITO electrode and ChOx-ChEt/PtNP/PPY/ITO nano bioelectrode at potential 0.06 V. It can be seen that R_{CT} value of for ChOxChEt/PtNP/PPY/ITO bioelectrode (264.98 Ω) is lower than that of the PtNP/PPY/ITO electrode (475.04 Ω) revealing that PtNP provides a desired micro-environment for immobilization of ChOx-ChEt where in the immobilized enzyme has better conformation and it retains its natural activity. But the values of double layer capacitance, C_{dl} are almost identical for the electrodes (**Table 2**).

Figure 12 shows the enzyme activities of the ChOxChEt/PtNP/PPY/ITO nano bioelectrodes, evaluated as a function of pH varying from 6.0 to 7.8 at room temperature (25°C) with 100 mg/dl cholesterol oleate solution. The least value of R_{CT} obtained at pH 7.0 (**Figure 12**) indicates that ChOx-ChEt/PtNP/ PPY/ITO bioelectrodes are more active at pH 7.0 at which ChOx and ChEt molecules retain the natural structure and do not get denatured. Thus all the experiments have been conducted at pH 7.0 at 25°C.

Figures 13A and B represents the response studies of the ChOx-ChEt/ PtNP/PPY/ITO nano bioelectrode which have been investigated as a function of cholesterol oleate concentration (100 μL of 2.5×10^{-4} to 6.5×10^{-3} M/l) at the bias potential of 0.06 V and pH 7.0. The observed experimental data of ChOx-ChEt/PtNP/PPY/ITO nanobioelectrode as a function of cholesterol oleate concentration data is fitted using EC_4 [R(Q[R(Q[RT])])]. The surface charge transfer resistance (R_{CT}) decreases with increase in cholesterol oleate concentration as shown in **Figure 13**A(a).

Table 2. The parameters of electrochemical impedance spectra [EIS] for (a) PtNP/ PPY/ITO and (b) ChOx-ChEt/ PtNP/PPY/ITO electrodes at 0.06 V

Curve (Figure 10)	Electrode	Rs-value	Cdl-value	RCT-value
a	PtNP/PPY/ITO	48.18	6.3665E−06	475.04
b	ChOx-ChEt/PtNP/PPY/ITO	41.44	5.9794E−06	264.98

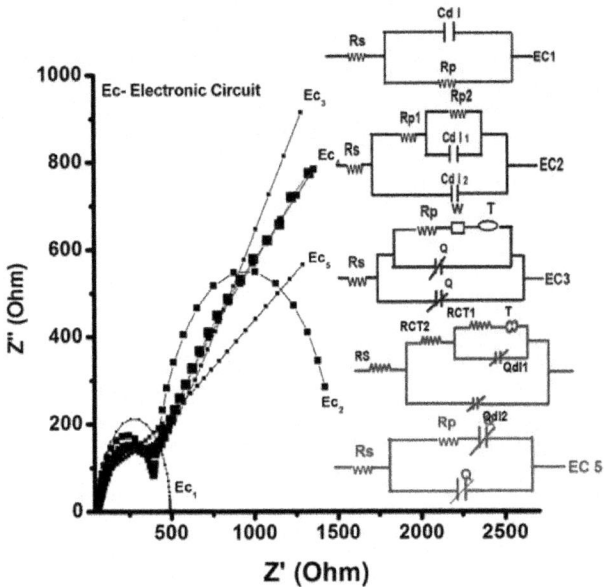

Figure 10. Electrochemical impedance spectra of PtNP/ PPY/ITO electrode fitted using different electronic circuits.

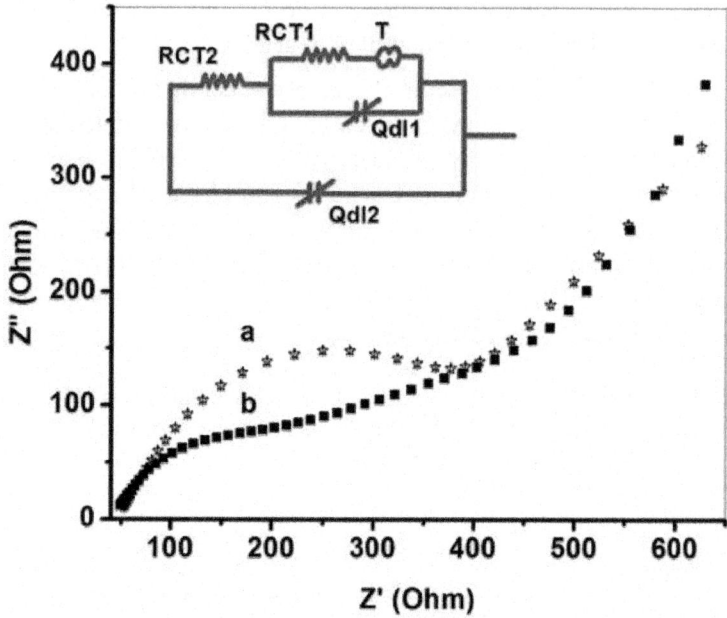

Figure 11: Electrochemical Impedance spectra of (a) PtNP/ PPY/ITO and (b) ChOx-ChEt/PtNP/PPY/ITO at 0.06 v.

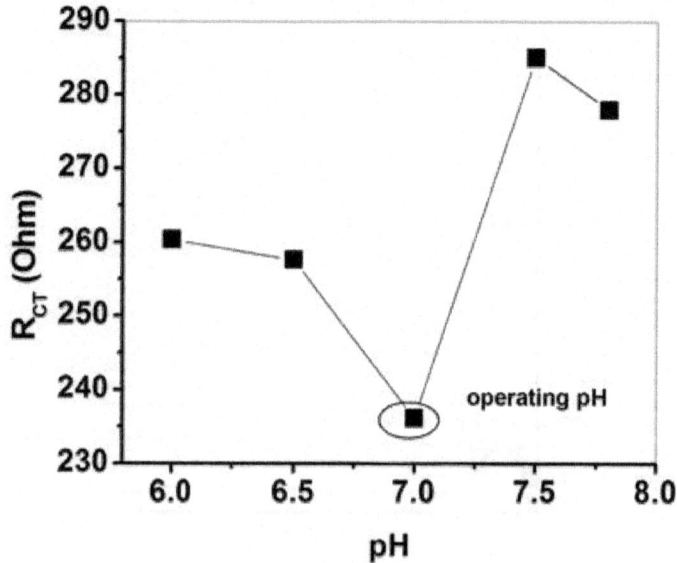

Figure 12: Electrochemical impedience spectra of ChOxChEt/PtNP/PPY/ITO nano-bioelectrode for pH studies.

At the same time double layer capacitance (C_{dl}) of nanobioelectrodes increases with increasing the concentration of cholesterol oleate as shown in **Figure 13**A(b). This can be attributed to the fact that ChOx and ChEt present on ChOx-ChEt/PtNP/PPY/ITO nanobioelectrodes convert cholesterol oleate to choleste- 4-ene-3-one. The generated electrons during the reoxidation of ChOx after enzymatic reaction are transferred to the ChEt-ChOx/PtNP/PANI/ITO electrode via Fe(III)/ Fe(IV) couples that helps in enhanced charge transfer rate leading to decreased in R_{ct} value resulting in increased sensitivity of the sensor. The biochemical reaction mechanism at ChOx-ChEt/PtNP/PPY/ITO bioelectrode is shown in Scheme 2. The overall biochemical reaction mechanism demonstrates the decrease in charge transfer resistance of the nano bioelectrode with increase in cholesterol oleate concentration as the ionic conductivity of the electrode increases.

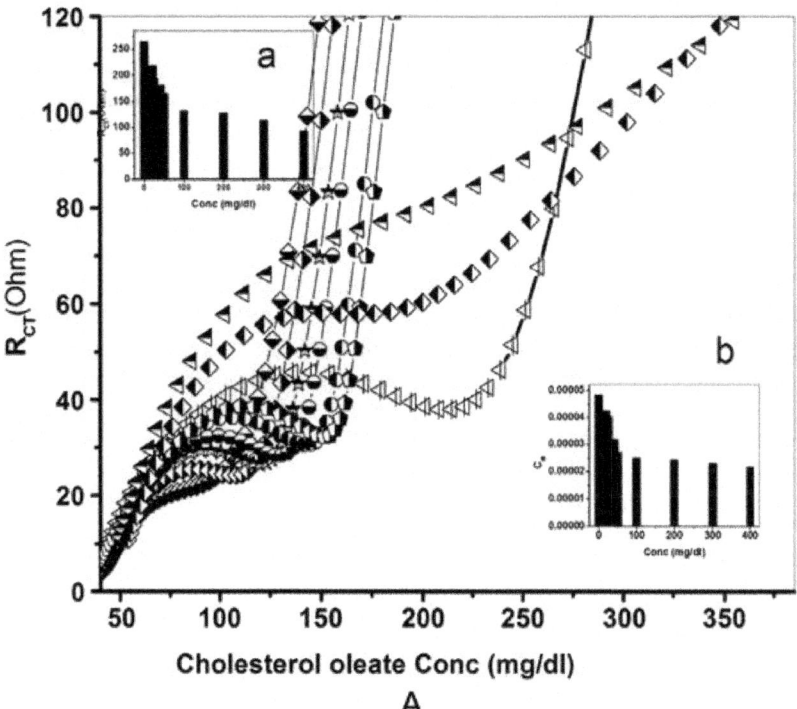

A

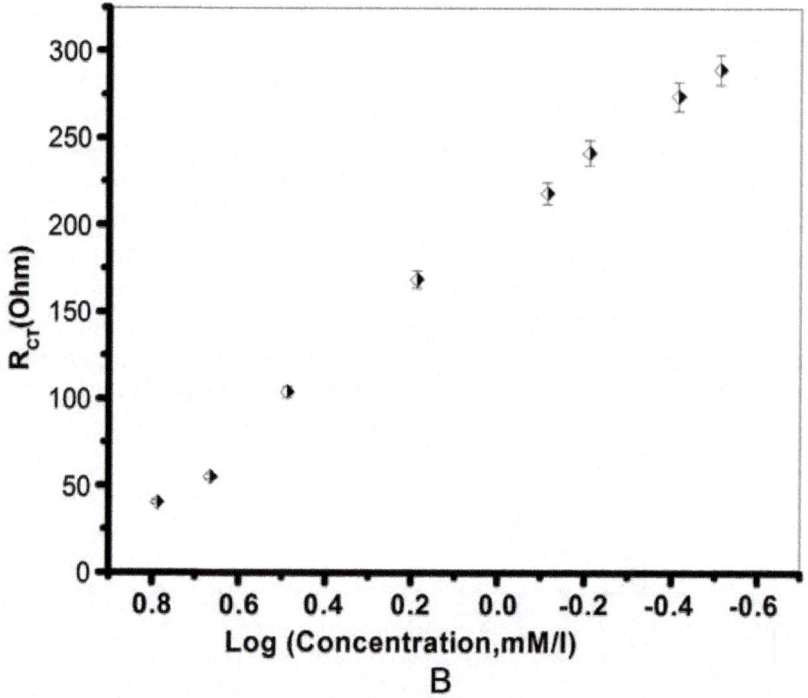

Figure 13: A. Electrochemical Impedance spectra [EIS] response of ChOx-ChEt/PtNP/PPY/ITO bioelectrode as a function of cholesterol oleate concentration (0.3 - 6.1 mM) at bias potential of 0.06 V and fit using EC4.13; B. Calibration curve between R_{CT} value and logarithm of cholesterol oleate concentration (0.3 - 6.1 mM).

The Equation (15) represents the linear regression for ChOx-ChEt/PtNP/ PPY/ ITO bioelectrode as obtained from the calibration curve by plotting R_{CT} vs log of cholesterol oleate concentration.

$$R_{CT} = 195 - 196 \times \log\left[\text{cholesterol concentration}\right]$$

(15)

with a correlation coefficient of 0.99.

The results of experiments carried out in triplicate sets reveal reproducibility of the system within 5%. It has been shown that the ChOx-ChEt/PtNP/PPY/ ITO bioelectrodes represents characteristics such as low detection limit (0.25 mM), fast response time (20 s), high sensitivity (196 Ω/mM/cm^2). The detection limit is determined by the minimum concentration of cholesterol oleate at which decrease in R_{CT} value as compared to bioelectrode is observed. As impedance measurement is carried out at 0.06 V, the developed nano bioelectrode shows excellent reusability (data not shown). The shelf-life of the ChOxChEt/PtNP/PPY/ITO nanobioelectrode measured after an interval of 1 week has been estimated as 7 weeks. The decrease in the value of R_{CT} has been found to be about 10% up to about 7 weeks after which the current decreases sharply resulting in about 70% loss in about 10 weeks (data not shown). The value of the MichaelisMenten constant (Km) obtained as 0.2 mM for ChOxChEt/PtNP/PPY/ITO using Lineweaver-Burke plot reveals that PtNP/PPY/ITO matrix facilitates enzymatic reaction and helps the immobilized enzyme to achieve better conformation for faster enzymatic reaction resulting in enhanced enzymatic activity. The conformational changes are known to affect a biochemical reaction at bioelectrode that in turn may be influenced by the surface morphology and the nature of immobilization matrix. The double layer capacitance (C_{dl}) generally increases with the increase of charge transfer resistance (R_{CT}). The same trend is observed with the C_{dl} value of the nanobioelectrode with the concentration of cholesterol oleate (**Figure 13**A(a) and (b)).

Figure 14 shows the selectivity of ChOx-ChEt/PtNP/ PPY/ITO nanobioelectrode, estimated by comparing magnitude of R_{CT} by adding normal concentration of interferents such as glucose (5 mM), ascorbic acid (0.05 mM), uric acid (0.1 mM), urea (5 mM), sodium pyruvate acid (0.5 mM) and sodium ascorbate (0.05 mM). The role of interferents has been examined by mixing equal volume of desired interferent with cholesterol. In **Figure 14**, the first bar (Cho-oleate) shows the value of charge transfer resistance (R_{CT}) obtained with 3.6 mM cholesterol oleate. The remaining bars show the variation of the R_{CT} corresponding to the mixture of cholesterol oleate and interferents in a 1:1 ratio. The straight line parallel to the X-axis shows the observed R_{CT} in the presence of desired interferents, revealing a maximum of 7.4% interference. The percentage interference (% inter) has been calculated using Equation (16) for various interferents:

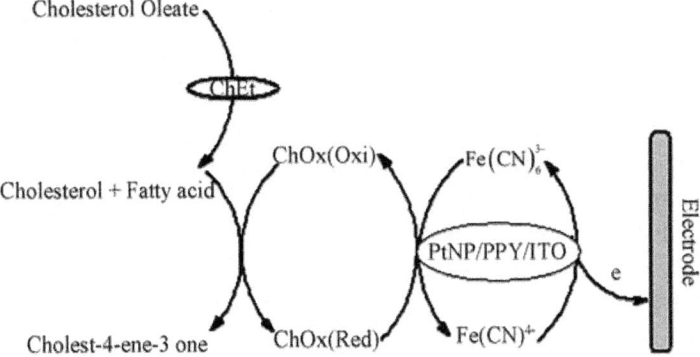

Scheme 2: Schematic diagram for biochemical reaction at ChOx-ChEt/PtNP/PPY/ITO bioelectrode for total cholesterol detection.

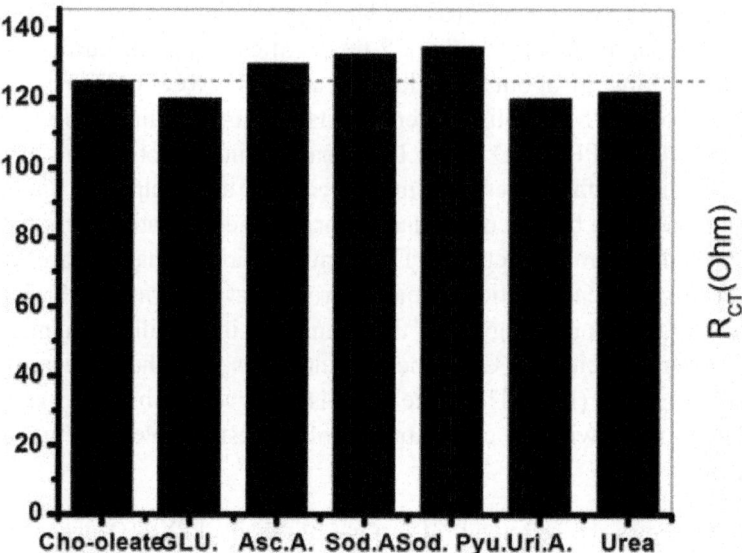

Figure 14: Electrochemical impedience spectral of ChOxChEt/PtNP/PPY/ITO nano-bioelectrode for interferent studies.

$$\% \text{ Interference} = \frac{\left[R_{CT_{cho}} - R_{CT_{Int}} \right]}{R_{CT_{cho}}} \times 100$$

(16)

where, $R_{CT_{cho}}$ is the charge transfer resistance of the nanobioelectrode obtained with 3.6 mM cholesterol oleate concentration and $R_{CT_{Int}}$ is the charge transfer resistance corresponding to the mixture of cholesterol oleate and interferents in a 1:1 ratio. It has been found that ChOx-ChEt/PtNP/PPY/ITO nano bioelectrode is not significantly affected by the presence of interferents The biosensing characteristics of ChEt-ChOx/PtNP/PPY/ITO bioelectrode has been compared with other total cholesterol biosenosrs, reported in literature have been summarized in **Table 3**. It can be noted that the Km value obtained by the impedance measurement reflects actual interaction between enzyme and the substrate as the impedance spectra depends on interfacial property.

CONCLUSION

The matrices of Pt-nanoparticle (PtNP) and PPY multilayer electrodes have been fabricated for the development of total cholesterol biosensor. The novelty of

Table 3: Shows characteristics of the ChEt-ChOx/PPYMWCNT/PTS/ITO bio-electrode including reported in the literature.

Sl No	Components of biosensor	Characteristics	Reference
1)	[Mat]: Ppy [E]: ChOx, ChEt [Mol]: Adsorption [M]: Ampero. vs Ag/AgCl	[L]: 1 - 8 mM/l [SL]: 4 weeks	Singh et al. [21]
2)	[Mat]: Polyaniline [E]: ChEt, ChOx [Mol]: Covalent [M]: Ampero. vs Ag/AgCl	[L]: 50 - 500 mg/dl [RT]: 40 s [SL]: 6 weeks [S]: 7.4×10^{-4} nA	Singh et al. [22]
3)	[Mat]: Polyaniline [E]: ChE, ChOx, HRP [Mol]: Covalent [M]: Ampero. vs Ag/AgCl	[L]: up tp 500 mg/dl [K_m]: 75 mg/dl [RT]: 240 s [SL]: 6 week [S]: 0.042 µA	Singh et al. [23]
4)	[Mat]: Dithiobissuccinimidyl propionate [E]: ChOx, ChEt [Mol]: Covalent [M]: Ampero. vs Ag/AgCl	[L]: 50 - 400 mg/dl [K_m]: 0.95 mM [Sl]: 7 weeks	Arya et al. [40]
5)	[Mat]: 3-aminopropyl-modified controlled-pore glass (APCPG) [E]: ChOx, ChEt, HRP [Mol]: Covalent [M]: Ampero. vs Ag/AgCl	[L]: 1.2 µl - 1 mM/l [K_m]: 2 mM [RT]: 120 s [SL]: 3 weeks	Arya et al. [42]
6)	[Mat]: AT cut quartz crystal [E]: ChOx, ChEt [M]: Impedence	-	Arya et al. [43]
7)	[Mat]: CHIT-SiO$_2$-MWCNT/ITO [E]: ChET, CHOx [Mol]: Covalent [M]: Ampero. (DPV) vs Ag/AgCl	[L]: 0.15 - 7.68 mM [K_m]: 0.52 mM	Pratima et al. [25]
8)	[Mat]: Polypyrrole/MWCNT nano composite [E]: ChET, CHOx [Mol]: Covalent [M]: Ampero. (DPV) vs Ag/AgClk	[L]: 0.4 mM to 6.5 mM/l [K_m]: 0.02 mM [RT]: 9 s [Sl]: 9 weeks	Basu et al. [29]
9)	[Mat]: Polypyrrole/PtNP composite [E]: ChET, CHOx [Mol]: Covalent [M]: Impedimetric vs Ag/AgClk	[L]: 0.4 mM to 6.5 mM/l [K_m]: 0.2 mM [RT]: 20 s [Sl]: 10 weeks	Present work

ChOx-ChEt/PtNP/PPY/ITO nanobioelectrode is the enhanced electronic conductivity over PtNP/PPY/ITO, in spite of insulating nature of ChOx and ChEt which demonstrates that the nanobioelctrode offers an enzyme friendly microenvironment to retain the bioactivity of the enzyme system. The above nano bioelectrodes ChOxChEt/PtNP/PPY/ITO offer an excellent performance in terms of linearity, sensitivity, detection limit, response time and shelf life. This is attributed to the presence of PtNP along with PPY to enhance the overall biochemical reaction. The unique features of the ChOx-ChEt/PtNP/ PPY/ ITO nanobioelectrode lie on the novelty of fabrication process, measurement technique, reusability, minimum interference and very low K_m value. The limitation of the impedance measurement is the semi-logarithm relationship of the calibration curve and selection of the representative circuit from the various Nyquist plots.

ACKNOWLEDGEMENTS

We thank Dr. Ashok Kumar Chauhan (Founder President, Amity University Uttar Pradesh) for providing the facilities. We are thankful to Dr. R. P. Singh, Director, AINT, AUUP for interesting discussions. We acknowledge the financial assistance received from the Department of Biotechnology, Govt. of India (Project No BTPR 11123/ MD/32/41/2008 DBT) for the financial support, India.

REFERENCES

1. M. Nauck, W. Marz, J. Jarausch, H. Cobbaert, A. Sagers, D. Bernard. O. Delanghe, G. Honauer, P. Lehmann, E. Oestrich, A. Von Eckardstein, S. Walch, H. Wieland and G. Assmann, "Multicenter Evaluation of a Homogeneous assay for HDL-Cholesterol without Sample Pretreatment," Clinical Chemistry, Vol. 43, No. 9, 1997, pp. 1622-1629.

2. P. Norouzi1, F. Faridbod, E. Nasli-Esfahan, B. Larijani and M. R. Ganjali, "Cholesterol Biosensor Based on MWCNTs-MnO$_2$ Nanoparticles Using FFT Continuous Cyclic Voltammetry," International Journal Electrochemistry, Vol. 5, No. 7, 2010, pp. 1008-1017.

3. Y. Chen and M. Gotoh, "Amperometric Needle-Type Glucose Sensor Based on a Modified Platinum Electrode with Diminished Response to Interfering Materials," Analytical Chimica Acta, Vol. 265, No. 1, 1992, pp. 5-14.

4. Z. Xu, X. Chen and S. Dong, "Electrochemical Biosensors Based on Advanced Bioimmobilization Matrices," Trends in Analytical Chemistry, Vol. 25, No. 9, 2006, pp. 898-908. http://dx.doi.org/10.1016/j.

trac.2006.04.008

5. M. Tian, G. Wu and A. Chen, "Unique Electrochemical Catalytic Behavior of Pt Nanoparticles Deposited on TiO_2 Nanotubes," American Chemical Society Catalyst, Vol. 2, 2012, pp. 425-432.

6. A. N. Hendji, P. Bataillard and N. Jaffrezic-Renault, "Covalent Immobilization of Glucose Oxidase on Silanized Platinum Microelectrode for the Monitoring of Glucose," Sensors Actuators B, Vol. 15, 1993, pp. 127-134. http://dx.doi.org/10.1016/0925-4005(93)85038-C

7. S. B. Hall, E. A. Khudaish and A. L. Hart, "Electrochemical Oxidation of Hydrogen Peroxide at Platinum Electrodes. Part V: Inhibition by Chloride," Electrochimica Acta, Vol. 45, No. 21, 2000, pp. 3573-3579. http://dx.doi.org/10.1016/S0013-4686(00)00481-3

8. M. Hepel, "The Electrocatalytic Oxidation of Methanol at Finely Dispersed Platinum Nanoparticles in Polypyrrole Films," Journal of Electrochemical Society, Vol. 145, No. 1, 1998, pp. 124-134. http://dx.doi.org/10.1149/1.1838224

9. S. Mokrane, L. Makhlouf and N. Alonso-Vante, "Electrochemical Behaviour of Platinum Nanoparticles Supported on Polypyrrole (PPy)/C Composite," ECS Transaction, Vol. 6, 2008, pp. 93-103.

10. S. S. Jeon, C. Kim, J. Ko and S. S. Im, "Pt Nanoparticles Supported on Polypyrrole Nanospheres as a Catalytic Counter Electrode for Dye-Sensitized Solar Cells," Journal of Physical Chemistry C, Vol. 115, No. 44, 2011, pp. 22035-22039.http://dx.doi.org/10.1021/jp206535c

11. M. Yang, Y. Yang, H. Yang, G. Shen and R. N. Yu, "LayerBy-Layer Self-Assembled Multilayer Films of Carbon Nanotubes and Platinum Nanoparticles with Polyelectrolyte for the Fabrication of Biosensors," Biomaterials, Vol. 27, 2006, pp. 246-255.http://dx.doi.org/10.1016/j.biomaterials.2005.05.077

12. H. Chen, R. Yuan, Y. Chai, J. Wang and W. Li, "Glucose Biosensor Based on Electrodeposited Platinum Nanoparticles and Three-Dimensional Porous Chitosan Membranes," Biotechnology Letter, Vol. 32, No. 10, 2010, pp. 1401-1404.http://dx.doi.org/10.1007/s10529-010-0303-z

13. R. Bittman, "Cholesterol: Its functional and Metabolism in Biology and Medicine," Plenum Press, New York, 1997. http://dx.doi.org/10.1007/978-1-4615-5901-6

14. P. L. Yeagle, "Biology of Cholesterol," CRC Press, Boca Raton, 1998.

15. C. F. Ana Maria, B. Oliveira, G. M. Helena and A. P. Piedade, "An Electrochemical Bienzyme Membrane Sensor for Free Cholesterol,"

Bioelectrochemistry Bioengineering, Vol. 28, 1992, pp. 105-115.

16. T. Siao, Y. C. Chen and C. A. Lee, "Amperometric Cholesterol Biosensors Based on Carbon Nanotube-Chitosanplatinum-Cholesterol Oxidase Nanobiocomposite," Sensors Actuators B, Vol. 135, No. 1, 2008, pp. 96-101. http://dx.doi.org/10.1016/j.snb.2008.07.025

17. S. Brahim, D. Narinesing and A. Guiseppi-Elie, "Amperometric Determination of Cholesterol in Serum Using a Biosensor of Cholesterol Oxidase Contained within a Polypyrrole-Hydrogel Membrane," Analytical Chimica Acta, Vol. 448, no. 1-2, 2001, pp. 27-36. http://dx.doi.org/10.1016/S0003-2670(01)01321-6

18. M. K. Ram, P. Bertoncello, H. Ding, S. Paddeu and C. Nicolini, "Cholesterol Biosensors Prepared by Layer-ByLayer Technique," Biosensors Bioelectronics, Vol. 16, No. 9, 2001, pp. 849-856. http://dx.doi.org/10.1016/S0956-5663(01)00208-1

19. S. Brahim, D. Narinesing and A. Guiseppi-Elie, "Amperometric Determination of Cholesterol in Serum Using a Biosensor of Cholesterol Oxidase Contained within a Polypyrrole-Hydrogel Membrane," Analytical Chimica Acta, Vol. 448, 2001, pp. 27-36.http://dx.doi.org/10.1016/S0003-2670(01)01321-6

20. M. L. Moraes, N. C. de Souza, O. H. Caio, F. Marystela, P. Ubirajara, F. Rodrigues, R. Antonio, Jr., Z. Valtencir and N. O. Osvaldo, Jr., "Immobilization of Cholesterol Oxidase in LBL Films and Detection of Cholesterol Using ac Measurements," Material Science Engineering C, Vol. 29, No. 2, 2009, pp. 442-447.http://dx.doi.org/10.1016/j.msec.2008.08.040

21. S. Singh, A. Chaubey and B. D. Malhotra, "Amperometric Cholesterol Biosensor Based on Immobilized Cholesterol Esterase and Cholesterol Oxidase on Conducting Polypyrrole Films," Analytical Chimica Acta, Vol. 502, No. 2, 2004, pp. 229-234.http://dx.doi.org/10.1016/j.aca.2003.09.064

22. S. Suman, P. R. Solanki, M. K. Pandey and B. D. Malhotra, "Covalent Immobilization of Cholesterol Esterase and Cholesterol Oxidase on Polyaniline Films for Application to Cholesterol Biosensor," Analytical Chimica Acta, Vol. 568, No. 1-2, 2006, pp. 126-132.http://dx.doi.org/10.1016/j.aca.2005.10.008

23. S. Singh, P. R. Solanki, M. K. Pandey and B. D. Malhotra, "Cholesterol Biosensor Based on Cholesterol Esterase, Cholesterol Oxidase and Peroxidase Immobilized onto Conducting Polyaniline Films," Sensors Actuators: B, Vol. 115, No. 1, 2006, pp. 534-541.http://dx.doi.

org/10.1016/j.snb.2005.10.025

24. G. Li, J. M. Liao, G. Q. Hu, N. Z. Ma and P. J. Wu, "Study of Carbon Nanotube Modified Biosensor for Monitoring Total Cholesterol in Blood," Biosenor and Bioelectronics, Vol. 20, No. 10, 2005, pp. 2140-2144. http://dx.doi.org/10.1016/j.bios.2004.09.005

25. P. R. Solanki, A. Kaushik, A. A. Ansari, A. Tiwari and B. D. Malhotra, "Multi-Walled Carbon Nanotubes/Sol-GelDerived Silica/Chitosan Nanobiocomposite for Total Cholesterol Sensor," Sensors Actuators B, Vol. 137, No. 2, 2009, pp. 727-735.http://dx.doi.org/10.1016/j.snb.2008.12.044

26. S. Aravamudhana, A. Kumarb, S. Mohapatra and S. Bhansali, "Sensitive Estimation of Total Cholesterol in Blood Using Au Nanowires Based Micro-Fluidic Platform," Biosensors and Bioelectronics, Vol. 22, No. 9-10, 2007, p. 2289.http://dx.doi.org/10.1016/j.bios.2006.11.027

27. S. K. Arya, P. Pandey, S. P. Singh, M. Datta and B. D. Malhotra, "Dithiobissuccinimidyl Propionate Self Assembled Monolayer Based Cholesterol Biosensor," Analyst, Vol. 132, No. 10, 2007, pp. 1005-109. http://dx.doi.org/10.1039/b707000d

28. S. K. Arya, M. Datta, S. P. Singh and B. D. Malhotra, "Biosensor for Total Cholesterol Estimation Using N-(2- Aminoethyl)-3-aminopropyltrimethoxysilane Self-Assembled Monolayer," Analytical Bioanalytical Chemistry, Vol. 389, 2007, pp. 2235-2242.http://dx.doi.org/10.1007/s00216-007-1655-7

29. K. Singh, P. R. Solanki, T. Basu and B. D. Malhotra, "Polypyrrole/Multiwalled Carbon Nanotubes Based Biosensor for Cholesterol Estimation," Polymers for Advanced Techniques, Vol. 23, No. 7, 2011, pp. 1084-1091. http://dx.doi.org/10.1002/pat.2020

30. R. Chauhan, Deepshikha and T. Basu, "Development of a Reusable Transducer Matrice Based on Nano Structured Based on Nano Structured Conducting Polyaniline and Its Application to Cholesterol Biosensor," Science for Advanced Material, Vol. 4, No. 1, 2012, pp. 96-125. http://dx.doi.org/10.1166/sam.2012.1256

31. J. R. Macdonald, "Impedance Spectroscopy," Wiley Interscience Publication, New York, 1987

32. M. I. Prodromidis, "Impedimetric Biosensors and Immunosensors," Pakistan Journal of Analytical Environmental Chemistry, Vol. 8, 2007, pp. 69-71.

33. K. Singh, T. Basu, P. R. Solanki and B. D. Malhotra, "Poly(pyrrole-co-N-methyl pyrrole) for Application to Cholesterol Sensor," Journal Material

Science, Vol. 44, No. 4, 2009, pp. 954-961. http://dx.doi.org/10.1007/s10853-008-3184-y

34. T. K. Vishnuvardhan, V. R. Kulkarni, C. Basavaraja and S. C. Raghavendra, "Synthesis, Characterization and a.c. Conductivity of Polypyrrole/Y_2O_3 Composites, Bulletin," Material Science, Vol. 29, No. 1, 2006, pp. 77-83.

35. C. Dhand, S. K. Arya, M. Datta and B. D. Malhotra, "Polyaniline-Carbon Nanotube Composite Film for Cholesterol Biosensor," Analytical Biochemistry, Vol. 383, No. 2, 2008, pp. 194-199. http://dx.doi.org/10.1016/j.ab.2008.08.039

36. V. Nandakumara, J. T. La Belle, J. Reed, M. Shah, D. Cochran, L. Joshi and T. L. Alford, "A Methodology for Rapid Detection of Salmonella typhimurium Using LabelFree Electrochemical Impedance Spectroscopy," Biosenors Bioelectronics, Vol. 24, No. 4, 2008, pp. 1039-1042. http://dx.doi.org/10.1016/j.bios.2008.06.036

37. J. G. Guan, Y. Q. Miao and Q. J. Zhang, "Impedimetric biosensors," Journal of Biosensor and Bioengineering, Vol. 97, No. 4, 2004, pp. 219-226,

38. J. S. Daniels and N. Pourmand, "Label-Free Impedance Biosensors: Opportunities and Challenges," Electroanalysis, Vol. 19, No. 12, 2007, pp. 1239-1257.http://dx.doi.org/10.1002/elan.200603855

39. Y. S. Fung and Y. Y. Wong, "Self-Assembled Monolayers as the Coating in a Quartz Piezoelectric Crystal Immunosensor to Detect Salmonella in Aqueous Solution," Analytical Chemistry, Vol. 73, No. 21, 2001, pp. 5302- 5309. http://dx.doi.org/10.1021/ac010655y

40. C. E. H. Berger and J. Greve, "Differential SPR Immunosensing," Sensors Actuators B, Vol. 63, No. 1, 2000, pp. 103-108. http://dx.doi.org/10.1016/S0925-4005(00)00307-5

41. A. Kaushik, P. R. Solanki, K. Kaneto, C. G. Kim, S. Ahmad and B. D. Malhotra, "Nanostructured Iron Oxide Platform for Impedimetric Cholesterol Detection," Electroanalysis, Vol. 22, No. 10, 2010, pp. 1045-1055.http://dx.doi.org/10.1002/elan.200900468

Chapter 12

HETEROCYCLIC SYNTHESIS VIA ENAMINONES: SYNTHESIS AND MOLECULAR DOCKING STUDIES OF SOME NOVEL HETEROCYCLIC COMPOUNDS CONTAINING SULFONAMIDE MOIETY

Ahmed Abd El-Hameed Hassan[1,2]

[1]Chemistry Department, Faculty of Science, Al-Azhar University, Nasr City, Egypt
[2]Chemistry Department, Faculty of Medicine, Jazan University, Jazan, KSA

ABSTRACT

4-Acetyl-N, N-diethylbenzenesulfonamide (1) was condensed with dimethylformamide dimethylacetal (DMF-DMA) to yield the enaminone, which reacts readily with different reagents to give the corresponding pyrazoles, triazolopyrimidine, imidazopyridine, pyrimidopyrimidine, pyrazolopyrimidine, benzofuran, pyranone, pyridine, pyrimidine and pyrazolopyridazine derivatives. These compounds were designed to comply with the general features of sulfonamide pharmacophore which act as Cyclooxygenase (COX-2) inhibitors. Virtual screening using molecular docking studies of the synthesized compounds was performed by (MOE), the molecular docking results indicate that, some synthesized compounds suitable inhibitor against (COX-2) with further modification.

INTRODUCTION

Enaminones are polydentate reagents that have been utilized extensively in this decade as building blocks in organic synthesis [1] -[6] . Also sulfonamides possess many types of biological activities and representatives of this class of pharmacological agents are widely used in clinic as antibacterial [7] , antithyroid, diuretic, hypoglycaemic and anti-cancer [8] -[12] . Moreover, non-steroidal anti-inflammatory drugs (NSAIDs) are widely employed in musculoskeletal disease, as well as their anti -inflammatory properties [13] .

After widely evaluation, NSAIDs are efficacy in different clinical setting, and act as a COX inhibitor (COX-1 and COX-2) through inhibiting the production of prostaglandins (PGs) [14] -[16] . Diclofenac is a one from famous available members of this drug's class under current clinical usage [17] , and suffers from a common toxicity of gastrointestinal drawback, due to inhibition non-selectivity of cyclooxygenase enzyme [18] -[20] , also, its display anti-microbial [21] -[23] , ulcerogenic, analgesic, anti-inflammatory, lipid peroxidation [24] [25] , antitumor [26] and inhibitor formation of transthyretin amyloid fibril properties [27] . In this paper, we have reported a variety of syntheses of heteroaromatics developed using functionally substituted enaminones as readily obtainable building blocks possessing multiple electrophilic and nucleophilic moieties.

MATERIAL AND METHODS

Experimental

All melting points, antioxidant and anticancer activities are uncorrected. IR spectra (KBr) were recorded on FT-IR 5300 spectrometer and Perkin Elmer spectrum RXIFT-IR system (v, cm^{-1}). The^{1}H-NMR spectra were recorded in (DMSO-d6) at 300 MHz on a Varian Mercury VX-300 NMR spectrometer (δ, ppm) using TMS as an internal standard. ^{13}C-NMR spectra were recorded on Varian Mercury VX 300 NMR using DMSO-d6 as solvent and TMS as an internal standard. Mass spectrum was obtained on GC MS-QP 1000 EX mass spectrometer at 70 eV. Elemental analyses were carried out by the Microanalytical Research Center, Faculty of Science, Cairo University and Al-Azhar University.

(E)-4-(3-(Dimethylamino)acryloyl)-N, N-diethylbenze-nesulfona-mide (2).

A mixture of 4-acetyl-N,N-diethylbenzenesulfonamide 1(0.01 mol) and DMF-DMA (0.012 mol) in dry xylene (50 ml) was heated under reflux for 4 hr. The separated solid was filtered off, washed with ethanol and recrystallized to give (2).Color: bright yellow; Yield: 83%; M.p.: 96°C - 98°C; FT-IR (KBr, v, cm^{-1}): 3100 (CH-arom.), 2920 (CH-aliph.), 1646 (CO), 1336, 1154 (SO$_2$); ^{1}HNMR (300 MHz, DMSO-d$_6$, δ, ppm): 1.01 (t, 6H, CH$_3$-CH$_2$), 2.90 and 3.13 (2s, 6H, N(CH$_3$)$_2$), 3.29 (q, 4H, CH$_2$-CH$_3$), 5.79 and 7.72 (dd, 2H, olefinic CH = CH; J = 12.25 Hz), 7.78 and 8.02 (dd, 4H, AB-ArH; J = 8.4 Hz); ^{13}C-NMR (300 MHz, DMSO-d$_6$, δ, ppm):14.6, 40.3, 42.3, 91.4, 127.1, 128.5, 141.8, 144.1, 155.4, 184.7; Anal. Calcd. for C$_{15}$H$_{22}$N$_2$O$_3$S: C, 58.04; H, 7.14; N, 9.02. Found: C, 57.91; H, 7.03; N, 8.90.

4-(1-Acetyl-1H-pyrazol-3-yl)-N, N-diethylbenzenesulfonamide (4).

A mixture of enaminone 2(0.01 mol) and hydrazine hydrate (0.01 mol) in ethanol/acetic acid (30 ml) (1:1) was heated under reflux for 5hrs. During the reflux period, a crystalline solid was separated. The separated solid was filtered off, washed with ethanol to give 4. Color: White; Yield: 73% ; M.p.: 187°C - 189°C ; FT-IR (KBr, v, cm^{-1}): 1684 (CO), 1380, 1152 (SO$_2$); ^{1}HNMR (300 MHz, DMSO-d$_6$, δ, ppm): 1.02 (t, 6H, CH$_3$-CH$_2$), 1.79 (s, 3H, CH$_3$), 3.31 (q, 4H, CH$_2$-CH$_3$), 6.81 and 8.85 (dd, 2H,H- 4, H-5 pyrazole), 7.77 and 7.98 (dd, 4H, AB-ArH; J = 8.4 Hz); MS (EI, m/z (%)): 321 (M$^+$)(8.6), 77 (100); Anal. calcd. for C$_{15}$H$_{19}$N$_3$O$_3$S: C, 56.06; H, 5.96; N, 13.07. Found: C, 55.92; H, 5.80; N, 12.93.

4-(1-(2-Cyanoacetyl)-1H-pyrazol-3-yl)-N,N-diethylbenzenesulfonamide (5).

A mixture of enaminone 2(0.01mol) and cyanoacetohydrazide (0.01 mol) in ethanol/acetic acid (30 ml) (1:1) was heated under reflux for 3hrs. During the reflux period, a crystalline solid was separated. The separated solid was filtered off, washed with ethanol to give 5. Color: yellow; Yield: 68%; M.p.: 215°C - 217°C; FT-IR (KBr, v, cm^{-1}): 2224 (CN), 1638 (CO), 1334, 1156 (SO$_2$);^{1}HNMR (300 MHz, DMSO-d$_6$, δ, ppm): 1.03 (t, 6H, CH$_3$- CH$_2$), 3.22 (q, 4H, CH$_2$-CH$_3$), 3.31 (s, 2H, CH$_2$), 7.47 and 8.23 (dd, 2H,H-4, H-5 pyrazole), 7.96 and 8.30 (dd, 4H, AB-ArH; J = 8.4 Hz); Anal. calcd. for C$_{16}$H$_{18}$N$_4$O$_3$S: C, 55.48; H, 5.24; N, 16.17. Found: C, 55.33; H, 5.11; N, 16.01.

4-(4-Cyanobenzo[4,5]imidazo[1,2-a]pyridin-1-yl)-N,N-diethylbenzenesulfonamide (7).

A mixture of enaminone 2 (0.01 mol) and 1H-benzo-imidazole-2-ylacetonitrile (0.01 mol) in glacial acetic acid (30 ml) was refluxed for 2 hr. The solid product which obtained after cooling was collected by filtration and recrystallized to give 7. Color: yellowish white; Yield: 65%; M.p.: 332 - 334°C; FT-IR (KBr, v, cm^{-1}): 2986 (CH-aliph.), 2228 (CN), 1334, 1156 (SO$_2$); ^{1}HNMR (300 MHz, DMSO-d$_6$, δ, ppm): 1.08 (t, 6H, CH$_3$-CH$_2$), 3.38 (q, 4H, CH$_2$-CH$_3$), 7.02 - 8.33 (m, 10H, ArH + 2H pyridine ring); MS, m/z (%): 404 (M$^+$)(20.4), 268 (100); ^{13}C-NMR (300 MHz, DMSO-d$_6$, δ, ppm): 14.3, 40.3, 103.2, 112.3, 114.7, 120.5, 122.0, 126.6, 128.0, 129.4, 130.4, 137.1, 138.1, 144.8, 160.9; Anal. calcd. for C$_{22}$H$_{20}$N$_4$O$_2$S: C, 65.33; H, 4.98; N, 13.85. Found: C, 65.20; H, 4.86; N, 13.71.

4-(8-Oxo-6-thioxo-7,8-dihydro-6H-pyrimido[1,6-a]pyrimidin-4-yl)-N,N-diethylbenzenesulfonamide (8).

A mixture of enaminone 2 (0.01mol) and 6-amino-2-thiouracil (0.01 mol) in glacial acetic acid (40 ml) was refluxed for 3 h. The solvent was removed by distillation under reduced pressure and the resulting solution was left to cool. The solid precipitate was collected by filtration; Color: yellow; Yield: 69%; M.p.: 315°C - 317°C; FT-IR (KBr, v, cm^{-1}): 3110 (NH), 2978 (CH-aliph.), 1686 (CO), 1336, 1158 (SO$_2$); ^{1}HNMR (300 MHz, DMSO-d$_6$, δ, ppm): 1.02 (t, 6H, CH$_3$-CH$_2$), 1.90 (s, 1H, SH), 3.18 (q, 4H, CH$_2$-CH$_3$), 7.93 - 8.97 (m, 7H, ArH + pyrimidine ring); MS (EI, m/z (%)):390 (M$^+$)(29.2), 254 (83.3), 56 (100); Anal. calcd. for C$_{17}$H$_{18}$N$_4$O$_3$S$_2$: C, 52.29; H, 4.65; N, 14.35. Found: C, 52.14; H, 4.51; N, 14.22.

4-([1,2,4]Triazolo[4,3-a]pyrimidin-5-yl)-N,N-diethylbenzenesulfonamide (10).

A mixture of enaminone 2 (0.01 mol) and 3-amino-1H-1, 2,4-triazole (0.01 mol) in acetic acid (30 ml) was refluxed for 5 hr. During the reflux period, a crystalline solid was separated. The separated solid was filtered off, washed with ethanol and recrystallized to give the compound 10. Color: yellow; Yield: 73%; M.p.: 238°C - 240°C; FT-IR (KBr, v, cm^{-1}): 2980 (CH-aliph.), 1336, 1160 (SO$_2$); ^{1}HNMR (300 MHz, DMSO-d$_6$, δ, ppm): 1.05 (t, 6H, CH$_3$-CH$_2$), 3.20 (q, 4H, CH$_2$-CH$_3$), 7.67 and 8.97 (dd, 2H, pyrimidine ring; J$_{6,7}$ = 4.6 Hz), 7.98 and 8.28 (dd, 4H, AB-ArH; J = 8.4 Hz), 8.72 (s, 1H, CH Triazole); Anal. calcd. for C$_{15}$H$_{17}$N$_5$O$_2$S: C, 54.36; H, 5.17; N, 21.13. Found: C, 54.22; H, 5.04; N, 21.04.

General procedure for preparation of (13a,b).

A mixture of enaminone 2 (0.01 mol) and 5-amino-3-(methylthio)-1H-pyrazole-4-carbonitrile or 3-phenyl- 1H-pyrazol-5-amine (0.01 mol) in ethanol (60 ml) was heated under reflux for 5 hr. During the reflux period, a crystalline solid was separated. The separated solid was filtered off, washed with ethanol and recrystallized from the appropriate solvents to give (13a,b).

4-(3-Cyano-2-(methylthio)pyrazolo[1,5-a]pyrimidin-7-yl)-N,N diethylbenzenesulfonamide (13a).

Color: brownish yellow; Yield: 75%; M.p.: 265 - 267°C; FT-IR (KBr, v, cm^{-1}): 2976 (CH-aliph.), 2220 (CN), 1352, 1154 (SO$_2$); ^{1}HNMR (300 MHz, DMSO-d$_6$, δ, ppm): 1.04 (t, 6H, CH$_3$-CH$_2$), 2.68 (s, 3H, SCH$_3$), 3.21 (q, 4H, CH$_2$-CH$_3$), 7.54 and 8.82 (dd, 2H, pyrimidine ring; J$_{5,6}$ = 4.6 Hz), 7.98 and 8.28

(dd, 4H, AB-ArH; J = 9.1 Hz); MS (EI, m/z (%)):401 (M$^+$)(26.4), 265 (100); Anal. calcd. for C$_{18}$H$_{19}$N$_5$O$_2$S$_2$: C, 53.85; H, 4.77; N, 17.44. Found: C, 53.71; H, 4.62; N, 17.31.

4-(2-Phenylpyrazolo[1,5-a]pyrimidin-7-yl)-N,N-diethylbenzenesulfon-amide (13b).

Color: yellow; Yield: 68%; M.p.: 150°C - 152°C; FT-IR (KBr, v, cm^{-1}): 2970 (CH-aliph.), 1352, 1154 (SO$_2$); MS (EI, m/z (%)):406 (M$^+$)(32.6), 270 (100); Anal. calcd. for C$_{22}$H$_{22}$N$_4$O$_2$S: C, 65.00; H, 5.46; N, 13.78. Found: C, 64.85; H, 5.32; N, 13.64.

4-(5-Hydroxybenzofuran-3-carbonyl)-N,N-diethyl-benzenesul-fonamide (17).

To a stirred solution of enaminone (1; 0.01 mol) in glacial acetic acid (30 ml), 1,4-benzoquionone (0.01 mol) was added, stirring was continued for 7 hr. At room temperature, the reaction mixture was evaporated in vacuo and the solid product was isolated by filtration and recrystallized to give (17). Color: white; Yield: 78%; M.p. 235°C - 237°C; FT-IR (KBr, v, cm^{-1}): 3320 (OH), 1614 (CO), 1340, 1156 (SO$_2$); ^{1}HNMR (300 MHz, DMSO-d$_6$, δ, ppm): 1.03 (t, 6H, CH$_3$-CH$_2$), 3.19 (q, 4H, CH$_2$-CH$_3$), 6.84 and 7.52 (dd, 2H, H-6,7 benzofuran), 7.49 (s, 1H, H-4 benzofuran), 7.93-8.02 (dd, 4H, AB-ArH; J = 7.65 Hz), 8.62 (s, 1H, H-2 benzofuran), 9.52 (s, 1H, OH exchangeable with D$_2$O); ^{13}C-NMR (DMSO-d$_6$) δ: 14.7, 40.3, 107.0, 112.6, 115.2, 125.8 127.6, 130.0, 142.3, 143.3, 149.5, 155.4, 156.1, 191.4. Anal. Calcd. for C$_{19}$H$_{19}$NO$_5$S: C, 61.11; H, 5.13; N, 3.75. Found: C, 61.05; H, 5.09; N, 3.70.

N-(6-(4-(N,N-diethylsulfamoyl)phenyl)-2-oxo-2H-pyran-3-yl) benzamide (19).

A mixture of enaminone 2 (0.01 mol) and benzoylglycine (0.01 mol) in acetic anhydride (30 ml) was heated under reflux for 2 hr. The reaction mixture was concentrated in vacuo. The solid product which formed upon cooling was filtered off then washed with ethanol. Color: white; Yield:73%; M.p.: 255°C - 257°C; FT-IR (KBr, v, cm^{-1}): 3336 (NH), 2972 (CH-aliph.), 1700 and 1660 (2CO), 1334, 1154 (SO$_2$); ^{1}HNMR (300 MHz, DMSO-d$_6$, δ, ppm): 1.02 (t, 6H, CH$_3$-CH$_2$), 3.15 (q, 4H, CH$_2$-CH$_3$), 7.32 and 8.22 (dd, 2H, pyranone; J = 7.6 Hz), 7.53 - 8.03 (m, 9H, ArH), 9.69 (s, 1H, NH); ^{13}C-NMR (300 MHz, DMSO-d$_6$, δ, ppm): 14.6, 40.2, 91.1, 109.1, 126.1, 128.2, 129.1, 129.3, 135.2, 140.1, 158.2, 163.2, 163.6; Anal. calcd. for C$_{22}$H$_{22}$N$_2$O$_4$S: C, 64.37; H, 5.40; N, 6.82. Found: C, 64.24; H, 5.25; N, 6.68.

General Procedure for the Reaction of 2 with Active Methylene Compounds to Form (20a,b) and (21).

Sodium ethoxide solution (0.23 g sodium metal in 25 ml absolute ethanol) was added with stirring to a mixture of 2 (0.01 mol) and cyanoacetamide, cyanothioacetamide and malononitrile dimer (0.01 mol) in absolute ethanol (25 ml). The reaction mixture was refluxed for 3 h, then poured into cooled water (50 ml) and neutrallized with diluted hydrochloric acid. The precipitate that formed was filtered off, dried, and crystallized from the appropriate solvents to give:

4-(5-Cyano-6-oxo-1,6-dihydropyridin-2-yl)-N,N-diethylbenzenesulfonamide (20a).

Color: yellow; Yield: 77%; M.p.: 220°C - 222°C; FT-IR (KBr, v, cm^{-1}): 2926(CH-aliph.), 2226 (CN), 1648 (CO), 1332, 1156 (SO$_2$); ^{1}HNMR (300 MHz, DMSO-d$_6$, δ, ppm): 1.02 (t, 6H, CH$_3$-CH$_2$), 3.19 (q, 4H, CH$_2$-CH$_3$), 7.87 - 8.47 (m, 7H, ArH + NH); MS (EI, m/z (%)): 331 (M$^+$)(13.8), 64 (100); Anal. calcd. for C$_{16}$H$_{17}$N$_3$O$_3$S: C, 57.99; H, 5.17; N, 12.68. Found: C, 57.84; H, 5.03; N, 12.55.

4-(5-Cyano-6-mercaptopyridin-2-yl)-N,N-diethylbenzenesulfonamide (20b).

Color: yellowish white; Yield: 65%; M.p.: 234°C - 236°C; FT-IR (KBr, v, cm^{-1}): 3114 (NH), 2938 (CHaliph.), 2224 (CN), 1334, 1154 (SO$_2$); MS (EI, m/z (%)): 347 (M$^+$)(24.1), 84 (100); Anal. calcd. for C$_{16}$H$_{17}$N$_3$O$_2$S$_2$: C, 55.31; H, 4.93; N, 12.09. Found: C, 55.17; H, 4.80; N, 11.95.

4-(5-Cyano-6-(dicyanomethylene)-1,6-dihydropyridin-2-yl)-N,N-diethylbenzenesulfonamide (21).

Color: yellow; Yield: 68%; M.p.: 296°C - 298°C; FT-IR (KBr, v, cm^{-1}): 3240 (NH), 2214 (CN), 1328, 1156 (SO$_2$); MS (EI, m/z (%)): 379 (M$^+$) (12.2), 55 (100); Anal. calcd. for C$_{19}$H$_{17}$N$_5$O$_2$S: C, 60.14; H, 4.52; N, 18.46. Found: C, 60.01; H, 4.37; N, 18.32.

General procedure for preparation of (23a,b).

To a mixture of enaminone 2 (0.01 mol) and thiourea or guanidine hydrochloride (0.01 mol) in ethanol (40 ml) was added a few drops of piperidine as catalyst. The reaction mixture was refluxed for 4 hr, then poured into cold water (50 ml) and neutralized with diluted hydrochloric acid. The precipitate that formed was filtered off, dried, and crystallized from the appropriate solvents to give:

4-(2-Thioxo-1,2-dihydropyrimidin-4-yl)-N,N-diethyl-benzenesulfonamide (23a).

Color: faint brown; Yield: 82%; M.p.: 190°C - 192°C; FT-IR (KBr, v, cm⁻¹): 3126 (NH), 2976 (CH-aliph.), 1336, 1164 (SO₂); ¹HNMR (300 MHz, DMSO-d₆, δ, ppm): 1.05 (t, 6H, CH₃-CH₂), 3.17 (q, 4H, CH₂-CH₃), 7.47 and 8.16 (dd, 2H, pyrimidine ring; J = 6.6 Hz), 7.93 and 8.31 (dd, 4H, AB-ArH; J = 8.4 Hz), 13.8 (s, 1H, NH); MS (EI, m/z (%)):323 (M⁺)(25.5), 308 (78.4), 187 (100); Anal. calcd. for C₁₅H₁₇N₅O₂S: C, 54.36; H, 5.17; N, 21.13. Found: C, 54.22; H, 5.04; N, 21.02.

4-(2-Aminopyrimidin-4-yl)-N,N-diethylbenzenesulfonamide (23b).

Color: yellow; Yield: 75%; M.p.: 175°C - 177°C; FT-IR (KBr, v, cm⁻¹): 3478, 3300 (NH₂), 3162 (CH-arom.), 2980 (CH-aliph.), 1332, 1152 (SO₂); MS (EI, m/z (%)): 306 (M⁺)(19.0), 291 (62.0), 170 (100); Anal. calcd. for C₁₄H₁₈N₄O₂S: C, 54.88; H, 5.92; N, 18.29. Found: C, 54.73; H, 5.80; N, 18.15.

General procedure for preparation of (24a,b).

A mixture of enaminone 2 (0.01 mol) and p-toluidine or p-phenitidine (0.01 mol) in a mixture of ethanol/acetic acid (50 ml) (1:1) was heated under reflux for 3 hr. During the reflux period, a crystalline solid was separated. The separated solid was filtered off, washed with ethanol and recrystallized from the appropriate solvents to give:

4-(3-(p-Tolylamino)acryloyl)-N,N-diethylbenzenesulfonamide (24a).

Color: yellow; Yield: 88%; M.p.: 265°C - 267°C; FT-IR (KBr, v, cm⁻¹): 3176 (NH), 2982 (CH-aliph.), 1644 (CO), 1334, 1154 (SO₂); ¹HNMR (300 MHz, DMSO-d₆, δ, ppm): 1.02 (t, 6H, CH₃-CH₂), 2.25 (s, 3H, CH₃), 3.16 (q, 4H, CH₂-CH₃), 6.09 (d, 1H, COCH, J = 7.6 Hz), 7.94 (m, 1H, CH-NH), 7.07-7.15 (dd, 4H, AB-ArH; J = 8.4 Hz), 7.85 - 7.96 (dd, 4H, AB-ArH; J = 8.4 Hz), 12.10 (d, 1H, NH exchangeable with D₂O); Anal. calcd. for C₂₀H₂₄N₂O₃S: C, 64.49; H, 6.49; N, 7.52. Found: C, 64.32; H, 6.35; N, 7.40.

4-(3-((4-Ethoxyphenyl)amino)acryloyl)-N,N-diethylbenzenesulfonamide (24b).

Color: reddish yellow; Yield: 85%; M.p.: 280°C - 282°C; FT-IR (KBr, v, cm⁻¹): 3170 (NH), 2980 (CH-aliph.), 1628 (CO), 1342, 1156 (SO$_2$); ¹HNMR (300 MHz, DMSO-d$_6$, δ, ppm): 1.01 (t, 6H, CH$_3$-CH$_2$), 1.27 (t, 3H, CH$_3$-CH$_2$O), 3.15 (q, 4H, CH$_2$-CH$_3$), 3.97 (q, 2H, OCH$_2$-CH$_3$), 6.06 (d, 1H, COCH, J = 7.8 Hz), 7.92 (m, 1H, CH-NH), 6.89 - 7.26 (dd, 4H, AB-ArH; J = 8.4 Hz), 7.83 - 8.06 (dd, 4H, AB-ArH; J = 8.4 Hz), 12.17 (d, 1H, NH exchangeable with D$_2$O); Anal. calcd. for C$_{21}$H$_{26}$N$_2$O$_4$S: C, 62.66; H, 6.51; N, 6.96. Found: C, 62.51; H, 6.36; N, 6.82.

4-(3-Benzoyl-1-phenyl-1H-pyrazole-4-carbonyl)-N,N-diethylbenzenesulfonamide (30).

To a mixture of enaminone 2 (0.01 mol) and the hydrazonoyl bromide (85; 0.01 mol) in benzene (40 ml) an equivalent amount of triethylamine was added. The reaction mixture was heated under reflux for 2 hr. the solvent was distilled at reduced pressure and the residual viscous liquid was taken in ethanol then the resulting solid was collected by filtration, washed thoroughly with ethanol, dried and finally recrystallized to give the compound 30. Color: yellow; Yield: 81%; M.p.: 224°C - 226°C; FT-IR (KBr, v, cm⁻¹): 2976 (CH-aliph.), 1660 (CO), 1336, 1154 (SO$_2$); ¹HNMR (300 MHz, DMSO-d$_6$, δ, ppm): 1.03 (t, 6H, CH$_3$-CH$_2$), 3.14 (q, 4H, CH$_2$-CH$_3$), 7.21 - 8.41 (m, 15H, ArH + CH pyrazole); MS (EI, m/z (%)): 478 (M⁺)(8.5), 351 (18.9.3), 76 (100); Anal. calcd. for C$_{27}$H$_{25}$N$_3$O$_4$S: C, 66.51; H, 5.17; N, 8.62. Found: C, 66.37; H, 5.04; N, 8.48.

4-(2,7-Diphenyl-2H-pyrazolo[3,4-d]pyridazin-4-yl)-N,N-diethylbenzenesulfonamide (31).

A mixture of pyrazole derivative 30 (0.01 mol) and hydrazine hydrate (0.012 mol) in ethanol (50 ml) was heated under reflux for 4 hr. The separated solid was filtered off, washed with ethanol and recrystallized to give 31. Color: brownish yellow; Yield: 65%; M.p.: 292°C - 294°C; FT-IR (KBr, v, cm⁻¹): 2974 (CH-aliph.), 1332, 1160 (SO$_2$); MS (EI, m/z (%)):483 (M⁺) (37.6), 347 (100); Anal. calcd. for C$_{27}$H$_{25}$N$_5$O$_2$S: C, 67.06; H, 5.21; N, 14.48. Found: C, 66.92; H, 5.08; N, 14.33.

Molecular Modeling Study

Generation of Ligand and Enzyme

Structures. Selection of COX structures.

Docking study was carried out for the target compounds into COX-1 (ID: 3N8Y) and COX-2 (ID: 1PXX) using MVD, 4.0 and MOE, 10. The crystal structure of the (COX) complexes with (1), which a selective inhibitor of COX-2 in co-crystallized form in the active site of the receptor. From X-ray crystal structure studies of the COX enzyme, the mouse enzyme is expected to be very similar to the human [28] , and can be used as a model for human COX enzyme.

Preparation of Small Molecule

Molecular modeling of the target compounds were built using MOE, and minimized their energy with PM3 through MOPAC. Our compounds were introduced into the (COX) binding site accordance the published crystal structures of (1) bound to the kinase.

Stepwise Docking Method

MOE Stepwise The crystal structure of the (COX) with a Diclofenac (1) as inhibitor molecule, was used in the receptor molecule, water and inhibitor molecules were removed, and hydrogen atoms were added. The parameters and charges were assigned to the MMFF94x force field. After alpha-site spheres were generated using the SITE FINDER module of MOE. The optimized 3D structures of molecules were subjected to generate different poses of ligands using triangular matcher placement method, which generating poses by aligning ligand triplets of atoms on triplets of alpha spheres representing in the receptor site points, a random triplet of alpha sphere centers is used to determine the pose during each iteration. The pose generated was rescored using London dG scoring function. The poses generated were refined with MMFF94x forcefield, also, the solvation effects were treated. The Born solvation model (GB/VI) was used to calculate the final energy, and the finally assigned poses were assigned a score based on the free energy in kJ/mol

RESULTS AND DISCUSSION.

Chemistry

Treatment   of   4-acetyl-N,N-diethylbenzenesulfonamide   (1)   with dimethylformamid-dimethylacetal (DMF-DMA) in dry dioxane afforded (E)-4-(3-(dimethylamino) acryloyl)-N,N-diethylbenzenesulfonamid (2) in high yield. (Scheme 1). The structure of the enaminone 2 was confirmed on the basis of elemental analysis and spectral data. Thus, the IR spectrum of compound 2 revealed absorption bands at (v cm^{-1}): 2910 (CH-aliph.) and 1646 (C = O), while its ^{1}H-NMR (δ ppm) spectrum (DMSO) indicated signals at: 2.90 and 3.13 (2s, 6H, N(CH$_3$)$_2$), 5.79 and 7.72 (dd, 2H, olefinic CH = CH; J = 12.25 Hz), which support that the structure in (E-form), not (Z-form). The reactivity of compound 2 towards some nitrogen nucleophiles was investigated. Thus, enaminone 2 was treated with hydrazine hydrate in refluxing ethanol/acetic, to produce intermediates 3 followed by acetylation to give 4-(1-acetyl-1H-pyrazol-3-yl)-N,N-diethylbenzenesulfonamide (4). In the same manner, the enaminone 2 reacted with 2-cyanoacetic acid hydrazide under the same experimental reaction conditions to afford 4-(1-(2- cyanoacetyl)-1H-pyrazol-3-yl)-N,N-diethylbenzenesulfonamide (5) (Scheme 1).

When compound 2 was allowed to react with active methylene reagent like: 2-cyanomethylbenzoimidazole in glacial acetic acid afforded 4-(4-cyanobenzo[4,5]imidazo [1,2-a] pyridin-1-yl)-N,N-diethylbenzenesulfonamide (7).

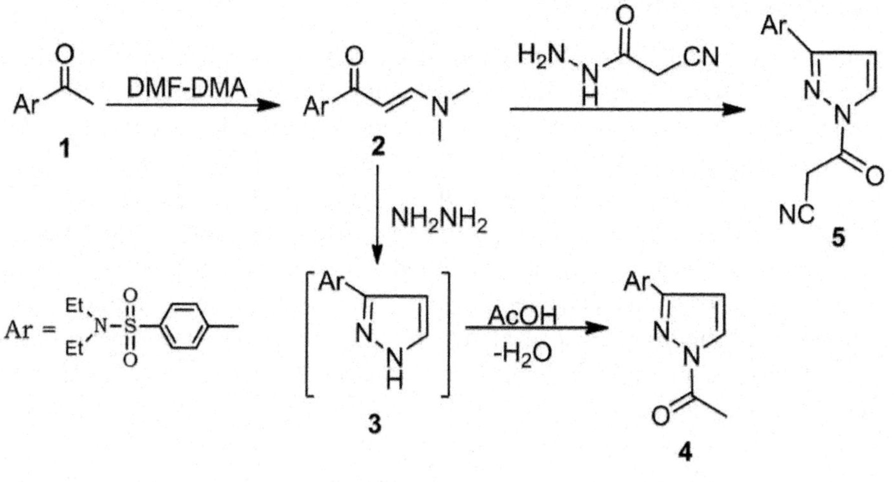

Scheme 1: SSynthesis of pyrazole derivatives.

Formation of 7 is assumed to proceed via the initial nucleophilic addition of the active methylene to enaminone double bond to afford the non-isolable intermediate 6 followed by elimination of water and dimethylamine (Scheme 2). On the other hand, the enaminone 2 condensed with 6-amino-2-thiouracil in glacial acetic acid at reflux temperature to afford N,N-diethyl-4-(8-oxo-6-

thioxo-7,8-dihydro-6H-pyrimido[1,6-a] pyrimidin-4-yl) benzenesulfonamide (8), (Scheme 2). Enaminone 2 reacted with 3-amino-1H-1,2,4-triazole in acetic acid under reflux to afford [1,2,4]triazolo[4,3-a]pyrimidine derivatives (10) via the addition of the exoamino group of aminotriazole to α,β-unsaturated moiety in compound 2 followed by elimination of dimethylamine molecule to yield the corresponding acyclic non-isolable intermediate , which undergoes intramolecular cyclization by elimination of water molecule to afford the final product 10 not 9 (Scheme 2).

In the same manner, the enaminone 2 reacted with another heterocyclic amine like: 5-amino-3-methyl thiopyrazole-4-carbonitrile [29] [30] and 5-Phenyl-2H-pyrazol -3-yl amine under the same experimental reaction conditions to afford pyrazolo [1,5-a]pyrimidine derivatives 13a,b. The formation of 13a,b was assumed to takes place via the addition of exoamino group of aminopyrazole to α,β-unsaturated moiety of enaminone 2 to yield the corresponding acyclic non-isolable intermediate 12a,b which undergoes intramolecular cyclization by the elimination of the dimethylamine and water molecules to afford the final product 13a,b not 11a,b (Scheme 3).

Furthermore, Compound 2 reacted with 1,4-benzoquin-one in glacial acetic acid at room temperature to yield a product which formulated as N,N-diethyl-4-(5-hydroxy-benzofuran-3-carbonyl)benzenesulfonamide (17). Elucidation of structure 17 and refusing of structure 15 was based on ^{1}H-NMR spectrum which indicates the disappearance of aldehydic signal and showed singlet signal at 9.52 ppm for H-2 benzofuran moiety. It's believed that electron rich (C-2) in the enaminone 2 initially adds to the activated double bond in the quinone yielding acyclic intermediate 16 which then cyclizes into 17 via dimethylamine elimination, and not afforded 15 (Scheme 4).

Enaminone 2 reacted with hippuric acid in acetic anhydride to give a product that was identified as N-(6-(4- (N,N-diethylsulfamoyl)phenyl)-2-oxo-2H-pyran-3-yl)benzamide (19), which confirmed on the basis of elemental analysis and spectral data, (Scheme 5). The formation of compound 19 was assumed to proceed via initial cyclization of hippuric acid into oxazolone derivative which then added to the activated double bond system of enaminone yielding 18 followed by intramolecular cyclization and rearrangement to give the final structure 19 (Scheme 5).

The reaction of 2 with some active nitriles such as cyanoacetamide, cyanothioacetamide and malononitrile dimer in sodium ethoxide was studied. Enaminones 2 were reacted with cyanoacetamide, cyanothioacetamide in refluxing ethanolic sodium ethoxide to give 4-(5-cyano-6-substituted pyridin-2-yl)-N,N-diethyl benzenesulfonamide 20a,b, (Scheme 6). Structures 20a,b were based on the correct elemental analyses and spectral data. Compound

2 was also reacting with malononitrile dimmer in refluxing ethanolic sodium ethoxide to give 4-(5-cyano-6-(dicyanomethylene)-1,6-dihydropyridin-2-yl)-N,N-diethylbenzenesulfonamide (21). All analytical and spectral data supported the suggested structure. The IR spectra showed an absorption band at 2214 (CN) cm⁻¹.

Enaminone 2 reacted with guanidine hydrochloride and thiourea in refluxing ethanol in the presence of pipridine to give 23a,b. A plausible mechanism for the formation of compounds 23a,b is outlined in (Scheme 7). Which then undergo intramolecular cyclization and subsequent aromatization via the elimination of dimethylamine and water molecules under the reaction conditions to give 23a,b as depicted in (Scheme 7). On the other hand, when enaminone (2) was treated with primary aromatic amines namely (p-toluidine and p-phenitidine) in a mixture of ethanol/acetic acid at reflux temperature afforded   4-(3-((p-substituted   amino)acryloyl)-N,N-diethylbenzene sulfonamides (24a,b) (Scheme 7).

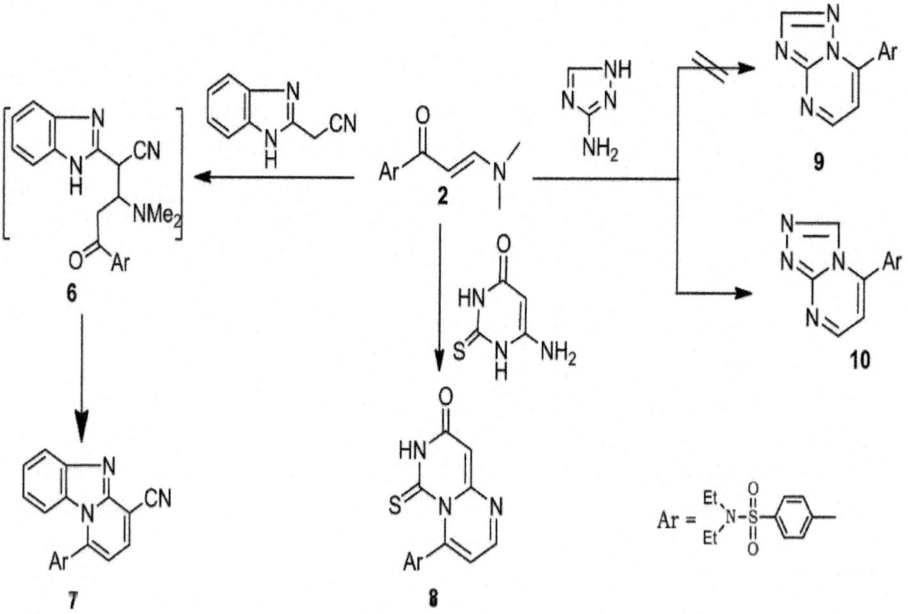

Scheme 2: Synthesis of benzoimidazo pyridine, pyrimido and triazolo pyrimidine derivatives.

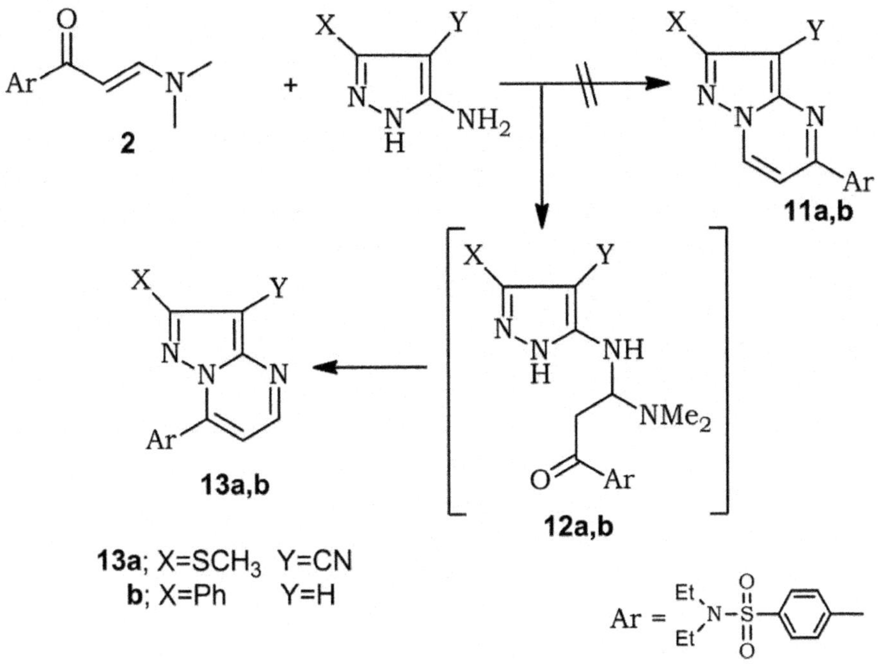

13a; X=SCH$_3$ Y=CN
b; X=Ph   Y=H

Scheme 3: Synthesis of pyrazolo pyrimidine derivatives.

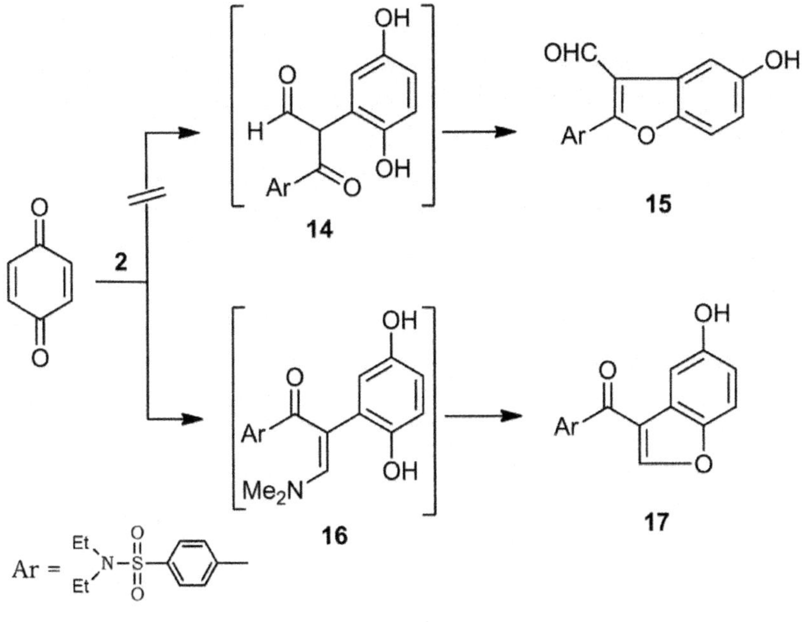

Scheme 4: Synthesis of benzofuran derivative.

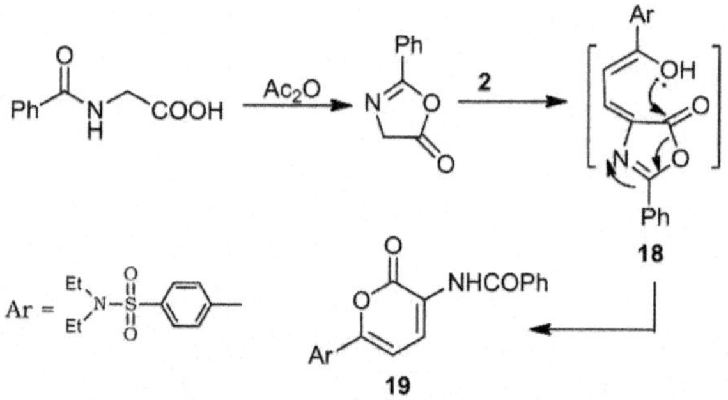

Scheme 5: Synthesis of pyranone derivative.

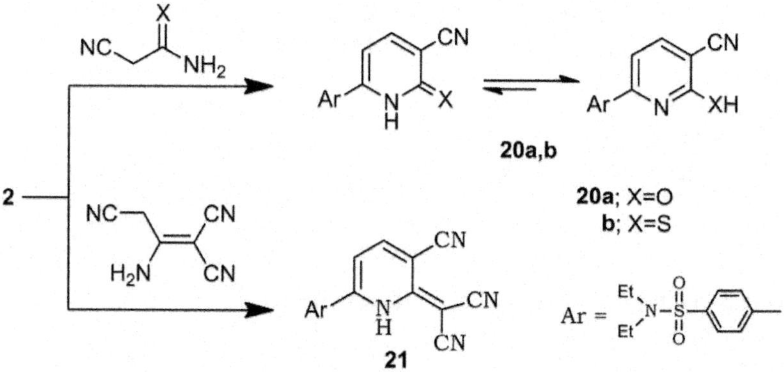

Scheme 6: Synthesis of pyridine derivative.

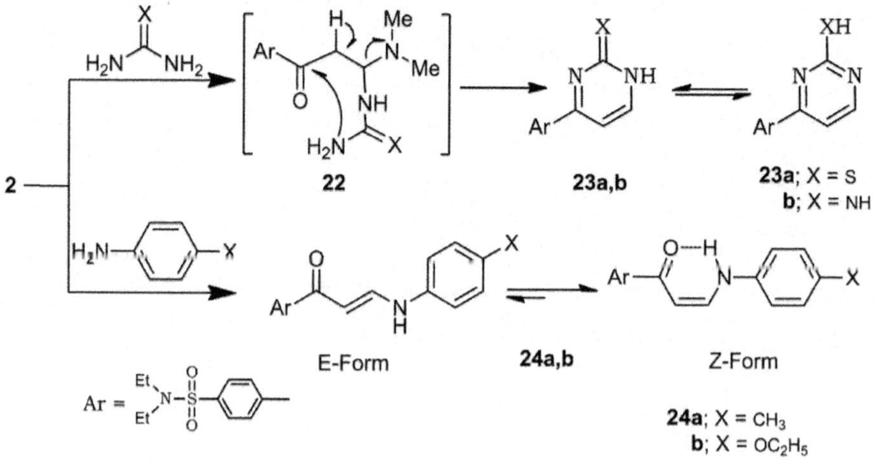

Scheme 7: Synthesis of pyrimidine derivatives.

The ^{1}H-NMR spectrum of compounds 24a,b support this structures is (Z-form) not (E-form), where the coupling constant of the doublet signals for olefinic protons equal to 7.6, 7.8 Hz, respectively. Stabilization of (Z-form) is achieved by intramolecular hydrogen bonding (Scheme 7).

Hydrazonyl halides [31]-[33] has been reported to add to α,β-unsaturated carbonyl compounds to yield a mixture of isomeric pyrazolines [34] [35] . In the present work the reaction of enaminone 2 with nitrileimine 26 (generated in situ from the treatment of the hydrazonoyl bromide 25 with triethylamine in refluxing m-xylene) gave only one isolable product (TLC). From which two proposed structures 28 or 30 seemed possible (Scheme 8). The other possible regioisomer 28 was excluded on the basis of the spectral data of the isolated products. For example, in the pyrazole ring system, C-4 is the most electron-rich carbon, thus H-4 is expected to appear in the ^{1}H-NMR spectra at higher field, typically near 6.0 ppm. On the other hand, H-5 is linked to the carbon attached to a nitrogen atom and thus it is deshielded to appear near 8.0 ppm. ^{1}H-NMR spectrum of 30 exhibits a singlet at 8.41 ppm, which indicates the presence of the pyrazole H-5 rather than H-4 [36]

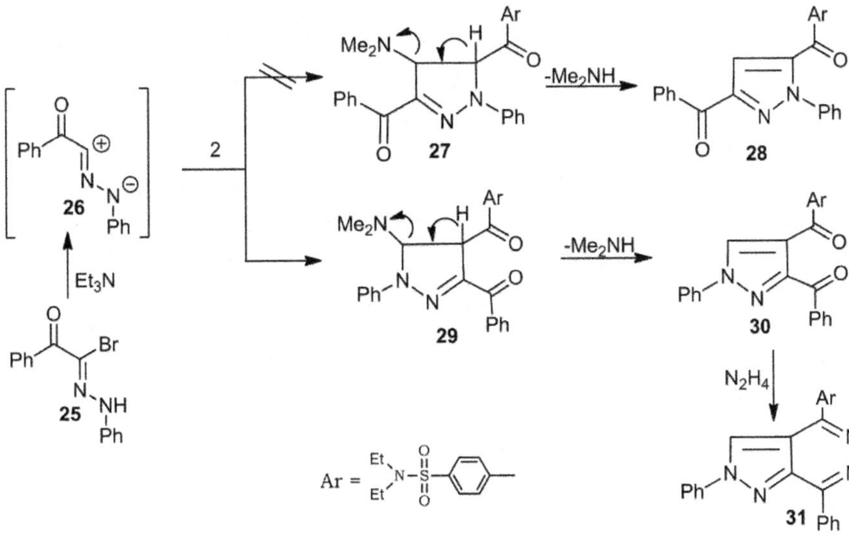

Scheme 8: Synthesis of pyrazolo pyridazine derivative.

Pyrazole derivative 30 was assumed to be formed via initial 1, 3-dipolar cycloaddition of nitrileimine 26 to the activated double bond in compound 2 forming non isolable intermediate 29 followed by the loss of dimethylamine (Scheme 8). Interaction of pyrazole derivative 30 with hydrazine hydrate in refluxing ethanol absolute afforded pyrazolo[3,4-d] pyridazine derivative (31), which confirmed by elemental analysis and spectral data (Scheme 8).

Docking Studies

In brief, two isoforms of COX protein are known: COX-1 is responsible for the physiological production of prostaglandins, which is expressed in most tissues; and COX-2, is responsible for the increasing production of prostaglandins during process of inflammation, which is induced by endotoxins, cytokines and mitogens in inflammatory cells [37] . Recently, from analysis of X-ray cocrystal of arachidonic acid with COX-2 showed that, carboxylate coordinated with Tyr-385 and Ser-530 [38] , as well as the action of NSAIDs, through the interaction carboxylate group with Tyr-385 and Ser-530, which stabilize the negative charge of the tetrahedral intermediate [39] , and demonstrated that, Tyr-385 and Ser-530 have a structural and functional evidence for the importance of them in the chelating of the ligands [40] . Molecular docking of the synthesized compounds into the active site of COX was performed, in order to obtain biological data on a structural basis, through rationalized ligand-protein interaction behavior. All calculations for docking experiment were performed with MOE 2008.10 [41] . The tested compounds were evaluated in silico (docking), using X-ray crystal structures of COX-2 (ID: 4COX). The tested compounds were docked into active sites of both enzymes COX-2. The active site of the enzyme was defined to include residues within a 10.0 Å radius to any of the inhibitor atoms. The scoring function of the most stable docking model for testing compounds was applied to evaluate the binding affinities of the inhibitors complexes with (COX) active site (Table 1). The complexes were energy-minimized with an MMFF94 force field [42] till the gradient convergence 0.05 kcal/mol was reached. The active compounds docked successfully into the COX-2 active site. The compounds 2.10 and 17 in COX-2 active site exhibited, binding scores (−98.09, −118.71 and −125.06) Kcal/mol, respectively.

Structure activity relationships:

In order to get a deeper insight into the nature and type of interactions of docked compounds, the complexes between each compound and COX-2 receptor were visualized, and depicted in (Figures 1-3). Since, the H bonds play an important role in the structure and function of biological molecules, the current ligand-receptor interactions were analyzed on the basis of H bonding. In order to reduce the complexity, hydrophobic and π-cation interactions (>6 Å) are not shown in the Figures 1-3. On COX-2 binding site (Figure 3); i-Compound 2 arranged in the binding pocket by adjusting phenyl ring

Table 1: Pharmacokinetic parameters important for good oral bioavailability of most active compounds.

CPD	2	4	7	8	10	17
HBD	0	0	1	0	0	0
HBA	7	5	6	7	6	6
CLogP	1.803	1.162	2.933	0.964	1.514	3.559
V	0	0	0	0	0	0
Vol.	185	176.25	181.875	201	195.875	196.25
TPSA	52.82	106.53	57.53	124.35	95.53	72.55
%ABS	90.7771	72.24715	89.15215	66.09925	76.04215	83.97025
Log S	−3.69206	−2.42823	−5.17052	−3.32469	−3.07276	−5.45284
HOMO	−9.46936	−8.95061	−9.14457	−9.95624	−9.57443	−8.68336
LUMO	−1.61311	−0.77546	−0.77307	−1.1573	−0.92765	−1.25907
mr	7.856251	8.17515	8.3715	8.79894	8.646781	7.42429
ΔE	3.928125	4.087575	4.18575	4.39947	4.32339	3.712145
η	0.254574	0.244644	0.238906	0.2273	0.2313	0.269386
S	−5.54124	−4.86304	−4.95882	−5.55677	−5.25104	−4.97122
χ	5.541235	4.863035	4.95882	5.55677	5.25104	4.971215
σ	3.90839	2.892804	2.937335	3.509251	3.188866	3.328666
ω	−9.46936	−8.95061	−9.14457	−9.95624	−9.57443	−8.68336

TPSA: Polar surface area (A^2); %ABS: Absorption percentage; Vol: Volume (A^3); HBA: Number of hydrogen bond acceptor; HBD: Number of hydrogen bond donor; V: Number of violation from Lipinski's rule of five; Log P: Calculated lipophilicity; Log S: Solubility parameter; mr: Molar Refractivity; ΔE: Energy Gaps (ev); η: Hardness (ev); S: Softness (ev); χ: Electronegativity (ev); σ: chemical potential (ev); ω: Electrophilicity (ev).

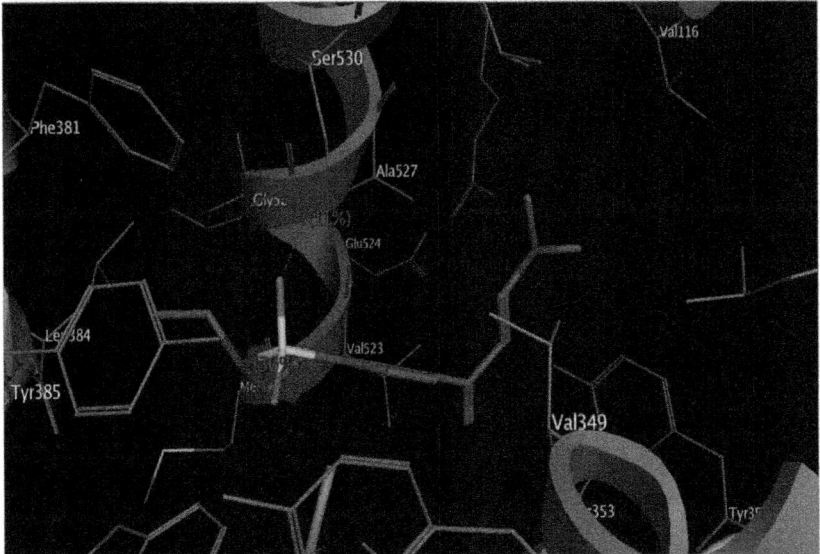

Figure 1: Docking compound 2 into the active site COX-2 with highest docking score H-bonds are in pink.

perpendicular with N-ethyl and formed an electrostatic bond with Ser-530 and Tyr-385; ii-Compound 10 and 17, were interacted with different modes with binding pocket, by forming two strong hydrogen bond interaction with important residue Arg-120, which explain increasing of the binding energy of its compounds. The results obtained clearly reveals that, the amino acid residues close to the reference molecules are mostly the same as those observed in the most active compounds complexes with protein (Figures 2 and 3). These results indicate that, the compounds 2, 10 and 17 act as selective inhibitors against COX-2.

ADMET factors profiling:

Oral bioavailability was considered to play an important role in the development of bioactive molecules as therapeutic agents. Many potential therapeutic agents fail to reach the clinic, because of their ADMET (absorption, distribution, metabolism, elimination and toxic) Factors.

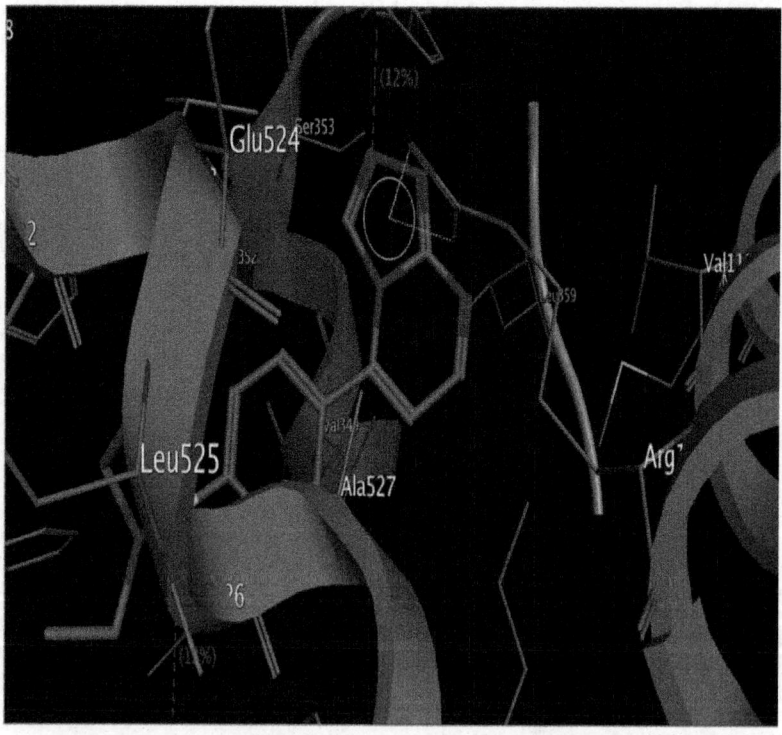

Figure 2: Docking compound 10 with a highest docking score into the active site COX-2.

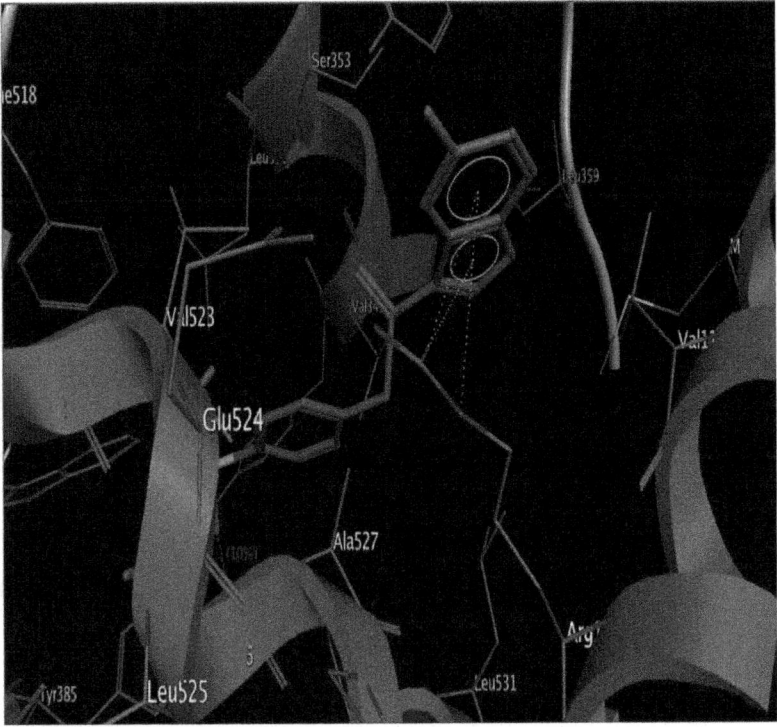

Figure 3: Docking compound 17 with a highest docking score into the active site COX-2.

Therefore, a computational study for prediction of ADMET properties of the molecules was performed for most active compounds, by the determination of topological polar surface area (TPSA), a calculated percent absorption (%ABS) which was estimated by Zhao et al. equation [29] , and "rule of five" formulated by Lipinski [30] , which established that, chemical compound could be an orally active drug in humans, if no more than one violation of the following rule:

1)   ClogP (partition coefficient between water and octanol) < 5.

2)   Number of hydrogen bond donor sites ≤ 5.

3)   Number of hydrogen bond acceptor sites ≤ 10, iv), molecular weight < 500.

In addition, the total polar surface area (TPSA) is another key property linked to drug bioavailability, the passively absorbed molecules with (TPSA > 140) have low oral bioavailability [43] . All calculated descriptors were performed using MOE Package, and their results were disclosed in Table 1.

Our results revealed that, the CLogP (factor of the lipophilicity) [44] was less than 5.0, hydrogen bond acceptors between (5 and 7), hydrogen bond donors (1), this data show these compounds fulfill Lipinski's rule. Also, the absorption percent is ranged between (~66% - 90%).

The HOMO and LUMO of a molecule play important roles in intermolecular interactions [33] , through the interaction between the HOMO of the drug with the LUMO of the receptor and vice versa. The interactions were stabilized inversely with energy gap between the interacting orbitals. Increasing HOMO energy and decreasing LUMO energy in the drug molecule lead to enhancement stabilizing interactions, and hence, binding to the receptor. Furthermore, the global and local chemical reactivity descriptors for molecules have been defined (Table 1), like softness (measures stability of molecules and chemical reactivity), hardness (reciprocal of softness), chemical potential, electronegativity (strength atom for attracting electrons to itself), electrophilicity index (measuring lowering energy due to maximal flowing electron between donor and acceptor) [45] -[51] . The results were shown in Table 1 and may explain the less toxicity and high affinity of its compounds against COX.

ACKNOWLEDGEMENTS

This study was supported by the Chemistry Department, Faculty of Science, Al-Azhar University. We are deeply thankful to Micro Analytical Center for making the IR, ^{1}H NMR, ^{13}C NMR and MS samples.

REFERENCES

1. Ferraz, H.M.C. and Goncalo, E.R.S. (2007) Recent Preparations and Synthetic Applications of Enaminones. Química Nova, 30, 957-964. http://dx.doi.org/10.1590/S0100-40422007000400035

2. Riyadh, S.M., Abdelhamid, I.A., Al-Matar, H.M., Hilmy, N.M. and Elnagdi, M.H. (2008) Enamines as Precursors to Polyfunctional Heteroaromatic Compounds; a Decade of Development. Heterocycles, 75, 1849-1905. http://dx.doi.org/10.3987/REV 07 625

3. Elassar, A.Z.A. and El-Khair, A.A. (2003) Recent Developments in the Chemistry of Enaminones. Tetrahedron, 59, 8463-8480. http://dx.doi.org/10.1016/S0040-4020(03)01201-8

4. Stanovnik, B. and Svete, J. (2004) Synthesis of Heterocycles from Alkyl 3-(Dimethylamino)Propenoates and Related Enaminones. Chemical Reviews, 104, 2433-2480. http://dx.doi.org/10.1021/cr020093y

5. Yermolayev, S.A., Gorobets, N.Y., Lukinova, E.V., Shishkin, O.V.,

Shishkina, S.V. and Desenko, S.M. (2008) An Efficient Synthesis of N1-Substituted 2,5-Dioxo-1,2,5,6,7,8-Hexahydro-3-Quinolinecarboxamide via Enolate Salts. Tetrahedron, 64, 4649-4655.http://dx.doi.org/10.1016/j.tet.2008.02.095

6.   Gorobets, N.Y., Yousefi, B.H., Belaj, F. and Kappe, C.O. (2004) Rapid Microwave-Assisted Solution Phase Synthesis of Substituted 2-Pyridone Libraries. Tetrahedron, 60, 8633-8644. http://dx.doi.org/10.1016/j.tet.2004.05.100

7.   Drews, J. (2000) Drug Discovery: A Historical Perspective. Science, 287, 1960-1964.http://dx.doi.org/10.1126/science.287.5460.1960

8.   Supuran, C.T., Casini, A. and Scozzafava, A. (2003) Protease Inhibitors of the Sulfonamide Type: Anticancer, Antiinflammatory, and Antiviral Agents. Medicinal Research Reviews, 23, 535. http://dx.doi.org/10.1002/med.10047

9.   Supuran, C.T. and Scozzafava, A. (2000) Carbonic Anhydrase Inhibitors and Their Therapeutic Potential. Expert Opinion on Therapeutic Patents, 10, 575-600.

10. Ghorab, M.M., Ragab, F.A., Heiba, H.I., Arafa, R.K. and El-Hossary, E.M. (2011) Docking Study, in Vitro Anticancer Screening and Radiosensitizing Evaluation of Some New Fluorine-Containing Quinoline and Pyrimidoquinoline Derivatives Bearing a Sulfonamide Moiety. Medicinal Chemistry Research, 20, 388-400. http://dx.doi.org/10.1007/s00044-010-9332-3

11. Boyd 3rd, A.E. (1988) Sulfonylurea Receptors, Ion Channels, and Fruit Flies. Diabetes, 37, 847-850. http://dx.doi.org/10.2337/diab.37.7.847

12. Kamel, M.M., Ali, H.I., Anwar, M.M., Mohamed, N.A. and Soliman, A.M. (2010) Synthesis, Antitumor Activity and Molecular Docking Study of Novel Sulfonamide-Schiff's Bases, Thiazolidinones, Benzothiazinones and Their CNucleoside Derivatives. European Journal of Medicinal Chemistry, 45, 572-580. http://dx.doi.org/10.1016/j.ejmech.2009.10.044

13. Nair, B. and Taylor-Gjevre, R. (2010) A Review of Topical Diclofenac Use in Musculoskeletal Disease. Pharmaceuticals, 3, 1892-1908. http://dx.doi.org/10.3390/ph3061892

14. Gabriel, S.E. and Matteson, E.L. (1995) Economic and Quality-of-Life Impact of NSAIDs in Rheumatoid Arthritis: A Conceptual Framework and Selected Literature Review. Pharmacoeconomics, 8, 479-490. http://dx.doi.org/10.2165/00019053-199508060-00004

15. Zochling, J., Bohl-Bühler, M.H.J., Baraliakos, X., Feldtkeller, E. and Braun, J. (2006) Nonsteroidal Anti-Inflammatory Drug Use in Ankylosing

Spondylitis—A Population-Based Survey. Clinical Rheumatology, 25, 794-800. http://dx.doi.org/10.1007/s10067-005-0132-y

16. Hochberg, M.C. (2005) COX-2 Selective Inhibitors in the Treatment of Arthritis: A Rheumatologist Perspective. Current Topics in Medicinal Chemistry, 5, 443-448.http://dx.doi.org/10.2174/1568026054201695

17. Warden, J.S. (2010) Prophylactic Use of NSAIDs by Athletes: A Risk/ Benefit Assessment. The Physician and Sports Medicine, 38, 132-138. http://dx.doi.org/10.3810/psm.2010.04.1770

18. Guyton, C.A. and Hall, J.E. (1998) Textbook of Medical Physiology. 9th Edition, Harcourt Asia Pte. Ltd.

19. Vane, J.R., Bakhle, Y.S. and Bolting, R.M. (1998) Cyclooxygenases 1 and 2. Annual Review of Pharmacology and Toxicology, 38, 97-120. http://dx.doi.org/10.1146/annurev.pharmtox.38.1.97

20. Guslandi, M. (1997) Gastric Toxicity of Antiplatelet Therapy with Low-Dose Aspirin. Drugs, 53, 1-5. http://dx.doi.org/10.2165/00003495-199753010-00001

21. Mazumdar, K., Dutta, N., Dastidar, S., Motohashi, N. and Shirataki, Y. (2006) Diclofenac in the Management of E. coli Urinary Tract Infections. In Vivo, 20, 613-619.

22. Dutta, N., Annadurai, S., Mazumdar, K., Dastidar, S.G., Kristiansen, J., Molnar, J., Martins, M. and Amaral, L. (2000) The Antibacterial Action of Diclofenac Shown by Inhibition of DNA Synthesis. International Journal of Antimicrobial Agents, 14, 249-251.http://dx.doi.org/10.1016/ S0924-8579(99)00159-4

23. Sriram, D., Yogeeswari, P. and Devakaram, R. (2006) Synthesis, in Vitro and in Vivo Antimycobacterial Activities of Diclofenac Acid Hydrazones and Amides. Bioorganic & Medicinal Chemistry, 14, 3113-3118. http:// dx.doi.org/10.1016/j.bmc.2005.12.042

24. Bhandari, S., Bothara, K., Raut, M., Patil, A., Sarkate, A. and Mokale, J. (2008) Design, Synthesis and Evaluation of Anti-Inflammatory, Analgesic and Ulcerogenicity Studies of Novel S-Substituted Phenacyl-1,3,4-Oxadiazole-2-Thiol and Schiff Bases of Diclofenac Acid as Nonulcerogenic Derivatives. Bioorganic & Medicinal Chemistry, 16, 1822- 1831.http://dx.doi.org/10.1016/j.bmc.2007.11.014

25. Amir, M. and Shikha, K. (2004) Synthesis and Anti-Inflammatory, Analgesic, Ulcerogenic and Lipid Peroxidation Activities of Some New 2-[(2, 6-Dichloroanilino) Phenyl]Acetic Acid Derivatives. European Journal of Medicinal Chemistry, 39, 535-545.http://dx.doi.org/10.1016/j. ejmech.2004.02.008

26. Barbaric, M., Kralj, M., Marjanovic, M., Husnjak, I., Pavelic, K., Filipovic Grcic, J., Zorc, D. and Zorc, B. (2007) Synthesis and in Vitro Antitumor Effect of Diclofenac and Fenoprofen Thiolated and Nonthiolated PolyaspartamideDrug Conjugates. European Journal of Medicinal Chemistry, 42, 20-29. http://dx.doi.org/10.1016/j.ejmech.2006.08.009

27. Oza, V., Smith, C., Raman, P., Koepf, E., Lashuel, H., Petrassi, H., Chiang, K., Powers, P., Sachettinni, J. and Kelly, J. (2002) Synthesis, Structure, and Activity of Diclofenac Analogues as Transthyretin Amyloid fibril Formation Inhibitors. Journal of Medicinal Chemistry, 45, 321-332. http://dx.doi.org/10.1021/jm010257n

28. Ramesh, K., Narayana Murthy, S., Karnakar, K. and Nageswar, Y.V.D. (2011) DABCO-Promoted Three-Component Reaction between Amines, Dialkyl Acetylenedicarboxylates, and Glyoxal. Tetrahedron Letters, 52, 3937-3941.http://dx.doi.org/10.1016/j.tetlet.2011.05.100

29. Hafiz, I.S. (2000) Enaminonitriles in Heterocyclic Synthesis: Synthesis of New 1, Dihydropyridine Pyrazolo [1, 5-a] Pyrimidine, Aminothiophene and Pyridine Derivatives. Zeitschrift fur Naturforschung, 55, 321.

30. Kiefer, J.R., Pawlitz, J.L., Moreland, K.T., Stegeman, R.A., Hood, W.F., Gierse, J.K., Steven, A.M., Goodwin, D.C., Rowlinson, S.W., Marnett, L.J., Stallings, W.C. and Kurumbail, R.G. (2000) Structural Insights into the Stereochemistry of the Cyclooxygenase Reaction. Nature, 405, 97-101.

31. Eweiss, N.F. and Osman, A. (1980) Synthesis of Heterocycles. Part II. New Routes to Acetylthiadiazolines and Alkylazothiazoles. Journal of Heterocyclic Chemistry, 17, 1713-1718. http://dx.doi.org/10.1002/jhet.5570170814

32. Nagakura, M., Ota, T., Shimadzu, N., Kawamura, K., Eto, Y. and Wada, Y. (1979) Syntheses and Antiinflammatory Actions of 4,5,6,7-Tetrahydroindazole-5-Carboxylic Acids. Journal of Medical Chemistry, 22, 48-52. http://dx.doi.org/10.1021/jm00187a012

33. Shawali, A.S. and Abdelhamide, A.O. (1976) Reaction of Dimethylphenacylsulfonium Bromide with N-Nitrosoacetarylamides and Reactions of the Products with Nucleophiles. Bulletin of the Chemical Society of Japan, 49, 321-324.http://dx.doi.org/10.1246/bcsj.49.321

34. Al-Zaydi, K.M. (2003) Microwave Assisted Synthesis, Part 1: Rapid Solventless Synthesis of 3-Substituted Coumarins and Benzocoumarins by Microwave Irradiation of the Corresponding Enaminones. Molecules, 8, 541-555. http://dx.doi.org/10.3390/80700541

35. Biere, H., Böttcher, I. and Kapp, J. (1983) Nonsteroidal Anti-Inflammatory

Agents. 11. Antiphlogistic Pyrazole Derivatives, III. Archiv der Pharmazie (Weinheim), 316, 608-616.

36. El-Taweel, F.M. and Elnagdi, M.H. (2001) Studies with Enaminones: Synthesis of New Coumarin-3-yl Azoles, Coumarin-3-yl Azines, Coumarin-3-yl Azoloazines, Coumarin-3-yl Pyrone and Coumarin-2-yl Benzo[b]Furans. Journal of Heterocyclic Chemistry, 38, 981-984. http://dx.doi.org/10.1002/jhet.5570380428

37. Hochgesang, G.P. and Marnett, L.J. (2000) Tyrosine-385 Is Critical for Acetylation of Cyclooxygenase-2 by Aspirin. Journal of the American Chemical Society, 122, 6514-6515.http://dx.doi.org/10.1021/ja0003932

38. Rowlinson, S.W., Kiefer, J.R., Prusakiewcz, J.J., Pawlitz, J.L., Kozak, K.R., Kalgutkar, A.S., Stallings, W.C., Kurumbail, R.G. and Marnett, L. (2003) A Novel Mechanism of Cyclooxygenase-2 Inhibition Involving Interactions with Ser-530 and Tyr-385. Journal of Biological Chemistry, 278, 45763-45769. http://dx.doi.org/10.1074/jbc.M305481200

39. Kurumbail, R.G., Stevens, A.M., Gierse, J.K., McDonald, J.J., Stegeman, R.A., Pak, J.Y., Gildehaus, D., Miyashiro, J.M., Penning, T.D., Seibert, K., Isakson, P.C. and Stallings, W.C. (1996) Structural Basis for Selective Inhibition of Cyclooxygenase-2 by Anti-Inflammatory Agents. Nature, 384, 644-648. http://dx.doi.org/10.1038/384644a0

40. Sidhu, R.S., Lee, J.Y., Yuan, C. and Smith, W.L. (2010) Comparison of Cyclooxygenase-1 Crystal Structures: Cross-Talk between Monomers Comprising Cyclooxygenase-1 Homodimers. Biochemistry, 49, 7069-7079. http://dx.doi.org/10.1021/bi1003298

41. Chemical Computing Group. Inc., MOE, 2009, 10.

42. Halgren, T.A. (1996) Merck Molecular Force Field I. Basis, Form, Scope, Parameterization, and Performance of MMFF94. Journal of Computational Chemistry, 17, 490-519.http://dx.doi.org/10.1002/(SICI)1096-987X(199604)17:5/6<490::AID-JCC1>3.0.CO;2-P

43. Clark, D.E. and Pickett, S.D. (2000) Computational Methods for the Prediction of 'Drug-Likeness'. Drug Discovery Today, 5, 49-58. http://dx.doi.org/10.1016/S1359-6446(99)01451-8

44. Wildman, S.A. and Crippen, G.M. (1999) Prediction of Physicochemical Parameters by Atomic Contribution. Journal of Chemical Information and Computer Sciences, 39, 868-873. http://dx.doi.org/10.1021/ci990307l

45. Fukui, K. (1982) Role of Frontier Orbitals in Chemical Reactions. Science, 218, 747-754.http://dx.doi.org/10.1126/science.218.4574.747

46. Jose, A.P. and Robert, R.S. (1991) Carbene/Anion Complexes. Unusual

Structural and Thermochemical Features of .Alpha.-Halocarbanions in the Gas Phase. Journal of the American Chemical Society, 113, 1845-1847.

47. Parr, R.G., Szentpaly, L.V. and Liu, S. (1999) Electrophilicity Index. Journal of the American Chemical Society, 121, 1922-1924. http://dx.doi.org/10.1021/ja983494x

48. Chattaraj, P.K., Maiti, B. and Sarkar, U. (2003) Philicity: A Unified Treatment of Chemical Reactivity and Selectivity. The Journal of Physical Chemistry A, 107, 4973-4975.http://dx.doi.org/10.1021/jp034707u

49. Parr, R.G., Donnelly, R.A., Levy, M. and Palke, W.E. (1978) Electronegativity: The Density Functional Viewpoint. The Journal of Chemical Physics, 68, 3801-3814.http://dx.doi.org/10.1063/1.436185

50. Parr, R.G. and Pearson, R.G. (1983) Absolute Hardness: Companion Parameter to Absolute Electronegativity. Journal of the American Chemical Society, 105, 7512-7516.http://dx.doi.org/10.1021/ja00364a005

51. Parr, R.G. and Yang, W. (1989) Density Functional Theory of Atoms and Molecules. Oxford University Press, Oxford.

Chapter 13

OPTIMIZATION OF GRAFTED FIBROUS POLYMER AS A SOLID BASIC CATALYST FOR BIODIESEL FUEL PRODUCTION

Yuji Ueki, Seiichi Saiki, Takuya Shibata, Hiroyuki Hoshina, Noboru Kasai, Noriaki Seko

Environment and Industrial Materials Research Division, Quantum Beam Science Center, Sector of Nuclear Science Research, Japan Atomic Energy Agency, Takasaki, Japan

ABSTRACT

Grafted fibrous polymer with quaternary amine groups could function as a highly-efficient catalyst for biodiesel fuel (BDF) production. In this study, the optimization of grafted fibrous polymer (catalyst) and transesterification conditions for the effective BDF production was attempted through a batch-wise transesterification of triglyceride (TG) with ethanol (EtOH) in the presence of a cosolvent. Trimethylamine was the optimal quaternary amine group for the grafted fibrous catalyst. The optimal degree of grafting of the grafted fibrous catalyst was greater than 170%. The optimal transesterification conditions were as follows: The optimal molar quantity of quaternary amine groups, transesterification temperature, molar ratio of TG and EtOH, and primary alkyl alcohol were 0.8 mmol, 80°C, 1:200, and 1-pentanol, respectively. The grafted fibrous catalyst could be applied to BDF production using natural oils. Furthermore, the grafted fibrous catalyst could be used repeatedly after regeneration involving three sequential processes, i.e., organic acid, alkali, and alcohol treatments, without any significant loss of catalytic activity.

INTRODUCTION

The global energy consumption has increased every year and has more than doubled between 1970 and 2008 (1970: 207 quadrillion British thermal units (Btu); 2008: 505 quadrillion Btu); experts predict that it will increase by another

53% by 2035 to become 770 quadrillion Btu [1] [2] . This growing global demand for energy will likely lead to the depletion of fossil fuels, unprecedented emissions of air pollutants and greenhouse gases, high energy-resource prices, and regional conflicts due to uneven distribution of fossil fuels. Therefore, to avoid and overcome these problems, many researchers worldwide are working on the development of methods to harvest energy from renewable sources such as solar, wind, hydraulic, geothermal, and biomass, together with innovative practical technologies. Among these, biomass energy is garnering attention because it has a high energy density and requires relatively facile handling and storage. Currently, studies on the production of biomethane from cellulose and organic waste [3] [4] , production of bioethanol from starch and nonfood crops [5] [6] , conversion of vegetable oils, animal fats, and waste oils into biodiesel fuel (BDF) [7] -[15] , and production of biofuels from algae [16] [17] are in progress.

BDF is defined as the monoalkyl esters of long-chain fatty acids (carbon chain lengths of 12 - 24), and is generally produced by transesterification of triglyceride (TG), which is the main constituent of vegetable oils and animal fats, with short-chain primary alcohols such as methanol or ethanol (EtOH). BDF has a number of important advantages: It is renewable, environmentally friendly, easily biodegradable, and compatible with current fuel infrastructure, and its production process, transport, and storage are very straightforward. Additionally, BDF can be used in compression-ignition engines instead of petroleum diesel with little or no modifications to the engine components. For these reasons, BDF has attracted significant attention as an alternative fossil fuel. BDF production can be classified into several methods according to the type of catalyst: homogeneous catalyst (alkaline and acid), heterogeneous catalyst (metal oxides, carbonates, zeolites, heteropoly acids, functionalized zirconia or silica, ion-exchange resins, hydrotalcites and alkaline salts) and enzymatic (lipases) catalysts and a noncatalytic supercritical method [7] -[15] . Some methods for the industrial production of BDF from oil and fat have already been developed; currently, an alkali catalyst method using a homogeneous catalyst such as NaOH or KOH is predominantly used because of its relatively fast reaction rates. However, this method has some disadvantages: The alkali catalyst corrodes equipment, the reaction requires a large amount of water during the neutralization and washing processes, undesirable byproducts are produced by a saponification side reaction, and it is difficult to separate the homogenized catalyst from the reaction mixture. These problems increase the BDF production costs.

To solve or minimize these problems, authors proposed the use of grafted fibrous polymer as a solid basic catalyst [18] . Grafted fibrous polymer (grafted fibrous catalyst) was synthesized by radiation-induced graft polymerization which could impart a desired functional group into pre-existing polymeric materials (i.e., trunk polymer) without altering their inherent properties, and the grafted fibrous polymer was used as adsorbents for environmental reclamation and obtaining resources [19] -[25] . In particular, when nonwoven fabric having both large specific surface area and high contact efficiency is selected as a trunk polymer for grafting, the BDF production speed (i.e., transesterification speed) of the grafted fibrous catalyst was more than twice that of commercial granular anion exchange resin, and the grafted fibrous catalyst could efficiently produce BDF within a shorter period of time [18] . This difference in the transesterification speed of the grafted fibrous catalyst and anion exchange resin was mostly attributed to their shape differences: The porous anion exchange resin had numerous micropores that increase the surface area. Thus, most of the reaction sites (i.e., functional groups) of the porous resin were located within its micropores, and hence the reactants were transported to the functional groups by diffusion (i.e., concentration gradient). Therefore, diffusional mass-transfer was the rate-determining factor for BDF production. In contrast, in the grafted fibrous catalyst, the functional groups were immobilized onto the graft chains, and the reactants were easily and immediately transported to the functional groups by convective flow of the reaction mixture. Therefore, the diffusional mass-transfer resistance of the reactants to the functional groups could be neglected. Additionally, the catalyst had further advantages: it was less corrosive than homogeneous alkali catalysts, did not produce soap, and was easily recovered from the BDF, which negated the need for neutralization and washing. For these advantages, authors think that the grafted fibrous catalyst will contribute to simplify and streamline the BDF production process, reduce the BDF production costs, reduce the waste product, and establish the large-scale BDF production process.

In our previous paper, we found that the grafted fibrous catalyst could function as a highly-efficient catalyst for BDF production [18] . The objective of this study is to characterize the catalytic properties of the grafted fibrous catalyst and determine the optimal BDF production conditions using the grafted fibrous catalyst. First, we demonstrated the effectiveness of type of immobilized quaternary amine groups onto the grafted fibrous catalyst, from both perspectives of the ease of amination and the catalytic performance. Second, the effects of the catalyst and transesterification conditions on the catalytic activity, transesterification speed, and production yield of BDF were elucidated by batch-wise transesterification of TG with EtOH in the presence of a cosolvent. Additionally, the effect of the type of natural oil on the transesterification yield and speed was investigated. Finally, the recyclability of the grafted fibrous catalyst and regeneration of the deactivated grafted fibrous catalyst were evaluated.

EXPERIMENTAL

Materials

Polyethylene-coated polypropylene (PE/PP) nonwoven fabric was supplied by Kurashiki Textile Manufacturing Co., Ltd. (Osaka, Japan) and used as a trunk polymer for BDF catalysts. The average fiber diameter of the PE/PP nonwoven fabric was 13 μm. 4-Chloromethylstyrene (CMS; purity > 95%) was purchased from AGC Seimi Chemical Co., Ltd. (Kanagawa, Japan). Polyoxyethylene (20) sorbitan monolaurate (Tween 20), which was used as a nonionic surfactant for preparing the monomer emulsion, was obtained from Kanto Chemical Co., Inc. (Tokyo, Japan), as were trimethylamine (TMA), triethylamine (TEA), tri-n-propylamine (TPA), tri-n-butylamine (TBA), triolein (purity > 60%), and oleic acid (purity > 80%). Sodium hydroxide, methanol, EtOH, 1-propanol, 2-propanol, 1-butanol, 1-pentanol, 1-hexanol, 1-octanol, 1-decanol, 1-dodecanol, acetonitrile, hexane, decane, linseed oil, beef tallow, and citric acid were supplied by Wako Pure Chemical Industries, Ltd. (Osaka, Japan). Rapeseed oil (51% erucic acid) and safflower oil (food-grade, 70% - 80% linoleic acid) were purchased from MP Biomedicals, LLC. (CA, USA). Palm oil (32% - 47% palmitic acid, 34% - 44% oleic acid) was obtained from Spectrum Chemical Mfg. Corp. (NJ, USA). EtOH, 2-propanol, acetonitrile, and hexane were of HPLC grade and all other chemicals were of reagent grade unless otherwise stated. Deionized water obtained from a Milli-Q deionization system (Nihon Millipore K.K., Tokyo, Japan) was used directly for preparing the monomer emulsions and aqueous solutions.

Synthesis of Grafted Fibrous Catalyst for Biodiesel Fuel Production

Grafted fibrous catalyst for BDF production was synthesized by radiation-induced emulsion grafting technique, as published previously [18] . PE/PP nonwoven fabric was irradiated with electron beam (100 kGy). Then, the irradiated PE/PP nonwoven fabric reacted with a deaerated monomer emulsion, which was composed of 3 wt% CMS, 0.3 wt% tween 20 and 96.7 wt% deionized water, in a glass ampoule and was kept in water bath at 40°C. Afterward, the CMS-grafted nonwoven fabric was recovered from the emulsion and washed repeatedly with water and then methanol to remove residual monomers and homopolymers. The amount of CMS grafted onto the PE/PP nonwoven fabric was evaluated by the degree of grafting (Dg [%]), which was calculated using the following equation:

$$Dg[\%] = 100 \times (W_1 - W_0)/W_0$$

where W_0 and W_1 are the dry weights of the PE/PP nonwoven fabrics before and after grafting, respectively.

To introduce quaternary amine groups into the CMS-graft chains of CMS-grafted nonwoven fabric (catalyst precursor), the catalyst precursor was treated with 0.25 and 2.5 M amine-ethanol solutions at 60°C for 24 h. To eliminate residual amine reagents, the aminated catalyst precursor was then washed with deionized water until it was neutral. The resultant aminated catalyst precursor is referred to as the grafted fibrous catalyst. The quaternary amine-group densities of the grafted fibrous catalysts were estimated by an analysis of their nitrogen content using an elemental analyzer (model: 2400 Series II CHNS/O elemental analyzer, PerkinElmer, Inc., MA, USA). The quaternary amine-group density of the grafted fibrous catalyst and degree of amination (Da [%]) of the CMS-graft chain are defined as follows:

Quaternary amine-group density [mmol-amine/g-catalyst] = (N/100)/14 × 1000, and

$$Da[\%] = 100 \times \left[152.6 \times (W_2 - W_0)/(W_1 - W_0) - 152.6\right]/M_{Amine}$$ where N, W_2, and M_{Amine} are the measured nitrogen content (%) of the grafted fibrous catalyst, weight of the grafted fibrous catalyst, and molecular weight of the introduced amine compound, respectively. The molecular weights of TMA, TEA, TPA, and TBA are 59.11, 101.19, 143.27 and 185.35 g/mol, respectively, while the values 14 and 152.6 refer to the molecular masses of N and CMS, respectively.

Transesterification Procedure with Grafted Fibrous Catalyst

The catalytic performance of the grafted fibrous catalyst was evaluated by the transesterification of TG and alcohols under the following standard conditions, as published previously [18] .

The transesterification tests were performed in a batch reactor equipped with a magnetic stirrer. The reactor was initially filled with 1.4 g (1.6 mmol) of triolein as a TG and 3.6 g (78 mmol) of EtOH as an alcohol, followed by the addition of 5.0 g of decane as a cosolvent in order to obtain a homogeneous phase. After stirring for 10 min, approximately 0.2 g of the pretreated grafted fibrous catalyst was immersed in the homogenous reaction solution, and the resultant mixture was heated to the reaction temperature (standard: 50°C) while stirring at 600 rpm. The reaction time is set at zero when the grafted fibrous catalyst was added to the reaction solution. In a typical test, the molar ratio of

TG to EtOH was 1:50, and the Dg, type of quaternary amine group, and amine group density of the grafted fibrous catalyst were ~300%, TMA, and 3.6 mmol-amine/g-catalyst, respectively. The molar ratio of TG to quaternary amine groups in the reaction mixture was fixed at 2:1, i.e., 0.8 mmol of quaternary amine group of the grafted fibrous catalyst was in the reaction mixture per test. At pertinent intervals, 0.1 mL aliquots were withdrawn from the reaction solution, quenched to room temperature, and diluted to 10 times the volume with 2-propanol/hexane (5:4, v/v) for compositional analysis. The effect of the molar ratio of TG to quaternary amine group (16:1 - 2:1), type of quaternary amine group, reaction temperature (20°C - 80°C), molar ratio of TG to EtOH (1:3 - 1:500), type of alcohol, and reaction time on the conversion ratio of TG to biodiesel fuel were investigated in detail.

Compositional Analysis

The compositions of the collected samples were analyzed using high-performance liquid chromatography (HPLC), according to the separation conditions described by Holčapek et al. [26] . The HPLC system (Shimadzu, Kyoto, Japan) consisted of a pump (model: LC-20AD) with a quaternary gradient system, degasser (model: DGU-20A5), system controller (model: CBM-20Alite), sample injector (model: 7725i; Rheodyne, CA, USA) with a 1.0 µL sample reservoir, column oven (model: CTO-20AC), and UV-Vis detector (model: SPD-20A). System control and data processing were carried out using Shimadzu LC solution software. A reversed-phase C18 column (model: L-column ODS, column size: 2.1 mm i.d. × 150 mm long, particle diameter: 5 µm, Chemicals Evaluation and Research Institute, Tokyo, Japan) was used for separation. The compositional analysis was conducted with a quaternary mobile phase consisting of deionized water (solvent A), acetonitrile (solvent B), 2-propanol (solvent C), and hexane (solvent D). The quaternary gradient elution program was as follows: 30% A + 70% B at 0 min, 100% B at 5 min, 50% B + 27.8% C + 22.2% D at 10 min, and isocratic elution with 50% B + 27.8% C + 22.2% D for the last 5 min. The flow rate was set at 1.0 mL/min, the effluent was monitored at 205 nm, and the column temperature was maintained at 40°C. The conversion ratio of TG to BDF was calculated from the rate of change of the total HPLC peak areas for TG before and after transesterification.

RESULTS AND DISCUSSION

Effect of Type of Quaternary Amine Groups on Catalytic Performance

To produce BDF using grafted fibrous catalyst, OH⁻ ions, which were chemically immobilized on the grafted fibrous catalyst, are the actual catalytic elements, while the polymer matrix acts as a scaffold to immobilize these ions. Although OH⁻ ions can be easily immobilized onto the grafted fibrous catalyst by interionic interaction with the cationic functional groups such as a quaternary amine group, it is considered that the type of quaternary amine group has possibility to have a significant influence on the catalytic performance. In order to select the best quaternary amine group of grafted fibrous catalyst, we discussed the effects of four types of amines with different alkyl chain lengths, such as TMA, TEA, TPA, and TBA, from both perspectives of the reactivity with CMS-grafted chains and the catalytic performance.

A CMS-grafted nonwoven fabric with 170% Dg was used as the catalyst precursor, and each amine concentration was fixed at 0.25 M. As expected, the quaternary amine-group density increased with increasing reaction time, and TMA, which has the shortest alkyl chain length, exhibited the highest reactivity toward the CMS-graft chains. As shown in Figure 1, amination with 0.25 M TMA began immediately and completed within 1 h; the quaternary amine-group density and Da reached 3.0 mmol-TMA/g-catalyst and 93%, respectively. However, other amine reagents with longer alkyl chain lengths, such as TEA, TPA, and TBA, were less reactive because of the increased steric hindrance of the amine compounds. Amination of the CMS-graft chains did not completely finish even after 24 h of amination with 0.25 M amine solution; the Da for the products of the reactions with TEA, TPA, and TBA were 81%, 63%, and 65%, respectively. To enhance these values, the CMS-grafted nonwoven fabric was treated with 2.5 M TEA-, TPA-, and TBA-EtOH solutions. After 24 h of amination with 2.5 M TEA, TPA, and TBA, Da values of 91%, 85%, and 88%, respectively, were achieved. These values are comparable to that for TMA. However, longer alkyl chain lengths of the amine compounds led to reduced quaternary amine-group densities per gram of grafted fibrous catalysts, even when the Da values for each amine were comparable to that of TMA. As denoted inFigure 1, the quaternary amine-group density for each amine reached 2.6 mmol-TEA/g-catalyst, 2.2 mmol-TPA/g-catalyst, and 2.0 mmol-TBA/g-catalyst after 24 h of amination. The effect of the Dg of the CMS-grafted nonwoven fabric on amination was also investigated, and it was found that the Dg value did not significantly influence the Da: The Da values after 1 h of amination with 0.25 M TMA were almost constant at 93%, regardless of the Dg. The quaternary amine-group densities were 2.6, 3.3, 3.6, and 3.8 mmol-TMA/g-catalyst for Dg values of 100%, 200%, 300%, and 400%, respectively.

Next, the effect of the type of immobilized quaternary amine groups on the catalytic activity was investigated; these results are given in Figure 2. As described in Section 2.3, transesterification was conducted in batch mode, and the molar quantity of quaternary amine groups of each catalyst in the reaction mixture was fixed at 0.8 mmol.

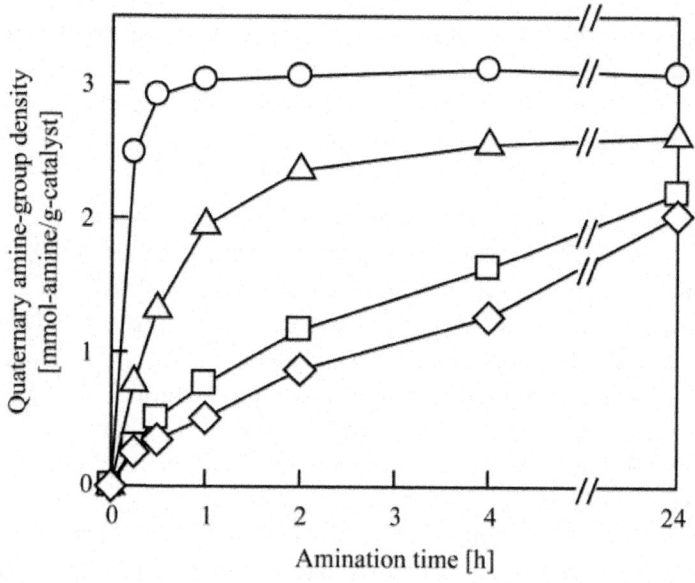

Figure 1: Effect of the type of amine reagents on the quaternary amine-group density of CMS-grafted nonwoven fabric.

As is evident from Figure 2, each catalyst exhibited similar catalytic activity toward BDF production regardless of the type of quaternary amine groups, and hence steric bulk, and transesterification proceeded with all the grafted fibrous catalysts. The conversion ratios of TG using TMA-, TEA-, TPA-, and TBA-type fibrous catalysts were 93%, 93%, 90%, and 90%, respectively, after 4 h of transesterification. The similarity of the transesterification speeds is attributed to the function of the quaternary amine groups, i.e., as a scaffold to immobilize the OH⁻ ions that are the actual catalytic elements; therefore, the quaternary amine groups were not directly involved in transesterification. From the results of the amination of CMS-grafted nonwoven fabric and the effect of the type of quaternary amine groups on the catalytic activity, we concluded that TMA is the optimal quaternary amine group for the grafted fibrous catalyst because it is the smallest trialkylamine and thus could easily be immobilized onto the CMS-graft chains. The order of reactivity of the trialkylamines for the amination of CMSgraft chains was TMA >> TEA > TPA > TBA.

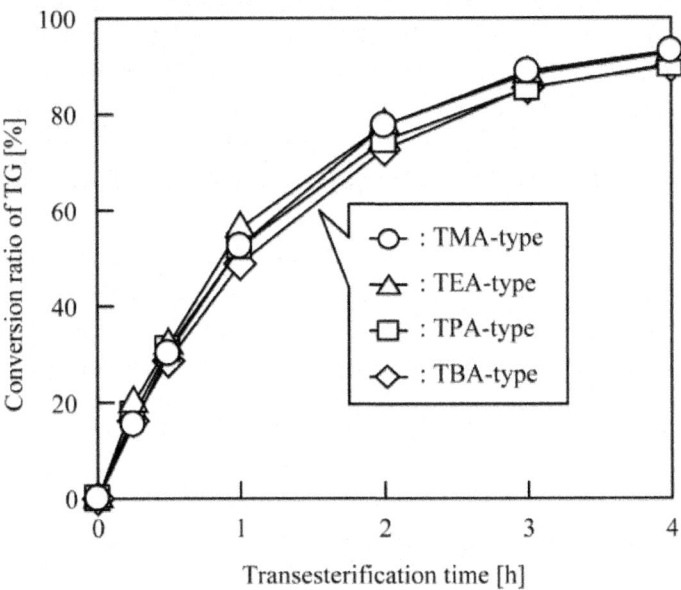

Figure 2: Effect of the type of immobilized quaternary amine group on the catalytic activity.

Effect of Catalyst Conditions on Catalytic Performance

The yield and rate of BDF production were greatly dependent on the catalyst conditions such as the Dg value, the shape and the fiber diameter of the grafted fibrous catalyst, and the transesterification conditions including the molar quantity of quaternary amine groups in the reaction mixture, transesterification temperature, molar ratio of TG to EtOH in the reaction mixture, type of alcohol, and so on. In this section, the effects of the catalyst conditions on catalytic performance were investigated.

First, the effects of the Dg value of the grafted fibrous catalyst on the catalytic activity and yield were investigated. In this experiment, the Dg of the grafted fibrous catalyst was controlled within a range up to 400%, and the molar quantity of TMA in the reaction mixture was 0.8 mmol per test. The other conditions were the same as the typical conditions described in Section 2.3. Regardless of the Dg value, each catalyst exhibited almost the same catalytic activity toward BDF production, except in the case of the grafted fibrous catalysts with 80% Dg or less. The TG conversion ratios using catalysts with 80%, 170%, 200%, 300%, and 400% Dg reached 70%, 87%, 91%, 93%, and 89%, respectively, after 4 h of transesterification. The decline in the transesterification efficiency of grafted fibrous catalyst with less than 80% Dg is attributed to the large volumes of these catalysts, which are sterically bulky, required to introduce 0.8 mmol of TMA into the reaction system. As a result, part of the grafted fibrous catalyst was not immersed in the reaction mixture during transesterification. Based on these results, we concluded that the optimal Dg value of the grafted fibrous catalyst to achieve effective transesterification was greater than 170%. The TMA group density of 170% Dg is 3.1 mmol-TMA/g-catalyst.

The fiber diameter of the grafted fibrous catalyst is also an important factor that affects the catalytic activity; accordingly, the relationship between the fiber diameter of the grafted fibrous catalyst and catalytic activity was examined. In this test, four types of grafted fibrous catalysts with different fiber diameters (i.e., 26, 39, 47, and 67 μm) were used. The grafted fibrous catalysts with different fiber diameters were prepared by controlling the fiber diameter of the nonwoven fabric to be 13, 19, 24, and 34 μm. The Dg and quaternary amine-group density of all grafted fibrous catalysts were fixed at approximately 300% and 3.6 mmol-TMA/g-catalyst, and the other conditions were the same as the typical conditions described in Section 2.3. Each catalyst exhibited almost the same catalytic activity toward BDF production regardless of the fiber diameter of the grafted fibrous catalyst, although the fiber diameter of the trunk polymer had a significant influence on the grafting efficiency. During emulsion grafting, trunk polymers with finer fibers had a larger specific surface area, which increased the speed and efficiency of the graft polymerization. A similar result was reported by Basuki et al. [27] . In contrast, during transesterification, the molar quantity of TMA in the reaction mixture was more important than the specific surface area because the reactants could be easily and immediately transported to the functional groups by the convective flow of the reaction mixture. Based on these results, the fiber diameter of the grafted fibrous catalyst did not significantly influence the catalytic activity. The conversion ratios of TG using the catalysts with fiber diameters of 26, 39, 47, and 67 μm were 93%, 92%, 88%, and 90%, respectively, after 4 h of transesterification.

Effect of Transesterification Conditions on Catalytic Performance

The effect of the molar quantity of quaternary amine group on the transesterification rate was investigated; these results are shown in Figure 3. In this experiment, the molar quantities of quaternary amine groups varied from

0.1 to 0.8 mmol TMA, and the other conditions were the same as the typical conditions described in Section 2.3. As expected, the transesterification rate increased with increasing molar quantity of TMA in the reaction mixture, and the conversion ratios of TG using 0.1, 0.2, 0.4, and 0.8 mmol TMA reached 27%, 48%, 76%, and 93%, respectively, after 4 h of transesterification. In this experiment, we concluded that the optimal molar quantity of quaternary amine groups (i.e., TMA) to achieve effective transesterification was 0.8 mmol per test.

Furthermore, the effect of the transesterification temperature on the transesterification rate of the grafted fibrous catalyst was studied in detail, and the results are shown in Figure 4(a). For comparison, the results for commercial granular anion exchange resin are provided in Figure 4(b). DIAION PA306S (Mitsubishi Chemical Co., Ltd., Tokyo, Japan) was used as the commercial granular anion exchange resin with TMA quaternary amine groups at a density of 4.2 mmol-TMA/g-dry resin with a particle diameter of 150 to 425 μm. In this experiment, the transesterification temperature was controlled within the range of 20°C to 80°C and TMA-type grafted fibrous catalyst was used. The other conditions were the same as the typical conditions described in Section 2.3. As indicated in Figure 4, although both the grafted fibrous catalyst and commercial granular resin exhibited good catalytic activities at all investigated transesterification temperatures, the transesterification rate of the grafted fibrous catalyst was faster than that of DIAION PA306S at all temperatures. As expected, higher transesterification temperatures led to faster transesterification rates for both catalysts. In particular, the transesterification was remarkably accelerated at high temperatures for the grafted fibrous catalyst. As seen in Figure 4(a), in the case of the grafted fibrous catalyst, about 3 h of transesterification was required to convert 90% of the initial TG into.

BDF at 50°C. In contrast, at 80°C, the required time to achieve a conversion ratio of 90% was only 40 min, which is less than one fourth the time required at 50°C. Based on these results, higher reaction temperatures resulted in more effective and efficient transesterification and, therefore, the grafted fibrous catalyst could dramatically reduce the time required for BDF production than the granular resin.

The ratio of TG to alcohol in the reaction mixture is also known to be a controlling factor for BDF production. Therefore, the effects of the molar ratio of TG to EtOH in the reaction mixture on the transesterification yield and rate were investigated; the results are given in Figure 5.

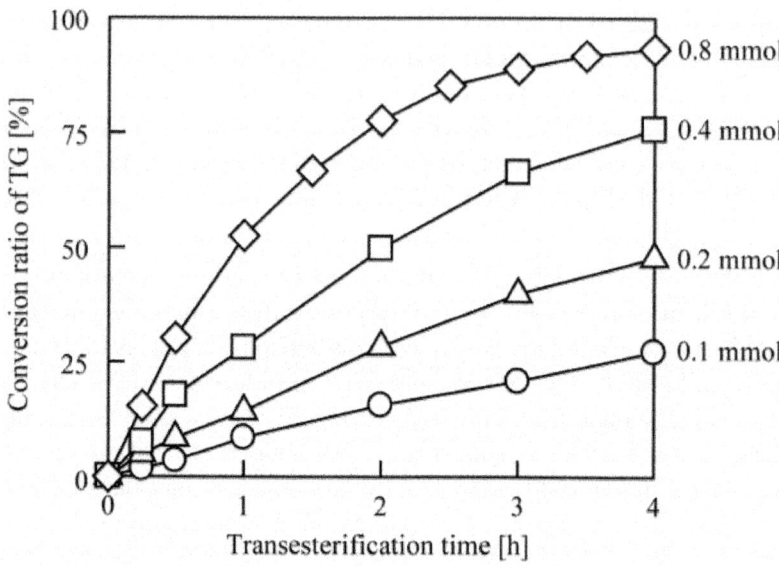

Figure 3: Effect of molar quantity of quaternary amine groups on transesterification speed.

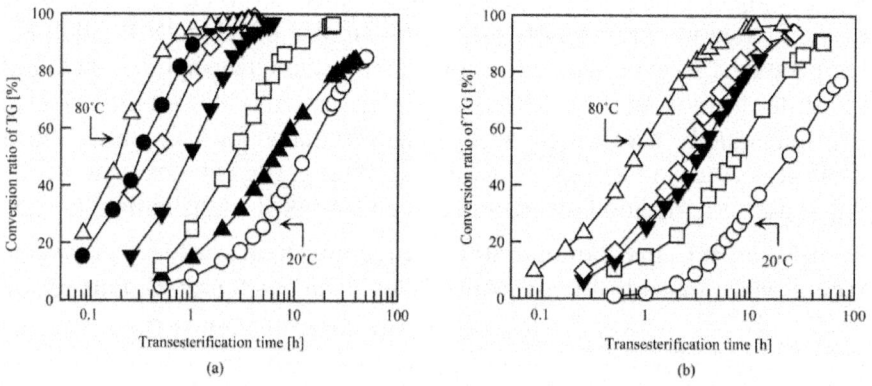

Figure 4: Effect of transesterification temperature on transesterification speed. Catalyst: (a) grafted fibrous catalyst and (b) commercial granular resin. Transesterification temperature: ○, 20°C; ▲, 30°C; □, 40°C; ▼, 50°C; ◇, 60°C; ●, 70°C; △, 80°C.

In this experiment, the molar ratios of TG to EtOH were adjusted to 1:3, 1:10, 1:25, 1:50, 1:100, 1:200, and 1:500, which corresponded to TG/EtOH volume ratios of 1:0.2, 1:0.6, 1:1.5, 1:3, 1:6, 1:12, and 1:30, respectively. The other conditions were the same as the typical conditions described in Section

2.3. As denoted in Figure 5, the molar ratio of TG to EtOH had a significant influence on the transesterification yield and rate: Transesterification was accelerated in the presence of excess EtOH. The transesterification times required to convert 90% of the initial TG into BDF were 4.47, 3.18, 2.09, 1.51 and 1.56 h for TG/EtOH molar ratios of 1:25, 1:50, 1:100, 1:200 and 1:500, respectively. TG/EtOH molar ratios of 1:3 and 1:10, the conversion ratio did not reach 90% even after 50 h. There are two potential reasons for these phenomena: One is that the EtOH in the reaction mixture was vaporized by heating during transesterification, and the other is that the relatively low volume proportion of EtOH to TG in the reaction mixture resulted in heterogeneous dispersion of the EtOH. The TG/EtOH molar ratios of 1:3 and 1:10 were corresponded to the TG/EtOH volume ratios of 1:0.2 and 1:0.6, respectively. In contrast, when the percentage of EtOH in the reaction mixture was significantly high, as at a TG/EtOH molar ratio of 1:500, EtOH diluted the reaction mixture and thereby hindered the rate of transesterification. Based on these results, the optimal molar ratio of TG and EtOH in the reaction mixture for the fastest BDF production was 1:200.

Finally, the effects of the type of alcohol on the transesterification yield and speed were investigated. In this test, nine primary alcohols with different linear alkyl chain lengths, i.e., 1 to 12 carbon atoms, were used; these were methanol (C1), EtOH (C2), 1-propanol (C3), 1-butanol (C4), 1-pentanol (C5), 1-hexanol (C6), 1-octanol (C8), 1-decanol (C10), and 1-dodecanol (C12). The molar ratio of TG to primary alkyl alcohol was fixed at 1:50, and the other conditions were the same as the typical conditions described in Section 2.3. Figure 6 shows HPLC chromatograms of the BDF samples produced from 30 min of transesterification of triolein with each alcohol, and the conversion ratio of TG with each alcohol is plotted versus the transesterification time in Figure 7. As shown inFigure 6, BDF was produced regardless of the primary alkyl alcohol used; therefore, the grafted fibrous catalyst is applicable to transesterification of TG with a variety of alcohols. Furthermore, the peaks corresponding to BDF gradually shifted to longer retention times as the alkyl chain length of the primary alkyl alcohol increased; this is attributed to the differences in the structures (and hydrophobicity) of the produced BDF, and demonstrated that differed BDFs were produced from different types of alcohol.

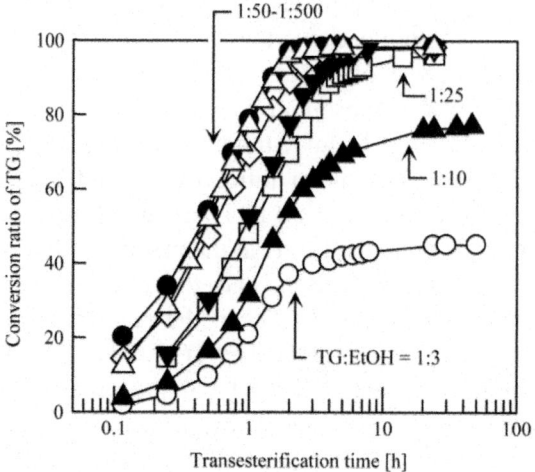

Figure 5: Effects of molar ratio of TG to EtOH in reaction mixture on transesterification yield and speed. Molar ratio of TG to EtOH: ○, 1:3; ▲, 1:10; □, 1:25; ▼, 1:50; ◇, 1:100; ●, 1:200; △, 1:500.

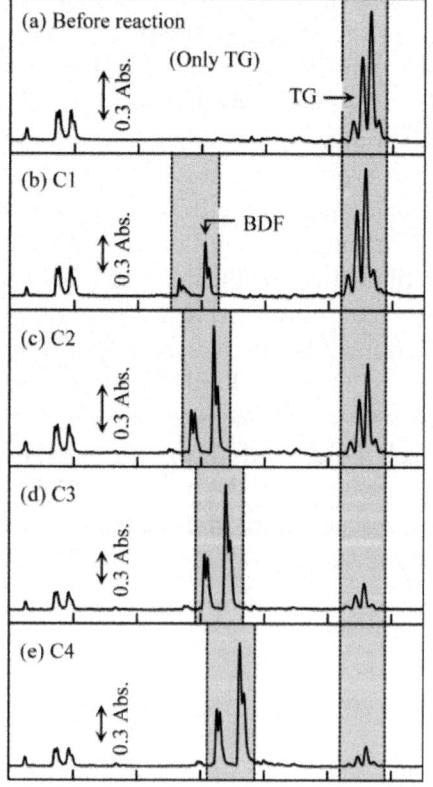

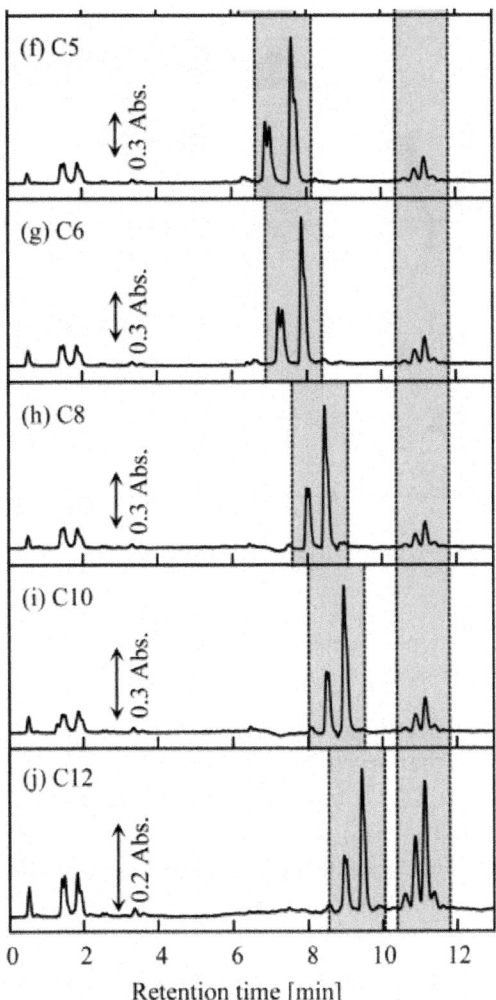

Figure 6: HPLC chromatograms of BDF samples produced by transesterification of triolein with different types of alcohol. Transesterification time: 30 min.

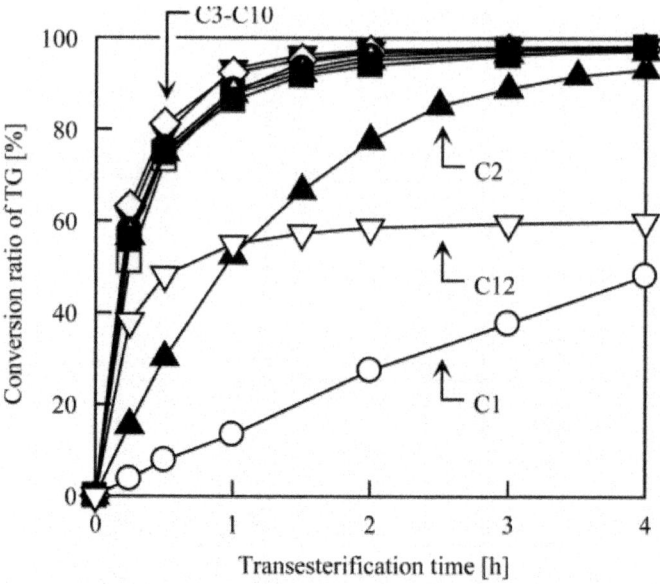

Figure 7: Effects of the type of alcohol on transesterification yield and speed. Alcohol: ○, C1; ▲, C2; □, C3; ▼, C4; ◇, C5; ●, C6; △, C8; ■, C10; ▽, C12.

As can be seen fromFigure 7, the yield and speed of BDF production varied depending on the type of primary alkyl alcohol: The conversion ratios of TG using C1, C2, C3, C4, C5, C6, C8, C10, and C12 were 8%, 30%, 73%, 79%, 81%, 76%, 75%, 74%, and 48%, respectively, after 30 min of transesterification. These results indicate that transesterification is acelerated when more hydrophobic (i.e., lipophilic) primary alkyl alcohols were used. However, as is evident from the results of using C12, very hydrophobic alcohols resulted in slower transesterification than primary alkyl alcohols with moderate hydrophobicity, such as C4, C5, and C6. Furthermore, transesterification using C12 slowed significantly at the conversion ratio of about 55% during the first 1 h; afterward, transesterification progressed gradually to reach a conversion ratio of 64% after 24 h. This decline in transesterification yield is attributed to the primary alkyl alcohol being too hydrophobic to react with the OH ions, which are highly hydrophilic, although highly hydrophobic alcohols are highly miscible with TG. In contrast, when primary alkyl alcohols with shorter alkyl chains, such as C1, were used, phase separation occurred before and after transesterification, even if a cosolvent such as decane was added to improve the uniformity of the reaction solution. This phase separation is attributed to the immiscibility of TG and C1. As a result, as shown in Figure 7, the trancesterification speed using methanol was very slow, because the reaction occurred only at the interface between the TG and C1 layer. The conversion ratio of TG using C1 reached 96% after 24 h of transesterification. To increase the miscibility of the two compounds, Tang et al. suggested that higher pressure and temperature are needed [28]. Based on the above results, the structure of BDF and transesterification speed could be controlled to some extent by the type of alcohol used. In this study, the primary alkyl alcohol that enabled the fastest production of BDF was C5, and the order of transesterification speed was C1 << C2 << C3 < C4 < C5 > C6 > C8 > C10 >> C12.

BDF Production from Vegetable Oils and Animal Fat

From the above experimental results, it was elucidated that the grafted fibrous catalyst shows good activity for BDF production from TG with an alcohol. To enable practical use of the grafted fibrous catalyst, it must adapt to a wide variety of feed oils such as natural vegetable oils, animal fats, and waste oils, which contain different types of TGs [29] [30], in addition to the reagent-grade triolein used in this study. Thus, the effect of the type of natural oil on the transesterification yield and speed was investigated.

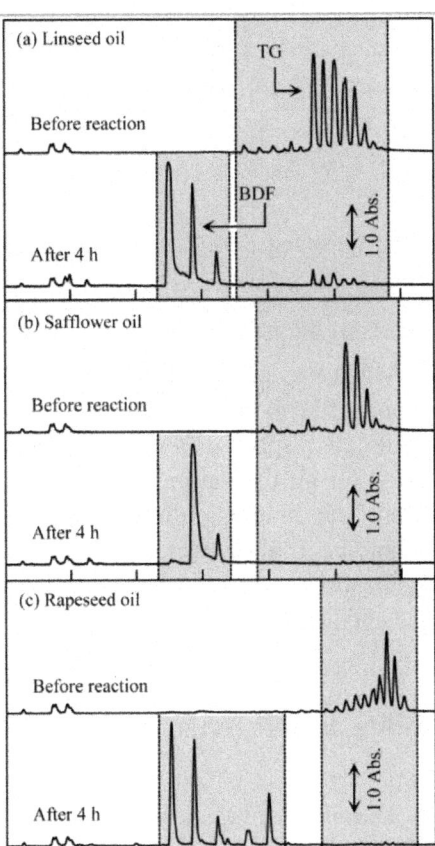

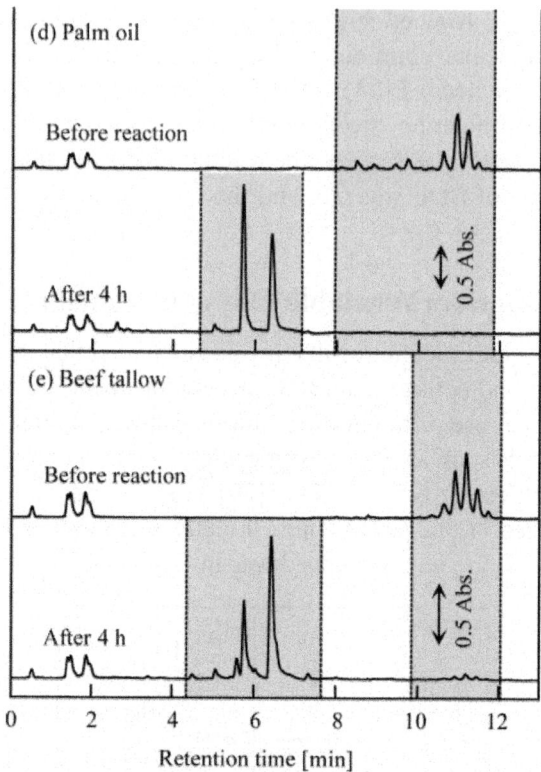

Figure 8: BDF production from vegetable oils and animal fat using grafted fibrous catalyst.

In the chromatograms of the natural oils, many peaks

derived from TG, which was contained in the natural oils, were observed before transesterification, unlike in the chromatogram of reagent-grade triolein. Although the transesterification speed differed according to the type of natural oil, the conversion ratios for all the natural oils were greater than 90% after 4 h of transesterification. The conversion ratios for linseed oil, safflower oil, rapeseed oil, palm oil, and beef tallow were 92%, 98%, 97%, 97%, and 93%, respectively, after 4 h of transesterification. Based on these results, it is evident that the grafted fibrous catalyst is applicable for BDF production from many types of natural oils.

Repeated-Use Stability and Regeneration of Grafted Fibrous Catalyst

Repeated-use stability and ease of regeneration are important aspects of the grafted fibrous catalyst for the production of large amounts of BDF and extended use of the catalyst. Firstly, the repeated-use stability of the grafted fibrous catalyst was evaluated by repeated transesterification of triolein with EtOH, and the immutability of the catalytic activity for each transesterification experiment was monitored. In this test, the grafted fibrous catalyst was only washed with EtOH after each transesterification. The molar ratio of TG to EtOH was fixed at 1:200, and the other conditions were the same as the typical conditions described in Section 2.3. As shown in Figure 9, the catalytic activity of the grafted fibrous catalyst gradually decreased with increasing number of transesterification reactions. There are two main reasons for this decay of catalytic activity: One is the removal of OH^- ions from the grafted fibrous catalyst, and the other is contamination of the quaternary amine groups by oleate ions, which are catalytically inactive. Oleate ions are generated during the transesterification process by a direct ion-exchange reaction of the OH^- ions, which were immobilized onto the graft chains, with the oleic acid group of triolein, diolein, or monoolein. In the presence of organic acid ions, the counterion of the quaternary amine group that was introduced into the graft chain is easily replaced with an organic acid ion instead of a OH^- ion. Additionally, rate of decrease of the catalytic activity was not steady: A rapid decrease was observed after the 8th and subsequent transesterifications. The transesterification times required to convert 90% of the initial TG into BDF were 0.99, 1.00, 1.04, 1.10, 1.18, 1.32, 1.44, 1.97, 2.90 and 12.00 h for the first ten reactions. At the 11th reaction, the catalytic activity of the grafted fibrous catalyst was almost negligible; thus, the conversion ratio only reached 9%, even after 24 h of transesterification.

Next, the deactivated grafted fibrous catalyst was regenerated, according to the procedure described by Shibasaki-kitakawa et al. [12] . The regeneration process consists of the following three steps: 1) Washing with 0.25 M citric acid solution (solvent: EtOH) to desorb the organic acid ions that cover the active sites, i.e., the quaternary amine groups, of the grafted fibrous catalyst, 2) regenerating with 1 M NaOH aqueous solution to replace the citric acid ions, which formed ionic bonds with the quaternary amine groups, with OH^- ions and washing with deionized water, and 3) washing with EtOH to restore the initial swelled condition. When the deactivated grafted fibrous catalyst was treated with the first and third steps and only the third step, the catalytic activities of each grafted fibrous catalyst did not recover their original state.

The conversion ratios of each grafted fibrous catalyst were almost zero, even after 24 h of transesterification. When the deactivated grafted fibrous catalyst was treated with the second and third steps, the catalytic activity of the grafted fibrous catalyst partially recovered: The conversion ratio after 1 h of transesterification was 25%, which is equivalent to about one-fourth of the original ratio (90% after 1 h of transesterification). Also, the transesterification reaction almost stopped after 4 h, and the conversion ratio after 4 h was about 43%. However, when the deactivated grafted fibrous catalyst was treated with all three steps, the catalytic activity of the grafted fibrous catalyst almost completely recovered to its original state:

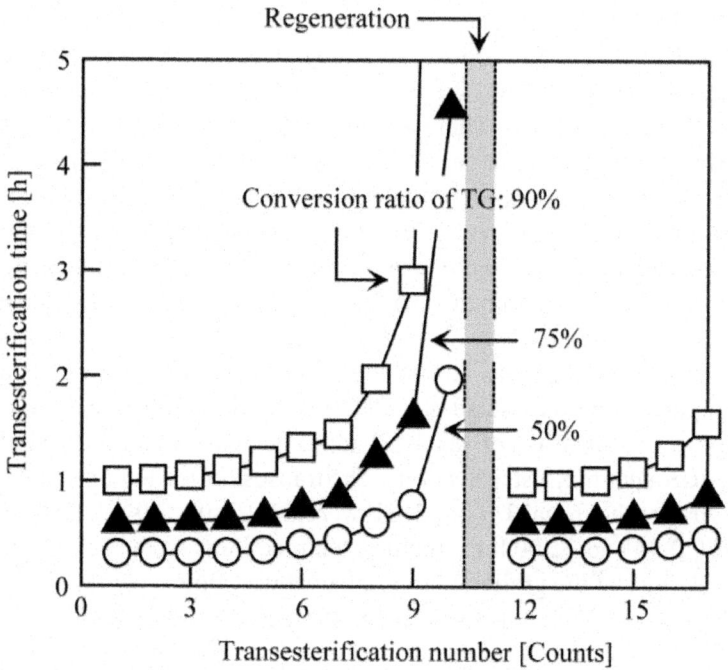

Figure 9: Repeated-use stability and regeneration of grafted fibrous catalyst.

The transesterification times required to convert 90% of the initial TG into BDF were 0.99 and 0.99 h for the 1st and 12th reactions, respectively. The conversion ratio and transesterification speed of the regenerated grafted fibrous catalyst almost overlapped with those of the unused grafted fibrous catalyst. As shown in Figure 9, the grafted fibrous catalyst could be used repeatedly, although the catalytic activity of the regenerated grafted fibrous catalyst gradually decreased with increasing number of transesterification reactions, as was the case with the unused grafted fibrous catalyst. Additionally, the regenerated grafted fibrous catalyst did not suffer significant physical or chemical damage, even after several transesterification-regeneration cycles. Based on these results, it is evident that the grafted fibrous catalyst could be used up to ten times without any proper regeneration and the BDF production capacity per gram of the grafted fibrous catalyst with 3.6 mmol-TMA/g-catalyst was 70 g because the weight ratio of triolein to grafted fibrous catalyst in the reaction mixture was fixed at ~7:1 (1.4 g/0.2 g) per test. Furthermore, it was confirmed that the three sequential regeneration processes, which involved organic acid, alkali and alcohol treatments, were required to completely recover the catalytic activity of the deactivated grafted fibrous catalyst. Also, the grafted fibrous catalysts are sufficiently stable for repeated use and sufficiently durable for long-term use.

CONCLUSION

TMA is the optimal quaternary amine group for the grafted fibrous catalyst from both perspectives of the ease of amination and the catalytic performance, although the various types of quaternary amine groups with different alkyl chain lengths could be introduced into the CMS-graft chains. The optimal Dg value of the grafted fibrous catalyst to achieve effective transesterification was greater than 170%, andthe quaternary amine-group density of 170% Dg is 3.1 mmol-TMA/g-catalyst. Additionally, the transesterification temperature, molar ratio of TG to EtOH, and type of alcohol significantly influenced the transesterification efficiency, as did the molar quantity of quaternary amine groups in the reaction mixture. The optimal molar quantity of quaternary amine groups, transesterification temperature, molar ratio of TG and EtOH, and primary alkyl alcohol were 0.8 mmol, 80°C, 1:200, and 1-pentanol, respectively. The grafted fibrous catalyst could be applied to BDF production using any type of natural oil as the starting oil. The grafted fibrous catalyst could be used ten times without any proper regeneration, and the BDF production capacity per gram of the grafted fibrous catalyst was ~70 g. The deactivated grafted fibrous catalyst could be recovered to its initial state by three sequential regeneration processes, including organic acid, alkali, and alcohol treatments, without any significant loss of the catalytic activity. Based on these results, the grafted fibrous catalyst, which has sufficient catalytic activity, efficient BDF production capacity, and high repeated-use stability for practical application in BDF production, will contribute to the feasibility of environmentally friendly energy resources, reduce the dependence on petroleum, and improve global environment preservation in the near future. Furthermore, we are firmly convinced that this new type of catalyst will encourage the dissemination of radiation processing and create new possibilities for radiation research.

REFERENCES

1.   US Department of Energy, Energy Information Administration (US DOE/EIA) (2001) International Energy Outlook 2001. US Department of Energy, Energy Information Administration, Washington DC.

2.   US Department of Energy, Energy Information Administration (US DOE/EIA) (2011) International Energy Outlook 2011. US Department of Energy, Energy Information Administration, Washington DC.

3.   Monlau, F., Sambusiti, C., Barakat, A., Guo, X.M., Latrille, E., Trably, E., Steyer, J.-P. and Carrere, H. (2012) Predictive Models of Biohydrogen and Biomethane Production Based on the Compositional and Structural Features of Lignocellulosic Materials. Environmental Science & Technology, 46, 12217-12225. http://dx.doi.org/10.1021/es303132t

4.   Li, Y., Zhang, R., Liu, X., Chen, C., Xiao, X., Feng, L., He, Y. and Liu, G. (2013) Evaluating Methane Production from Anaerobic Monoand Co-Digestion of Kitchen Waste, Corn Stover, and Chicken Manure. Energy & Fuels, 27, 2085-2091.http://dx.doi.org/10.1021/ef400117f

5.   Yangcheng, H., Jiang, H., Blanco, M. and Jane, J.-L. (2013) Characterization of Normal and Waxy Corn Starch for Bioethanol Production. Journal of Agricultural and Food Chemistry, 61, 379-386. http://dx.doi.org/10.1021/jf305100n

6.   Limayem, A. and Ricke, S.C. (2012) Lignocellulosic Biomass for Bioethanol Production: Current Perspectives, Potential Issues and Future Prospects. Progress in Energy and Combustion Science, 38, 449-467. http://dx.doi.org/10.1016/j.pecs.2012.03.002

7.   Tran, H.-L., Ryu, Y.-J., Seong, D.H., Lim, S.-M. and Lee, C.-G. (2013) An Effective Acid Catalyst for Biodiesel Production from Impure Raw Feedstocks. Biotechnology and Bioprocess Engineering, 18, 242-247. http://dx.doi.org/10.1007/s12257-012-0674-1

8.   Sagiroglu, A., Ozcan, H.M., Isbilir, S.S., Paluzar, H. and Toprakkiran, N.M. (2013) Alkali Catalysis of Different Vegetable Oils for Comparisons of Their Biodiesel Productivity. Journal of Sustainable Bioenergy Systems, 3, 79-85.http://dx.doi.org/10.4236/jsbs.2013.31011

9.   Mata, T.M., Sousa, I.R.B.G., Vieira, S.S. and Caetano, N.S. (2012) Biodiesel Production from Corn Oil via Enzymatic Catalysis with Ethanol. Energy & Fuels, 26, 3034-3041.http://dx.doi.org/10.1021/ef300319f

10. Babajide, O., Musyoka, N., Petrik, L. and Ameer, F. (2012) Novel Zeolite Na-X Synthesized from Fly Ash as a Heterogeneous Catalyst in Biodiesel Production. Catalysis Today, 190, 54-60.http://dx.doi.org/10.1016/j.cattod.2012.04.044

11. Ilham, Z. and Saka, S. (2012) Optimization of Supercritical Dimethyl Carbonate Method for Biodiesel Production. Fuel, 97, 670-677. http://dx.doi.org/10.1016/j.fuel.2012.02.066

12. Shibasaki-Kitakawa, N., Honda, H., Kuribayashi, H., Toda, T., Fukumura, T. and Yonemoto, T. (2007) Biodiesel Production Using Anionic Ion-Exchange Resin as Heterogeneous Catalyst. Bioresource Technology, 98, 416-421.http://dx.doi.org/10.1016/j.biortech.2005.12.010

13. Tsuji, T., Kubo, M., Shibasaki-Kitakawa, N. and Yonemoto, T. (2009) Is Excess Methanol Addition Required to Drive Transesterification of Triglyceride toward Complete Conversion? Energy & Fuels, 23, 6163-6167.http://dx.doi.org/10.1021/ef900622d

14. Demirbas, A. (2008) Comparison of Transesterification Methods for Production of Biodiesel from Vegetable Oils and Fats. Energy Conversion and Management, 49, 125-130.http://dx.doi.org/10.1016/j. enconman.2007.05.002

15. Helwani, Z., Othman, M.R., Aziz, N., Kim, J. and Fernando, W.J.N. (2009) Solid Heterogeneous Catalysts for Transesterification of Triglycerides with Methanol: A Review. Applied Catalysis A: General, 363, 1-10.http://dx.doi.org/10.1016/j.apcata.2009.05.021

16. Zhou, D., Zhang, S., Fu, H. and Chen, J. (2012) Liquefaction of Macroalgae Enteromorpha Prolifera in Sub-/Super-critical Alcohols: Direct Production of Ester Compounds. Energy & Fuels, 26, 2342-2351. http://dx.doi.org/10.1021/ef201966w

17. Menetrez, M.Y. (2012) An Overview of Algae Biofuel Production and Potential Environmental Impact. Environmental Science & Technology, 46, 7073-7085.http://dx.doi.org/10.1021/es300917r

18. Ueki, Y., Mohamed, N.H., Seko, N. and Tamada, M. (2011) Rapid Biodiesel Fuel Production Using Novel Fibrous Catalyst Synthesized by Radiation-Induced Graft Polymerization. International Journal of Organic Chemistry, 1, 20-25.http://dx.doi.org/10.4236/ijoc.2011.12004

19. Madrid, J.F., Ueki, Y. and Seko, N. (2013) Abaca/Polyester Nonwoven Fabric Functionalization for Metal Ion Adsorbent Synthesis via Electron Beam-Induced Emulsion Grafting. Radiation Physics and Chemistry, 90, 104-110.http://dx.doi.org/10.1016/j.radphyschem.2013.05.004

20. Ueki, Y., Dafader, N.C., Hoshina, H., Seko, N. and Tamada, M. (2012) Study and Optimization on Graft Polymerization under Normal Pressure and Air Atmospheric Conditions, and Its Application to Metal Adsorbent. Radiation Physics and Chemistry, 81, 889-898. http://dx.doi. org/10.1016/j.radphyschem.2012.02.031

21. Iwanade, A., Kasai, N., Hoshina, H., Ueki, Y., Saiki, S. and Seko, N. (2012) Hybrid Grafted Ion Exchanger for Decontamination of Radioactive Cesium in Fukushima Prefecture and Other Contaminated Areas. Journal

of Radioanalytical and Nuclear Chemistry, 293, 703-709. http://dx.doi.
org/10.1007/s10967-012-1721-2

22. Hoshina, H., Kasai, N., Shibata, T., Aketagawa, Y., Takahashi, M.,
Yoshii, A., Tsunoda, Y. and Seko, N. (2012) Synthesis of Arsenic Graft
Adsorbents in Pilot Scale. Radiation Physics and Chemistry, 81, 1033-
1035.http://dx.doi.org/10.1016/j.radphyschem.2012.02.018

23. Hoshina, H., Seko, N., Ueki, Y., Iyatomi, Y. and Tamada, M. (2010)
Evaluation of Graft Adsorbent with N-MethylD-Glucamine for Boron
Removal from Groundwater. Journal of Ion Exchange, 21, 153-156.

24. Seko, N., Hoshina, H., Kasai, N., Ueki, Y., Tamada, M., Kiryu, T., Tanaka,
K. and Takahashi, M. (2010) Novel System for Recovering Scandium
from Hot Spring Water with Fibrous Graft Adsorbent. Journal of Ion
Exchange, 21, 117-122.

25. Seko, N., Katakai, A., Hasegawa, S., Tamada, M., Kasai, N., Takeda, H.,
Sugo, T. and Saito, K. (2003) Aquaculture of Uranium in Seawater by
a Fabric-Adsorbent Submerged System. Nuclear Technology, 144, 274-
278.

26. Holcapek, M., Jandera, P., Fischer, J. and Prokes, B. (1999) Analytical
Monitoring of the Production of Biodiesel by High-Performance
Liquid Chromatography with Various Detection Methods. Journal
of Chromatography A, 858, 13-31.http://dx.doi.org/10.1016/S0021-
9673(99)00790-6

27. Basuki, F., Seko, N. and Tamada, M. (2010) Recovery of Scandium with
Phosphoric Chelating Adsorbent Prepared by Direct Radiation Graft
Polymerization. Journal of Ion Exchange, 21, 127-130.

28. Tang, Z., Du, Z., Min, E., Gao, L., Jiang, T. and Han, B. (2006) Phase
Equilibria of Methanol-Triolein System at Elevated Temperature and
Pressure. Fluid Phase Equilibria, 239, 8-11. http://dx.doi.org/10.1016/j.
fluid.2005.10.010

29. Gunstone, F.D., Hamilton, R.J., Padley, F.B. and Qureshi, M.I. (1965)
Glyceride Studies. V. The Distribution of Un-saturated Acyl Groups in
Vegetable Triglycerides. Journal of the American Oil Chemists Society,
42, 965-970.http://dx.doi.org/10.1007/BF02632456

30. Ramos, M.J., Fernández, C.M., Casas, A., Rodríguez, L. and Pérez,
á. (2009) Influence of Fatty Acid Composition of Raw Materials on
Biodiesel Properties. Bioresource Technology, 100, 261-268.http://
dx.doi.org/10.1016/j.biortech.2008.06.039

CITATION

CHAPTER 1

Okada, Y. , Huruya, H. and Imori, Y. (2015) Ligand Exchange Reaction of Ferrocene with Heterocycles. International Journal of Organic Chemistry, 5, 282-290. doi: 10.4236/ijoc.2015.54028.

CHAPTER 2

Saravanan, T. , Chadha, A. , Dinesh, T. , Palani, N. and Balasubramanian, S. (2015) Chemoenzymatic Synthesis of an Enantiomerically Enriched Bicyclic Carbocycle Using Candida parapsilosis ATCC 7330 Mediated Enantioselective Hydrolysis. International Journal of Organic Chemistry, 5, 271-281. doi: 10.4236/ijoc.2015.54027.

CHAPTER 3

Habibi, A. , Valizadeh, Y. , Mollazadeh, M. and Alizadeh, A. (2015) Green and High Efficient Synthesis of 2-Aryl Benzimidazoles: Reaction of Arylidene Malononitrile and 1,2-Phenylenediamine Derivatives in Water or Solvent-Free Conditions. International Journal of Organic Chemistry, 5, 256-263. doi: 10.4236/ijoc.2015.54025.

CHAPTER 4

Sanada, S. , Sumimoto, M. and Hori, K. (2015) Theoretical Study on Pd-Catalyzed Acylation of Allylic Esters with Acylsilanes and Acylstannanes. *International Journal of Organic Chemistry*, 5, 246-255. doi:10.4236/ijoc.2015.54024.

CHAPTER 5

Kawsar, S. , Ara, H. , Uddin, S. , Hossain, M. , Chowdhury, S. , Sanaullah, A. , Manchur, M. , Hasan, I. , Ogawa, Y. , Fujii, Y. , Koide, Y. and Ozeki, Y. (2015) Chemically Modified Uridine Molecules Incorporating Acyl Residues to Enhance Antibacterial and Cytotoxic Activities. *International Journal of Organic Chemistry*, **5**, 232-245. doi:10.4236/ijoc.2015.54023.

CHAPTER 6

Mousa, B. , Bayoumi, A. , Korraa, M. , Assy, M. and El-Kalyoubi, S. (2015) A Novel One-Pot and Efficient Procedure for Synthesis of New Fused Uracil Derivatives for DNA Binding. *International Journal of Organic Chemistry*, **5**, 37-47. doi: 10.4236/ijoc.2015.51005.

CHAPTER 7

Salman, A. , Abdel-Aziem, A. and Alkubbat, M. (2015) Synthesis, Spectroscopic Characterization and Antimicrobial Activity of Some New 2-Substituted Imidazole Derivatives. *International Journal of Organic Chemistry*, **5**, 15-28. doi: 10.4236/ijoc.2015.51003.

CHAPTER 8

Abu-Shanab, F. , Mousa, S. , Sherif, S. and Hassan, M. (2014) Preparation of Polyfunctionally Substituted Pyridine-2(1*H*)-Thione Derivatives as Precursors to Bicycles and Polycycles. *International Journal of Organic Chemistry*, **4**, 319-330. doi: 10.4236/ijoc.2014.45035.

CHAPTER 9

Bouterfas, K. , Mehdadi, Z. , Benmansour, D. , Khaled, M. , Bouterfas, M. and Latreche, A. (2014) Optimization of Extraction Conditions of Some Phenolic Compounds from White Horehound (Marrubium vulgare L.) Leaves. International Journal of Organic Chemistry, 4, 292-308. doi: 10.4236/ijoc.2014.45032.

CHAPTER 10

A. Alshammari and A. El-Gazzar, "Novel Synthesis Approach and Antiplatelet Activity Evaluation of 6-Arylmethyleneamino-2-Alkylsulfonylpyrimidin-4(3*H*)-one Derivatives and Its Nucleosides," *International Journal of Organic*

Chemistry, Vol. 3 No. 3A, 2013, pp. 28-40. doi: 10.4236/ijoc.2013.33A004.

CHAPTER 11

K. Singh, R. Chauhan, P. Solanki and T. Basu, "Development of Impedimetric Biosensor for Total Cholesterol Estimation Based on Polypyrrole and Platinum Nanoparticle Multi Layer Nanocomposite," *International Journal of Organic Chemistry*, Vol. 3 No. 4, 2013, pp. 262-274. doi: 10.4236/ijoc.2013.34038.

CHAPTER 12

El-Hameed Hassan, A. (2014) Heterocyclic Synthesis via Enaminones: Synthesis and Molecular Docking Studies of Some Novel Heterocyclic Compounds Containing Sulfonamide Moiety. *International Journal of Organic Chemistry*,4, 68-81. doi: 10.4236/ijoc.2014.41009.

CHAPTER 13

Ueki, Y. , Saiki, S. , Shibata, T. , Hoshina, H. , Kasai, N. and Seko, N. (2014) Optimization of Grafted Fibrous Polymer as a Solid Basic Catalyst for Biodiesel Fuel Production. *International Journal of Organic Chemistry*, 4, 91-105. doi: 10.4236/ijoc.2014.42011.

INDEX